Conservation Tourism

Conservation Tourism

Editor

Kornelia Marshall

scitus academics

1. INTRODUCTION

With the ongoing economic development, people's demands regarding quality of life, entertainment, and leisure are substantially increasing. To develop and upgrade the tourism industry, the Taiwanese government has considerably increased the convenience for foreign tourists to visit Taiwan by actively promoting policies such as opening up to visitors from China, liberalizing visa policies, simplifying entry and exit procedures, and expanding air routes. According to the World Tourism Barometer [1], Taiwan's international tourism revenue increased by two digits annually between 2009 and 2011, with an average growth rate of 23.2%, exceeding that of Hong Kong (22.4%), Singapore (16.2%), and South Korea (8.3%). Accordingly, Taiwan was ranked first among the four Asian Tigers and among the main Asia-Pacific countries and regions. The average growth rate for the number of foreign tourists visiting Taiwan between 2009 and 2012 was 17.6%, exceeding the global rate by 2.8% and the Asia-Pacific rate by 6.2%. In the *Travel and Tourism Competitiveness Report 2013* published by the World Economic Forum, Taiwan was ranked 33rd among 140 countries. Compared with its ranking in 2009 and 2011, Taiwan has advanced 10 and four places, respectively. These data demonstrate the substantial potential of Taiwan's tourism industry and the rapid demands for related growth.

Regarding changing travel behavior, in recent years, tourists have exhibited a preference for ecotourism. Additionally, travelling to Taiwan has become a popular trend among foreign tourists. Consequently, the number of tourists who choose Taiwan's national parks as recreation destinations has rapidly increased. According to data released by the Construction and Planning Agency, Ministry of the Interior [2], the number of tourists to Taiwan's national parks increased by 77% from 9,750,000 in 1999 to 17,300,000 in 2011. In addition, both the Battle of Guningtou and the 823 Artillery War, which ultimately enhanced the stability of the Taiwan Strait, occurred in the Kinmen area. Therefore, the Kinmen area played a unique role and has significance in contemporary history. To preserve the battle relics, cultural heritage, and natural resources of this area appropriately, the Kinmen National Park was founded in 1995. This was the first war memorial-themed historical and cultural heritage and natural resource conservation site established in Taiwan. After the Taiwanese government trialed the "mini three links" proposal, the number of Chinese tourists who visited Kinmen National Park increased from 202,138 in 2004 to 735,218 in 2011. The total amount of tourists who visited Kinmen National Park increased from 1,095,236 in 2004 to 2,164,248 in 2011. However, a rapid growth in the tourist industry is accompanied by recurring environmental impacts in the process of tourism, such as traffic jams, overexploitation of natural resources and other problems arising from tourists' improper behavior, which not only affect human life, the natural environment, and cultural heritage, but also accordingly give rise to a lot of pollution problems. Therefore, going by the premise that tourism resource development must ensure the sustainable development of ecology, economy, and society and meanwhile reduce the recreational impact, it is an urgent topic for this paper to (1) discuss how to ensure that tourism develops according to the principles of sustainable operation and in a way that aids the conservation of the environmental and ecological system; and (2) to consider the issues of environmental protection, such as biodiversity and climate change.

As the tourism industry continues to flourish, tourism-related environmental issues are becoming increasingly apparent every day. Tourism ecological capacity (TEC) has become the focus of tourism research. TEC refers to the maximum sum of productive land supplied for sustainable human use that has no harm on related ecosystem productive forces or the whole ecosystem. Tourist ecological capacity may be understood as the maximum ecological footprint in some natural and social conditions. Current domestic and foreign studies of TEC typically emphasize methods for evaluating and applying TEC, specifically, using quantified analysis approaches and directly or indirectly measuring TEC [3,4,5]. Under the wave of sustainable development, international society began to develop tools or indicators that can evaluate sustainable development one by one. They want to reasonably reflect the ecological environment, meanwhile analyzing resource consumption effectively and exploring the relationship among different kinds of environmental impact [6]. Generally speaking, the current evaluation indicators or measurement models of sustainable development established or developed internationally or domestically have their own features. Most of them can manage to include various sustainable development factors such as society, economy, ecology, and the environment [7]. However, when analyzing the aforementioned evaluation indicators and measurement models, the following concerns arose: (a) Certain evaluation indicators and measurement models are excessively complex to adequately reflect the connotations of sustainable development, and the dynamic indicators established for sustainable development are insufficient; (b) Several evaluation indicators or measurement models were developed based on comprehensive systems; thus, quantifying these indicators is difficult and even impossible, yielding low operability; (c) Some evaluation indicators and measurement models exhibit data accessibility problems and, thus, are challenging to apply. Zhang *et al.* [8] stated that although most existing sustainability evaluation methods can provide insight into the influence that human activities exert on various ecosystem functions, their applicability for evaluating relevant issues on a social and economic level is limited. In addition, most previous studies have not explored dynamic development trends. Hence, relevant literature has scope for improvement. Among the existing research, the ecological footprint (EF) concept proposed by Wackernagel and Rees [9] examines the index established for sustainability issues under the notion that human consumption behaviors depend on natural environments. The uniqueness of EF is its use of carrying capacity as the theoretical foundation and evaluation of environment sustainability with the assumption that all types of energy sources, material consumption, and waste production require the assimilation of productivity or absorption of land or water areas to transform human consumption behaviors and waste in certain areas into land size measurements of each person's consumption. Rees [10] asserted that the size of EF is directly proportional to environmental impacts, implying that environmental impact increases in correlation to EF.

Since the EF concept and computation method were proposed, EF has become a vital indicator of sustainable development for quantitative evaluation research. Additionally, EF has been widely employed in various fields as a simple, comprehensive indicator that conforms to sustainable development rationales. Regarding the application of EF to tourism and travel, Wackernagel and Yount [11] conducted a preliminary analysis of international tourism EF and reported that tourism EF (TEF) accounted for 10% of global EF. Gössling *et al.* [12] adopted Seychelles, Africa, as an example to establish an EF computation model for tourist destinations. Hunter [13] proposed the concept of a touristic ecological footprint, as well as its classification and application in sustainable tourism development. Cole and Sinclair [14] analyzed the touristic ecological footprint of tourists visiting the Indian Himalayas and recommended several strategies for sustainable development, such as treating waste materials, reducing the use of fossil fuels, developing ecotourism, and cultivating tourists' environmental protection awareness. Bagliani *et al.* [15] adopted EF to explore the influence that tourism activities in Venice, Italy, had on the local ecological environment. Patterson *et al.* [16] examined the differences between TEF and local biodiversity in Siena, Italy, to establish environment management improvement indicators. Kytzia *et al.* [17] considered the Alps resort Davos in Europe as an example, adopting a regional input-output model as an ecological footprint index (EFI) to examine how ecological efficiencies can be used to evaluate travel strategies. Li and Hou [18] calculated the TEF and TEC in the scenic zone of the Yellow Crane Tower on China for 2008. Their results indicated that the per capita TEF (PTEF) measured 0.0570 hm^2; of this, the contributions from transportation (55.89%) and waste (33.20%) accounted for comparatively high proportions.

With global environmental changes and frequent natural disasters, the international community has started to recognize the threat that the environment poses to human survival and the urgency of this issue. The International Institute for Applied System Analysis (IASA) officially proposed the concept of ecological security in 1989. The IASA defined ecological security as the condition where people's lives, health, wellbeing, basic rights, living necessities, essential resources, social order, and adaptability to environmental changes are not threatened. Ponsioen *et al.* [19] described ecological security as a state where the ecological environment required for the survival and development of a country is not or barely threatened. In other words, ecological security is when the natural ecological environment can satisfy the sustainable development requirements of individuals and communities, without damaging the natural ecological environment.

With numerous studies conducted on ecological security, the research methods employed vary. Scholars have investigated ecological security regarding the aspects of ecological risk assessments [20,21], ecological health [22,23], ecological models [24,25], and indicator systems [26,27]. However, most extant ecological security studies only provide quantitative descriptions based on literature reviews without implementing quantitative methods or introducing innovative strategies. For the studies that did conduct indicator system evaluations, the majority were static evaluations. Accordingly, ecological security management policies have remained passive for a long time and cannot be used to predict relevant trends. Warhurst [28] asserted that simplifying complex information and examining the factors influencing issues by using quantifying indicators can increase the objectivity of such indicators [29]. Rasul and Thapa [30] selected 12 indicators for evaluating the sustainable development of traditional agriculture and ecological security in Bangladesh. Bhandari and Grant [31] established an indicator system from three dimensions (*i.e.*, economy, ecology, and society) to evaluate ecological security in Western Nepal. Siche *et al.* [32] used EF and the environmental sustainability index (ESI) to establish ecological security evaluation indicators. Liu and Borthwick [33] adopted EFI and the carrying capacity of the environment to investigate ecological security evaluations. Yuan [34] employed the pressure-state-response model to establish a land ecological security evaluation index system for Hangzhou in Zhejiang Province, China, based on the dimensions of nature, economy, and society. In conclusion, this paper seeks to apply EF to national park ecological security evaluation and construct a tourism biocapacity evaluation model that is applicable to national parks. We first adopted TEF and TEC to evaluate the ecological conditions of Kinmen National Park between 2002 and 2011. Subsequently, environmental sustainability indicators such as tourism ecological deficit (TED), tourism ecological remainder (TER), EFI, ESI, and EF per capita and per NT$10,000 gross domestic product (GDP) were employed to evaluate the ecological security and resource utilization efficiency of Kinmen National Park. Finally, the issues reflected in various indicator values were analyzed to establish a systematic measurement instrument for promoting sustainable development and assessing the progress trends of sustainable development.

2. METHODS

2.1. Study Design

This study adopted the TEF concept proposed by Gössling *et al.* [10] and employed by Martin-Cejas and Sanchez [35] as the research framework for evaluating the EF of Kinmen National Park between 2002 and 2011. The evaluation items were divided into five categories: transportation ecological footprint (TREF), accommodation ecological footprint (ACCEF), activities ecological footprint (ACTEF), food and fiber consumption ecological footprint (FEF), and wastewater ecological footprint (WWEF). These EF evaluation items were then categorized into six types of biologically productive land to investigate the influence that EF exerts on the environment. The six types of productive lands comprised the ecological footprint of crop land (EF_{CL}), ecological footprint of grazing land (EF_{GL}), ecological footprint of forest land (EF_{FL}), ecological footprint of fishing grounds (EF_{FG}), ecological footprint of built-up land (EF_{BU}), and ecological footprint of carbon uptake land (EF_{CU}). The main evaluation items of each category and data sources are shown in Table 1.

2.2. Methods for Calculating Yield and Equivalence Factors

The Global Footprint Network has developed a national footprint account classifying biologically productive land into six types: crop, grazing, forest, fishing, carbon uptake, and built-up land. These land types have differing biological productivities, hence their areas are weighted to represent an equivalent area with the same biological productivity, *i.e.*, the global hectare. Abbreviated as "gha",the global hectare quantifies the biocapacity of the earth in a given year, where one global hectare measures the average productivity of biologically productive areas. The conversion calculation mainly adopts equivalence factor (EQF) and yield factor (YF).

EQF is the ratio of the potential biological productivity of a certain land type to the average potential biological productivity of all global lands and it is used to evaluate the difference between the six types of productive lands on the globe. As shown by Equation (1), the equivalence factor γ_k of type-k biologically productive land is the average productivity $\overline{Y_k}$ of such a type of land on the globe divided by the average productivity $\overline{Y}$ of all types of land on the globe:

$$\gamma_k = \frac{\overline{Y_k}}{\overline{Y}} \quad k = 1, 2, \ldots, 6 \tag{1}$$

Because different countries or regions have different resource endowments, the biological productivity varies according to different land types and even that of the same type of land varies from region to region. Therefore, in order for comparability and accumulativity between regions, it is required that the area of each type of land of research object be converted into an equivalent area with corresponding global average productivity and

Table 1. Evaluation items and data sources.

EF Category	Evaluation Indicators	Evaluation Items	Evaluation Content	Data Sources
TREF	Built-up land	Road use area	Road use area	Gössling *et al.* [12]; Kinmen National Park Administration Office [36]
		Parking lot area	Large vehicles, small vehicles, motorcycles, and bicycles	Gössling *et al.* [12], Kinmen National Park Administration Office [36]; Equipment Management System of Taiwan National Parks [37]
	Fossil energy	Resource usage	Transportation energy consumption	Gössling *et al.* [12]; Visitations and Revenues of National Parks [38]
ACCEF	Built-up area	Accommodation area	Hostel areas in national parks	Gössling *et al.* [12], Kinmen National Park Administration Office [36]; Monthly Report of Home Stay Facilities [39]
	Fossil energy	Accommodation energy consumption	Hostel energy consumption	
ACTEF	Built-up area	Recreation area	Recreation area	Kinmen National Park Administration Office [36]
	Fossil energy	Recreation energy consumption	Recreation energy expense	
FEF	Crop land	Food and fiber consumption when traveling	Grains, potatoes, sugar and honey, seeds and oilseeds, vegetables, fruits, fats, tobacco, and cotton	Food Supply and Utilization, Council of Agriculture [40]
	Grazing land		Meat, eggs, and diary	
	Carbon land		Coniferous trees, broad-leaved trees, fuel wood, and faggots of wood	
	Fishing grounds		Aquatic products	
WWEF	Built-up area	Wastewater equipment area	Treatment plant area	Kinmen National Park Administration Office [34]
	Fossil energy	Water purification energy consumption	Water purification energy expense	

Data source: Compiled in this study.

conversion factor, the YF. The YF λ_k of type-k land in a certain region is the ratio of the average productivity $\overline{y_k}$ of this type of land in this region to the global average productivity $\overline{Y_k}$ of the same type of land; the computational formula is Equation (2):

$$\lambda_k = \frac{\overline{y_k}}{\overline{Y_k}} \quad k=1, 2, \ldots, 6 \tag{2}$$

As for a given region, the physical area of its type-k land multiplied by λ_k is the area with the global average productivity of such a type of land and, multiplied by r_k, is the equivalent area with global average productivity, which has global comparability and the measurement unit of which is known as global hectare (gha). This work refers to the EQF and YF from the Ecological Footprint Atlas [41], as summarized in Table 2.

Table 2. Equivalence factors and yield factors for a given land type.

Land Type	Equivalence Factor	Yield Factor
Carbon uptake land	1.26	1.2
Crop land	2.51	1.15
Forestland	1.26	1.2
Grazing land	0.46	1.6
Built-up land	2.51	1.15
Fishing ground	0.37	0.9

Source: Global Footprint Network, Ecological Footprint Atlas (2010).

2.3. Model Computation Method

2.3.1. TEF Computation Model

TEF is composed of five elements: TREF, ACCEF, ACTEF, FEF, and WWEF. Relevant explanations are provided below:

$$TEF = TREF + ACCEF + ACTEF + FEF + WWEF \tag{3}$$

(a) TREF Computation

The computation of TREF is divided into two aspects: (a) the built-up area of transportation facilities used to travel (*i.e.*, road area and parking lot area); and (b) the transportation energy consumed during travel activities. The computation formula is shown below:

$$TREF = \left(\sum S_{transport} + \sum E_{transport}\right) \times F_v \quad (4)$$

where *TREF* represents the transport ecological foot print; *S transport* represents the built-up area of transportation facilities; *E transport* represents the fossil energy area transformed through transportation energy consumption; and F_v (v = 1, 2, ..., 6) represents the yield factor (YF) and equivalence factor (EQF) for the six types of biologically productive lands. Equation (4) was rewritten as Equation (5) according to the actual tourist traffic situation:

$$TREF = \left[\sum (s_i \times K_i) + \sum (N_j \times D_j \times e_j / r)\right] \times F_v \quad (5)$$

where s_i represents the built-up area of the ith type of transportation facility; K_i represents the tourist utilization rate of the ith type of transportation facility; N_j represents the number of tourists in the jth type of vehicle; D_j represents the average travel distance for tourists using the jth type vehicle; e_j represents the per capita unit energy consumption of the jth type vehicle; r represents the conversion factor of unit fossil fuel productive land area worldwide; and F_v represents the YF and EQF for the six types of biologically productive lands.

(b) ACCEF Computation

The computation of ACCEF involves two parts: (a) the accommodation construction land area provided to tourists; and (b) tourists' energy consumption during residence (e.g., energy consumed by air conditioners and lighting):

$$ACCEF = \left(\sum S_{accommodation} + \sum E_{accommodation}\right) \times F_v \quad (6)$$

where *ACCEF* represents the accommodation ecological footprint; $S_{accommodation}$ represents the construction land area of accommodation facilities; $E_{accommodation}$ represents the fossil energy area transformed through accommodation energy consumption; and F_v represents the YF and EQF for the six types of biologically productive lands.

Because energy consumption approaches and items are complex, difficult to calculate, and vary between regions and accommodation types, this study referred to the global residential land usable area and energy usage statistics provided by the UNWTO [42] as the standard for evaluating the energy consumption per bed every night in Kinmen National Park. Thus, Equation (7) can be rewritten as:

$$ACCEF = \left[\sum (S_i \times N_i) + \sum \left(365 \times N_i^{'} \times K_i \times e_i / r\right)\right] \times F_v \tag{7}$$

where S_i represents the construction land area of the ith type of accommodation facility bed; N_i represents the number of beds possessed by the ith type of accommodation facility; N'_i represents the number of beds actually used in the ith type of accommodation facility; K_i represents the average annual guest room rental rate for the ith type of accommodation facility;e_i represents the daily energy consumption for the ith accommodation facility; r represents the conversion factor of unit fossil fuel productive land area worldwide; and F_v represents the YF and EQF for the six types of biologically productive lands.

(c) ACTEF Computation

The computation of ACTEF involves two aspects: (a) the built-up land areas (e.g., tourist trails, highways, and scenic view spaces) within various types of scenic areas; and (b) the fossil energy area transformed through energy consumption, such as touring scenic sites by vehicle:

$$ACTEF = \left(\sum S_{visiting} + \sum E_{visiting}\right) \times F_v \tag{8}$$

where *ACTEF* represents the activities ecological footprint; $S_{visiting}$ represents the built-up land area of tourism and sightseeing facilities; $E_{visiting}$ represents the fossil energy area transformed through tourism and sightseeing energy consumption; and F_v represents the YF and EQF for the six types of biologically productive lands.

Because of the unique layout of Kinmen National Park, and the fact that the vehicles used to travel between subsidiary parks might have been included in the TREF, energy consumption was excluded from the calculation of ACTEF. Thus, Equation (9) can be rewritten as:

$$ACTEF = s_i \times F_v \tag{9}$$

where s_i represents the built-up land area of scenic sightseeing facilities; and F_v represents the YF and EQF for the six types of biologically productive lands.

(d) FEF Computation

The computation of FEF involves three aspects: (a) the building land area of food and beverage service facilities (e.g., local cuisine, buffet, and beverages); (b) biologically productive land area transformed through the consumption of various foods by tourists; and (c) biologically productive land area transformed through the consumption of fiber by tourists:

$$FEF = (\sum S_{food} + \sum C_{food} + \sum F_{food}) \times F_v \tag{10}$$

where *FEF* represents the food and fiber consumption ecological footprint; S_{food} represents the building land area of food services; C_{food} represents the biologically productive land area transformed through food consumption; F_{food} represents the fossil energy land area transformed through fiber consumption; and F_v represents the YF and EQF for the six types of biologically productive lands.

According to the actual food consumption situation, Equation (11) can be rewritten as:

$$TEF_{food} = \sum (N \times D \times c_i / p_i) \times F_v \tag{11}$$

where *N* represents the number of tourists; *D* represents the average days per trip; c_i represents the daily consumption of the ith type of food by tourists; P_i represents the average annual productivity of the ith type of food for biologically productive lands; and F_v represents the YF and EQF for the six types of biologically productive lands.

(e) WWEF Computation

The computation of WWEF primarily involves calculating the wastewater purification energy consumption resulting from various tourist activities conducted in the park. In this study, the electricity consumed in wastewater treatment plant operations was transformed into a carbon footprint to facilitate the inclusion of the environmental impact of wastewater in EF computations. Because the building land areas of wastewater plant facilities are designated to regular control areas and included as an item of ACTEF, only the wastewater treatment carbon footprints established based on electricity consumption were incorporated in the WWEF calculation:

$$WWEF = EC \times CF / FCS \times F_v \tag{12}$$

where *WWEF* represents the wastewater ecological footprint; *EC* represents the total electricity consumed by wastewater treatment plants; *CF* represents carbon dioxide conversion factors; *FCS* represents the CO_2 absorption rate of forest land, which was 3.6666(tCO2/ha/year); and F_v represents the YF and EQF for the six types of biologically productive lands.

2.3.2. The TEC Computation Model

The computation of TEC mainly relied on data (e.g., region partition and land utilization plans) published by the Kinmen National Park Administration Office [36] to estimate the capacity areas of the six types of biologically productive land in Kinmen National Park:

$$TEC = Ntec = N\sum_{i=1}^{6}(a_i r_i / y_i) \tag{13}$$

where *TEC* represents the tourism ecological capacity; *N* represents the number of tourists; *i* represents the types of biologically productive land; *tec* represents the per capita TEC; a_i represents the per capita biologically productive land area; r_i represents EQF; and y_i represents YF.

2.3.3. The Establishment of Sustainable Tourism Environment Evaluation Indicators

This study employed multiple quantitative indicators (e.g., TED, TER, EFI, ESI, and EF per capita and per NT$10,000 GDP) to establish a set of evaluation indicators regarding the tourism environment sustainability of national parks and provide the criteria for national parks to evaluate ecological security. The evaluation indicators employed in this study are introduced below.

(a) TED or TER

When the environmental carrying capacity of a region is less than necessitated by EF demands, an ecological deficit (ED) occurs, which indicates that the ecological carrying capacity of the region exceeds the ecological capacity. Consequently, the corresponding development model is comparatively less sustainable. When the environment carrying capacity of a region is greater than required by EF demands, an ecological remainder (ER) occurs, which indicates that the ecological carrying capacity of the region is sufficient to satisfy the corresponding carrying capacity and that the development model is comparatively more sustainable. Rees [43] stated that ED is caused by humans placing excessive demands on the ecosystem. Therefore, to maintain sustainable ecological development, ecological demands must be reduced. Moore *et al.* [44] adopted EF to examine ED/ER in Vancouver; the results indicated a severe deficit. The formulas for TED and TER are:

$$TER = TEC - TEF \tag{14}$$

$$TED = TEF - TEC \tag{15}$$

where *TER* represents tourism ecological remainder; *TED* represents tourism ecological deficit; *TEC* represents tourism ecological capacity; and *TEF* represents tourism ecological footprint.

(b) EFI

EFI involves comparing resource and energy expenditures with the ecological carrying capacity of a region to evaluate the resource utilization of the region or country and determine whether the resource and environment condition exhibits sustainable development characteristics. Xiao *et al.* [45] adopted EF as the criterion and employed EFI and the ecological occupancy index in ecological security evaluations and analysis to explore the corresponding ecological environment conditions. The EFI computation formula is expressed as Equation (16), and the EFI levels are shown in Table 3.

$$EFI = TEF / TEC \tag{16}$$

where *EFI* represents the ecological footprint index; *TEF* represents tourism ecological footprint; and *TEC* represents tourism ecological capacity.

Table 3. EFI levels and the corresponding conditions.

Level	EFI	EFI conditions
1	0.5	Safe
2	0.5~0.8	Moderately safe
3	0.8~1.0	Threshold
4	>1.0	Unsafe

(c) ESI

ESI is an environmental sustainability evaluation index developed by the Yale Center for Environmental Law and Policy (YCELP), the Center for International Earth Science Information Network (CIESIN), and the World Economic Forum [46]. ESI primarily evaluates the extent to which the ecology of a region can satisfy humans' ecological demands to assess whether the region can be developed sustainably. Cui *et al.* [47] adopted ESI to explore the development conditions of Shandong Province, China at that time. The ESI results indicated that the development conditions of Shandong Province were unsustainable. Siche *et al.* [32] claimed that both EF and ESI can be used as ecological security evaluation indicators. The ESI formula is presented as Equation (17), and the ESI levels are shown in Table 4.

$$ESI = TEC / (TEC + TEF) \quad (17)$$

where *ESI* represents the environmental sustainability index; *TEF* represents tourism ecological footprint; and *TEC*represents tourism ecological capacity.

Table 4. ESI levels.

Level	ESI	Regional ecological sustainability extent
1	>0.7	High sustainability
2	0.50–0.70	Low sustainability
3	0.30–0.50	Low unsustainability
4	<0.30	High unsustainability

Resource: YCELP and CIESIN [46].

(d) EF Per Capita and Per NT$10,000 GDP

EF per capita and per NT$10,000 GD refers to the ecological space occupied by NT$10,000 GDP; in other words, the ratio of total EF to NT$10,000 GDP. High NT$10,000 GDP indicates low regional resource utilization efficiency. Conversely, low NT$10,000 GDP indicates high regional resource utilization efficiency. Meyfroidt *et al.* [48] used ecological footprints per NT $10,000 GDP to inspect the resource utilization conditions in forests. Their results indicated that the resource utilization efficiency in the region exhibited a declining trend due to the increase in EF and stagnation of GDP. The EF per capita and per NT$10,000 GDP computation formula is:

$$\text{Ecological footprint per NT\$10,000 GDP} = \text{TEF/GDP} \tag{18}$$

where TEF represents tourism ecological footprint and GDP reflects the average incomes in the region.

3. RESULTS AND DISCUSSION

3.1. TEF Computation and Analysis Results

Table 5 lists the five types of activities in Kinmen National Park and the TEF computation results. TEF decreased from 7747.17535 gha in 2002 to 7071.86588 gha in 2005 before gradually increasing to 8369.85782 gha between 2007 and 2011. Among the five types of activity EF, ACTEF accounted for the largest proportion at an average of 80.395%, followed by FEF at an average of 13.263%, then TREF (5.4%), ACCEF (0.8%), and WWEF (0.2%), which accounted for the smallest proportion.

According to the empirical results, TREF increased from 337.70005 gha in 2007 to 588.21509 gha in 2011 because the increased number of tourists resulted in increased demand for vehicles, which further increased demands for liquefied fuel, thereby increasing TREF. ACCEF gradually increased from 62.63099 gha in 2007 to 100.65721 gha in 2011 primarily because changes in accommodation facilities and the number of tourists seeking accommodation influenced accommodation rates.

Table 5. EF for the five types of activities and total EF (unit: gha).

Year	TREF	ACCEF	ACTEF	FEF	WWEF	TEF
2002	394.58731	21.21110	6188.07870	1128.50208	14.79615	7747.17535
2003	332.82163	21.21110	6188.07870	744.50910	11.37336	7297.99389
2004	363.32671	21.21110	6188.07870	703.58981	11.47078	7287.67711
2005	332.84297	53.20533	6040.11671	635.09422	10.60665	7071.86588
2006	324.67955	65.50837	6040.11671	651.63703	10.23611	7092.17777
2007	337.70005	62.63099	6040.11671	595.89318	10.28346	7046.62440
2008	357.10648	72.90312	6040.11671	898.14068	10.88576	7379.15276
2009	462.48191	89.44121	6040.11671	1193.80303	14.64477	7800.48764
2010	550.13200	91.95318	6040.11671	1566.88595	18.82418	8267.91202
2011	588.21509	100.65721	5781.16880	1877.38212	22.43461	8369.85782
Average proportion	5.4%	0.8%	80.4%	13.2 %	0.2%	100.000%

Because this study assumed that all recreational activities for tourists in Kinmen National Park were within the range of regular control zones, recreation areas, and heritage areas, the combination of these three area types were considered ACTEF. Based on construction statistics released by CPAMI (2012), the gross area of Kinmen National Park did not change substantially between 2002 and 2010; significant changes only occurred in 2011. ACTEF decreased from 6188.079 gha in 2002 to 5781.169 gha in 2011. FEF exhibited a decreasing trend from 2002 to 2007 before increasing from 595.8931833 gha in 2007 to 1877.3821159 gha in 2011, with an average annual growth rate of 20.8%. These changes in FEF were related to the increased number of tourists. WWEF increased annually from 2007 to 2011, reaching 22.43460596 gha (Figure 1).

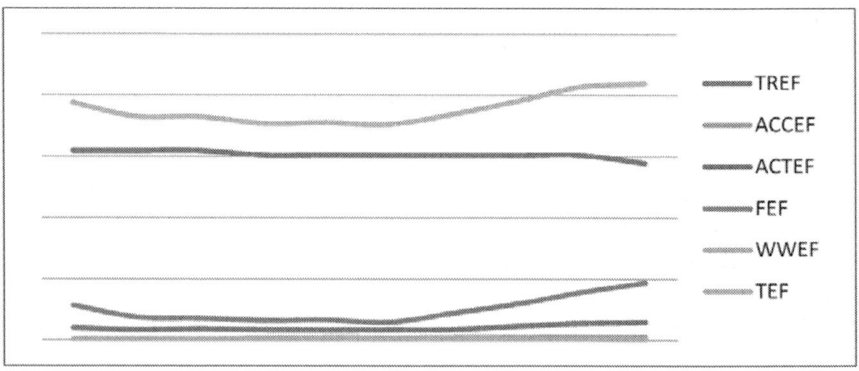

Figure 1. EF for the five types of activities and total EF between 2002 and 2011 (unit: gha).

The results presented in Table 5 were divided by the number of tourists who visited Kinmen National Park during the research period to obtain values for the five types of activities and PTEF (Table 6). Generally, PTEF has exhibited a declining trend for nearly 10 years, decreasing by 28.27% from 0.005391 gha in 2002 to 0.003867 gha in 2011. Among the five activities, the per capita activities ecological footprint (PACTEF) was substantially influenced by the number of tourists. PACTEF decreased by 57.6% from 0.006304 gha in 2007 to 0.002671 gha in 2011. The per capita transport ecological footprint (PTREF) has exhibited a year-by-year declining trend of 22.7% since 2007. The per capita food and fiber consumption ecological footprint (PFEF) increased from 0.000622 gha in 2007 to 0.000867 gha in 2011, with an average annual growth rate of 39.4%. The most substantial factor influencing PFEF was the number of tourists; PFEF increased in correlation to the number of tourists.

Table 6. The five types of activities and PTEF (unit: gha per capita).

Year	Number of tourists visiting Kinmen National Park	PTREF	PACCEF	PACTEF	PFEF	PWWEF	PTEF
2002	1,436,953	0.000275	0.000015	0.004306	0.000785	0.000010	0.005391
2003	1,088,860	0.000306	0.000019	0.005683	0.000684	0.000010	0.006702
2004	1,095,236	0.000332	0.000019	0.005650	0.000642	0.000010	0.006654
2005	1,002,065	0.000332	0.000053	0.006028	0.000634	0.000011	0.007057
2006	958,376	0.000339	0.000068	0.006302	0.000680	0.000011	0.007400
2007	958,107	0.000352	0.000065	0.006304	0.000622	0.000011	0.007355
2008	1,015,977	0.000351	0.000072	0.005945	0.000884	0.000011	0.007263
2009	1,386,778	0.000333	0.000064	0.004356	0.000861	0.000011	0.005625
2010	1,805,754	0.000305	0.000051	0.003345	0.000868	0.000010	0.004579
2011	2,164,248	0.000272	0.000047	0.002671	0.000867	0.000010	0.003867
Average proportion		5.2%	0.8%	81.6%	12.2%	0.2%	100.000%

Although the number of tourists visiting the park increases annually, the per capita accommodation ecological footprint (PACCEF) exhibited a declining trend from 2008 to 2011. The primary reason for this phenomenon could be that the average duration of trips to Kinmen National Park was short. Taiwan has been open to travel for mainland Chinese tourists in recent years, and their visits to Kinmen National Park are typically scheduled as day trips. Hence, the influence exerted by PACCEF on Kinmen National Park was less than that of the other three items (e.g., PTREF, PACTEF, and PFEF). The per capita wastewater treatment ecological footprint (PWWEF) decreased from 0.000011 gha in 2007 to 0.000010 gha in 2011. Consequently, the proportion of PWWEF in PTEF was relatively small (Figure 2).

According to the aforementioned analysis, the primary resource consumption during trips was PACTEF consumption. Because Kinmen National Park is a national historic battlefield park, to maintain the historic battle culture, the reserved building land area in Kinmen National Park considerably exceeds that of other parks, resulting in comparatively higher ACTEF consumption. However, the per capita total EF began to decrease from 2007, primarily because of the increased number of tourists.

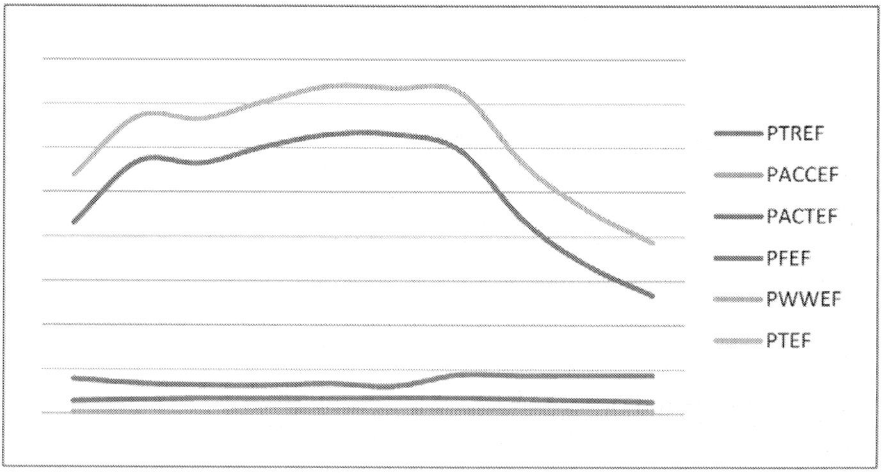

Figure 2. The five types of activities and PTEF between 2002 and 2011(unit: gha per capita).

3.2. Computation and Analysis Results for Sustainable Tourism Environment Evaluation Indicators

3.2.1. PTES/PTED

Table 7 shows the computation results for PTES/PTED. The per capita tourism ecological deficit (PTED) of Kinmen National Park increased by 45.37% from −0.000364 gha in 2002 to −0.000530 gha in 2011 primarily because of the increased number of tourists. Following the promotion of Project Vanguard for Excellence in Tourism starting in 2002, Taiwan implemented the Doubling Tourist Arrivals Plan and direct cross-Strait transportation in 2008. Additionally, mainland Chinese tourists have been allowed to travel independently in Taiwan since 2011. Consequently, Taiwan has attracted a greater number of tourists from mainland China, Southeast Asia, Europe, and the United States, increasing the number of tourist visitors to Kinmen National Park from 1,436,953 in 2002 to the peak value of

2,164,248 in 2011. Therefore, this study recommends that park authorities actively control the number of tourists in an effort to reduce TED.

Table 7. PTES/PTED and EFI (unit: gha per capita).

Year	PTEC	PTEF	PTES/PTED	EFI		
				Index	Level	Representation condition
2002	0.005027	0.005391	−0.000364	1.07	4	Unsafe
2003	0.006634	0.006702	−0.000068	1.01	4	Unsafe
2004	0.006596	0.006654	−0.000058	1.01	4	Unsafe
2005	0.007209	0.007057	0.000152	0.98	3	Barely safe
2006	0.007537	0.007400	0.000137	0.98	3	Barely safe
2007	0.007540	0.007355	0.000185	0.98	3	Barely safe
2008	0.007110	0.007263	−0.000153	1.02	4	Unsafe
2009	0.005209	0.005625	−0.000416	1.08	4	Unsafe
2010	0.004000	0.004579	−0.000578	1.14	4	Unsafe
2011	0.003338	0.003867	−0.000530	1.16	4	Unsafe

3.2.2. EFI

This study used EFI to measure the ecological security of the park; the results are shown in Table 7. EFI exhibited a declining trend from 1.07 in 2002 to 0.98 in 2007 before increasing from 1.02 in 2008 to 1.16 in 2011. This indicated that during that period, the level of ecological security in Kinmen National Park was unsafe, park development was deviating from sustainable development, and controls were required to improve the situation.

3.2.3. ESI

Table 7 shows the computation results of ESI. Between 2002 and 2009, ESI remained at Level 2, indicating low sustainability. However, since 2010, ESI has declined to Level 3, indicating unsustainability. If not controlled and improved, sustainable ecological development cannot be achieved.

3.2.4. Ecological Footprint Per Capita and Per NT$10,000 GDP

The analysis results of resource utilization efficiency in Kinmen National Park according to EF per NT$10,000 GDP are presented in Table 8. The EF per NT$10,000 GDP decreased from 7434.86 in 2002 to 5842.76 in 2009, indicating an increasing trend in resource utilization efficiency. However, the EF per NT$10,000 GDP increased from 5842.76 in 2009 to 6306.69 in 2011, indicating a decline in resource utilization efficiency.

Table 8. ESI and EF per NT$10,000 GDP.

Year	TEF	TEC	Per capita recurrent Income (NT$10,000)	ESI			EF per NT$10,000 GDP
				Index	Level	Representational State	
2002	7747.17535	8208.739	1.042008	0.514	2	Low sustainability	7434.86
2003	7297.99389	8208.739	1.063047	0.529	2	Low sustainability	6865.17
2004	7287.67711	8208.739	1.046024	0.530	2	Low sustainability	6967.03
2005	7071.86588	8208.739	1.109019	0.537	2	Low sustainability	6376.69
2006	7092.17777	8208.739	1.191444	0.536	2	Low sustainability	5952.59
2007	7046.62440	8208.739	1.306885	0.538	2	Low sustainability	5391.92
2008	7379.15276	8208.739	1.292542	0.527	2	Low sustainability	5709.02
2009	7800.48764	8208.739	1.335068	0.513	2	Low sustainability	5842.76
2010	8267.91202	8208.739	1.315137	0.498	3	Low unsustainability	6286.73
2011	8369.85782	8208.739	1.327140	0.495	3	Low unsustainability	6306.69

4. CONCLUSIONS AND RECOMMENDATIONS

4.1. Conclusions

This study employed EF, ecological capacity, and environmental sustainability evaluation indicators to examine the ecological security and resource use efficiency of Kinmen National Park. The empirical results were as follows: (a) TEF increased by 8.03% over 10 years from 7747.175 gha in 2002 to 8369.858 gha in 2011. Among the five activity EFs, ACTEF accounted for the highest proportion (80.4%), followed by FEF (13.26%), TREF (5.37%), ACCEF (0.8%), and WWEF (0.18%), which accounted for the smallest proportion; (b) Regarding environmental sustainability evaluation indicators, PTED increased by approximately 45.37% from −0.000364 gha in 2002 to −0.000530 gha in 2011. In 2011, EFI was ranked Level 4 at 1.16, and ESI was ranked Level 3 at 0.495, indicating that the level of ecological security for Kinmen National Park during that period was unsafe. The EF per NT$10,000 GDP decreased from 7434.86 (gha/NT$10,000) in 2002 to 6306.69 (gha/NT$10,000) in 2011, indicating a decline in resource utilization efficiency. Based on the aforementioned results, with the expanded scale of tourism to Kinmen National Park, the pressure that ecological occupancy exerts on the national ecosystem has exceeded the ecological capacity. The development of Kinmen National Park is likely to deviate from sustainable development if the ecosystem is not improved.

According to the empirical analysis results, the primary factors influencing various types of activity EF are presented below.

4.1.1. Number of Tourists

The number of tourists exerts a positive influence on the total EF from all activities. When tourist numbers increased, EF increased, as did the impact on the environment. From the perspective of per capita EF, the space resource allocated to each person declined with the increase in tourist numbers.

4.1.2. Energy Utilization Efficiency

The increase in fossil energy utilization efficiency effectively reduced the influence that the number of tourists exerted on TREF. However, this influential factor cannot be improved by park managers or decision makers. Thus, an effective method for reducing carbon footprint is to reduce indirect influences and the use of fossil energies.

4.2. Recommendations

Based on the primary research findings, several recommendations were proposed as a reference for managers and relevant organizations. These recommendations are listed below.

1) Kinmen National Park authorities should closely monitor the negative influence that tourism development exerts on sustainable development of the ecosystem. The environmental consciousness of tourists should be enhanced to prevent damage to the ecological environment of the park resulting from an excessive number of tourists.

2) Kinmen National Park authorities should conduct statistical analysis regarding the number of tourists to estimate the influences that tourists exert on EF; the results can serve as a reference for the sustainable development of Kinmen National Park.

3) The empirical analysis results indicated that the fossil energy consumed for transportation EF of Kinmen National Park was the key factor contributing to TREF. Therefore, energy-saving and carbon-reduction approaches to tourism should be promoted. Tourists should be encouraged to use public transportation with low energy intensities and vehicles with low carbon consumption, low energy consumption, and low pollution emissions (e.g., by providing bicycle and electric motorcycle rental services). In addition, global positioning systems should be installed in rental vehicles to enable national parks to effectively monitor the proportion of tourists who engage in recreational activities. Relevant data can be employed to adjust the collection and drop-off schedules at public transportation stations, effectively reducing transportation carbon footprints and the overall amount of fossil energy consumed by transportation.

4.3. Future Suggestion

When the ecological footprint method is used to analyze and evaluate the sustainable development of a tourist area, as ecological footprint is calculated by the year, the environment problems as a result of uneven distribution of tourists in time and space are ignored. Being influenced by climate, holidays, celebrations, *etc.*, tourists are characterized by seasonal fluctuations and the frequency of tourist activities and the concentration of tourists in a tourist area can both trigger special changes in some ecological resources of the tourist area (e.g., concentrated excessive emission of pollutants may cause permanent harm to flora and fauna in tourist areas) and cause permanent damage and such possible effects cannot be manifested in the process of ecological footprint calculation.

Water is one of the most consumed resources in human activities as it is involved in accommodation, catering, sanitation facilities, activities, *etc.* in the process of tourism. What is more, the ecological footprint of electricity consumption during wastewater treatment in Kinmen Park should also be taken into account. Different from previous research, this paper seeks to include the discharge and disposal of sewage and wastes into ecological footprint calculation. However, as relevant data are hard to obtain, this paper fails to include garbage disposal into the calculation; as a result, this paper may have underestimated the actual biocapacity of Kinmen National Park and it is suggested that follow-up studies incorporate sewage discharge and garbage disposal into the scope of discussion.

ACKNOWLEDGMENTS

We thank the Chung Shan Medical University for providing financial support for this research project under grant number CSMU-INT-101-26.

REFERENCES

1. United Nations World Tourism Organization (UNWTO) World Tourism Barometer (English Verison). Available online: http://www.e-unwto.org/content/h51h370763/?p=e612873b0ad64992b6d7d8921ba9a519&pi=0 (accessed on 1 February 2012).
2. Construction and Planning Agency, Ministry of the Interior (CPAMI). The Statistical Yearbook of Construction and Planning Agency Ministry of Interior. Available online: http://www.cpami.gov.tw/chinese/index.php?option=com_content&view=article&id=7716&Itemid=103 (accessed on 21 March 2012).

3. Mark, T.; Brown, S.U. Energy measures of carrying capacity to evaluate economic investments. *Popul. Environ.* **2001**,*22*, 471–472.
4. Xu, L.; Kang, P.; Wei, J. Evaluation of urban ecological carrying capacity: A case study of Beijing, China. *Energy Policy* **2010**, *2*, 1873–1880.
5. Zeng, C.; Liu, Y.; Liu, Y.; Hu, J.; Bai, X.; Yang, X. An integrated approach for assessing aquatic ecological carrying capacity: A case study of Wujin District in the Tai Lake Basin, China. *Int. J. Environ. Res. Public Health* **2011**, *8*, 264–280.
6. Chen, Y.; Chen, C.Y.; Hsieh, T. Establishment and applied research on environmental sustainability assessment indicators in Taiwan. *Environ. Monit. Assess.* **2009**, *155*, 407–417.
7. Chen, Y.; Chen, C.Y.; Hsieh, T. Exploration of sustainable development by applying green economy indicators.*Environ. Monit. Assess.* **2011**, *182*, 279–289.
8. Zhang, Y.; Yang, Z.; Yu, X. Measurement and evaluation of interactions in complex urban ecosystem. *Ecol. Model.***2006**, *96*, 77–89.
9. Wackernagel, M.; Rees, W.E. *Our Ecological Footprin: Reducing Human Impact on the Earth*; New Society Publishers: Gabriola Island, BC, Canada, 1996.
10. Rees, W.E. Eco-footprint analysis: Merits and brickbats. *Ecol. Econ.* **2000**, *32*, 371–374.
11. Wackernagel, M.; Yount, J.D. Footprints for sustainability: The next steps. *Environ. Dev. Sustain.* **2000**, *2*, 23–44.
12. Gössling, S.; Hansson, C.B.; Hörstmeier, O.; Saggel, S. Ecological footprint analysis as a tool to assess tourism sustainability. *Ecol. Econ.* **2002**, *43*, 199–211.
13. Hunter, C. Sustainable tourism and the touristic ecological footprint. *Environ. Dev. Sustain.* **2002**, *4*, 7–20.
14. Cole, V.; Sinclair, A.J. Measuring the ecological footprint of a Himalayan Tourist Center. *Mt. Res. Dev.* **2002**, *22*, 132–141.
15. Bagliani, M.; Da Villa, E.; Gattolin, M.; Niccolucci, V.; Patterson, T.; Tiezzi, E. The ecological footprint analysis for the Province of Venice and the relevance of tourism. In *The Sustainable City III. Urban Regeneration and Sustainability*; Marchettini, N., Brebbia, C.A., Tiezzi, E., Wadhwa, L.C., Eds.; WIT Press: Southampthon, UK, 2004; pp. 123–131.
16. Patterson, T.M.; Niccolucci, V.; Bastianoni, S. Beyond "more is better": Ecological footprint accounting for tourism and consumption in Val di Merse, Italy. *Ecol. Econ.* **2007**, *62*, 747–756.

17. Kytzia, S.; Walz, A.; Wegmann, M. How can tourism use land more efficiently? A model–based approach to land–use efficiency for tourist destinations. *Tour. Manag.* **2011**, *32*, 629–640. Li, H.; Hou, L. Evaluation on sustainable development of scenic zone based on tourism ecological footprint: Case study of Yellow Crane Tower in Hubei Province, China. *Energy Procedia* **2011**, *5*, 145–151.
18. Ponsioen, T.C.; Hengsdijk, H.; Wolf, J.; Ittersum, M.K.; Rötter, R.P.; Son, T.T.; Laborte, A.G. TechnoGIN, a tool for exploring and evaluating resource use efficiency of cropping systems in East and Southeast Asia. *Agric. Syst.* **2006**, *87*, 80–100.
19. Fernándeza, M.D.; Cagigal, E.; Vega, M.M. Ecological risk assessment of contaminated soils through direct toxicity assessment. *Ecotoxicol. Environ. Safety* **2005**, *62*, 174–184.
20. Critto, A.; Torresan, S.; Semenzin, E. Development of a site–specific ecological risk assessment for contaminated sites: Part I. A multi–criteria based system for the selection of ecotoxicological tests and ecological observations. *Sci. Total Environ.* **2007**, *379*, 16–33.
21. Bouyer, J.; Sana, Y.; Samandoulgou, Y. Identification of ecological indicators for monitoring ecosystem health in the trans–boundary W Regional Park: A pilot study. *Biol. Conserv.* **2007**, *138*, 73–88.
22. Pollino, C.A.; Woodberry, O.; Nicholson, A. Parameterisation and evaluation of a Bayesian network for use in an ecological risk assessment. *Environ. Model. Softw.* **2007**, *22*, 1140–1152.
23. Luijten, J.C.; Knappb, E.B.; Jonesa, J.W. A tool for community based assessment of the implications of development on water security in hillside watersheds. *Agric. Syst.* **2001**, *70*, 603–622.
24. Voinov, A.; Fitz, C.; Boumans, R. Modular ecosystem modeling. *Environ. Model. Softw.* **2004**, *19*, 285–304.
25. Bousquet, F.; Le Page, C. Multiagent simulations and ecosystem management: A review. *Ecol. Model.* **2004**, *176*, 313–332.
26. Robinson, J.; Bradley, M.; Busby, P.; Connor, D.; Murray, A.; Sampson, B.; Soper, B.W. Climate change and sustainable development: Realizing the opportunity. *Ambio* **2006**, *35*, 2–8.
27. Warhurst, A. Mining, mineral processing, and extractive metallurgy: An overview of the technologies and their impact on the physical environment. In *Environmental Policy in Mining: Corporate Strategy and Planning for Closure*; Warhurst, A., Noronha, L., Eds.; CRC Press LLC: Boca Raton, FL, USA, 2000.

28. Singh, R.K.; Murty, H.R.; Gupta, S.K.; Dikshit, A.K. An overview of sustainability assessment methodologies. *Ecol. Indic.* **2012**, *15*, 281–299.
29. Rasul, G.; Thapa, G.B. Sustainability analysis of ecological and conventional agricultural systems in Bangladesh. *World Dev.* **2003**, *31*, 1721–1741.
30. Bhandari, B.S.; Grant, M. Analysis of livelihood security: A case study in the Kali–Khola watershed of Nepal. *J. Environ. Manag.* **2007**, *85*, 17–26.
31. Siche, J.R.; Agostinho, F.; Ortega, E.; Romeiro, A. Sustainability of nations by indices: Comparative study between environmental sustainability index, ecological footprint and the energy performance indices. *Ecol. Econ.* **2008**, *66*, 628–637.
32. Liu, R.Z.; Borthwick, A.G.L. Measurement and assessment of carrying capacity of the environment in Ningbo, China. *J. Environ. Manag.* **2011**, *92*, 2047–2053.
33. Yuan, S.F. Study on regional land ecological security evolvement: A case of Hangzhou. *J. Camb. Stud.* **2011**, *6*, 30–40.
34. Martin-Cejas, R.R.; Sanchez, P.P.R. Ecological footprint analysis of road transport related to tourism activity: The case for Lanzarote Island. *Tour. Manag.* **2010**, *31*, 98–103.
35. Kinmen National Park Administration Office. Visitor Number Statistics. Available online: http://www.kmnp.gov.tw/ct/index.php?option =com_openinfo&view=openinfo&Itemid=313 (accessed on 5 February 2013).
36. The Equipment Management System of Taiwan National Park. The Statistical Table of Public Facility Classification, Kinmen National Park Administration Office. Available online: http://nppw.cpami.gov.tw /index.php (accessed on 10 February 2013).
37. Construction and Planning Agency, Ministry of the Interior. Visitations and Revenues of National Parks. Available online: http://np.cpami.gov.tw/chinese/index.php?option=com_statistic s&view=statistics&Itemid=182&gp=1 (accessed on 12 February 2013).
38. Tourism Bureau, M.O.T.C. Republic of China. Monthly Report of Home Stay Facilities. Available online: http://admin.taiwan.net.tw/statistics /month2.aspx?no=194 (accessed on 10 March 2013).
39. Council of Agriculture. Food Supply and Utilization. Available online: http://agrstat.coa.gov.tw/sdweb/public/book/Book.aspx (access ed on 6 March 2013).

40. Global Footprint Network. Ecological Footprint Atlas 2010. Available online: http://www.footprintnetwork.org/en/index.php/GFN/page/ecological_footprint_atlas_2010 (accessed on 18 March 2012).
41. United Nations World Tourism Organization (UNWTO). *Compendium of Tourism Statistics*; UNWTO: Madrid, Spain, 2001.
42. Rees, W.E. Getting Serious about Urban Sustainability: Eco-Footprints and the Vulnerability of 21st Century Cities. In*Canadian Cities in Transition: New Directions in the Twenty–first Century*; Oxford University Press: Toronto, ON, Canada, 2011; pp. 70–86.
43. Moore, J.; Kissinger, M.; Rees, W.E. An urban metabolism and ecological footprint assessment of Metro Vancouver. *J. Environ. Manag.* **2013**, *124*, 51–61.
44. Xiao, L.; Dong, L.L.; Lan, Y.X.; Zhao, X.G.; Wang, Y.M. Ecological security assessment in Jiangxi Province based on Ecological Tension Index. *Areal Res. Dev.* **2008**, *27*, 117–120.
45. World Economic Forum; Yale Center for Environmental Law and Policy; Center for International Earth Science Information Network (CIESIN) of Columbia University; Joint Research Centre of the European Commission. Environmental Sustainability Index, 2005. Available online: http://www.ciesin.columbia.edu/indicators/ESI/ (accessed on 9 August 2010).
46. Cui, Y.; Hens, L.; Zhu, Y.; Zhao, J. Environmental sustainability index of Shandong Province, China. *Int. J. Sustain. Dev. World Ecol.* **2004**, *11*, 227–233.
47. Meyfroidt, P.; Rudel, T.K.; Lambina, E.F. Forest transitions, trade, and the global displacement of land use. *Proc. Natl. Acad. Sci. USA* **2010**, *107*, 20917–20922.

Chapter 2

Preservation and Conservation through Cultural Heritage Tourism. Case Study: Musi Riverside Palembang

Tutur Lussetyowati

Department of Architecture, Sriwijaya University, Palembang, Indonesia

ABSTRACT

Palembang is the center of political and economic power in Southeast Asia in the past and was visited by foreign traders from China, Middle East and Persia. This paper reveals the potential of cultural heritage tourism in Palembang and shows the role of tourism in urban heritage preservation and conservation. Cultural tourism creates jobs and new business opportunities and strengthens local economy. Cultural heritage tourism helps also to protect cultural heritage and to improve the quality of life of residents and visitors. Linking tourism with heritage and culture will benefit the local economy. The main idea in cultural heritage tourism is to save urban heritage and culture, to share it with visitors, and to reach economic benefits.

1. INTRODUCTION

Palembang is the center for commerce, education and culture during the Sriwijaya Kingdom and Palembang Darussalam Sultanate until now. In the past during the Sriwijaya Kingdom in the 6-12th century, Palembang was the largest maritime empire, and the city had been visited by many foreign traders from China, Middle East and Persia. Many old buildings with various architectural style are located along the Musi River, which are clearly visible until today, such as Kampung Arab, Kampung Kapiten, Chinese Temple 10 Ulu, Sekanak Region, Pasar 16 Ilir, and Kampung Al Munawar. Unfortunately some heritage buildings are decaying although they are quite worth to become cultural heritage tourism.

The National Trust for Historic Preservation in the United States defines heritage tourism as "travelling to experience places and activities that authentically represent the stories and people of the past", and cultural heritage tourism is defined as "travelling to experience the places and activities that authentically represent the stories and people of the past and present." Cultural heritage tourism in Palembang is important because it brings positive impacts to the economic and social life of the city, and reinforces the identity of the city. The tourism will help preserve the cultural heritage and facilitate the harmony and understanding among people (Richards, 1996). The concept of tourism development in Palembang should support the maximizing of heritage potentials and the diversification of touristic objects and activities. The main priority should be providing new facilities for tourism and preserving the heritage by involving local people.

2. METHODOLOGY

This study is developed through explorative descriptive methods in which field study and interview with residents are included. The samples for interview are representative of local people who live in the heritage area. The in-depth interviews are conducted with adult members of the resident covering a wide range of issues regarding resident's view on heritage buildings and the benefit of cultural tourism to local people. Through the interview data on personal history, experiences and perspectives are collected. The research is focused on the contribution of cultural heritage tourism to the preservation and conservation of urban heritage.

3. FINDINGS

3.1 Cultural Heritage in Palembang

Culture heritage tourism in Palembang could not be separated from the existence of Musi River as a source of people's life and the starting point for the city growth. Musi River means a lot to local people as they rely on it for transportation system connecting one area to the other between Seberang Ulu and Seberang Ilir. In Seberang Ulu, most of heritage buildings are residential and public facilities such as mosques and temples. The area is characterized as multi-ethnic settlements of Malay, Arabic and Chinese. Some ethnic neighbourhood such as Kampung Arab, Kampung Kapiten, and 3-4 Ulu are identified as the origin of settlements in Palembang. The architectural style and authenticity of settlements along the Musi riverbank made this area unique and important to be preserved as a cultural heritage. The existing artefacts provide an ideal setting for conveying story of the past to the younger generation.

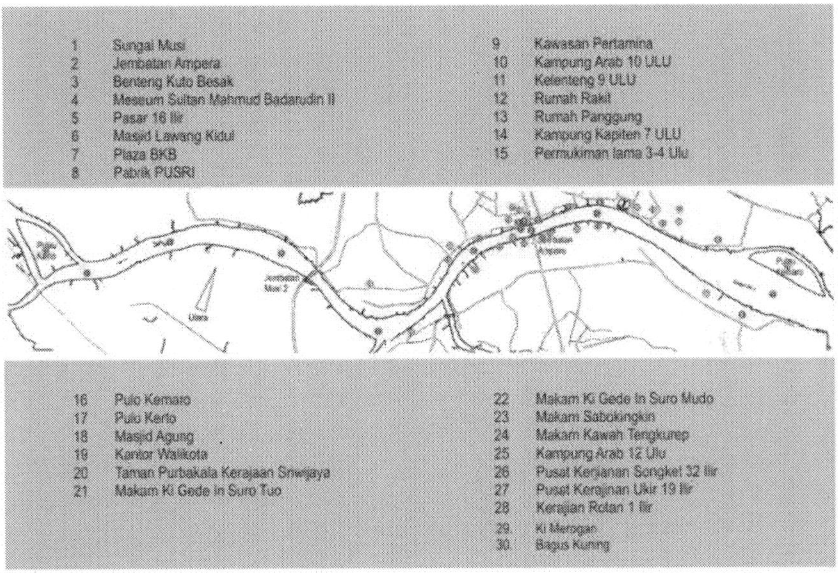

Figure 1. Map of Palembang heritage tourism. Source: Riverside Tourism Masterplan.

At first, the 9-10 Ulu area is habitated not by indigenous people but by Arabic, Persian and Chinese migrants. They settled down as the trade was getting increased in quantity and intensity in that they must stay longer in land and finally get married to local people. People built at first shelters on raft, and later on improved the raft house to a stilt house. At that time people tend to live in a group maintaining their origin culture and tradition, leaving the traces of two types of settlement: the Arabian and Chinese.

Houses in Kampung Arab can be differentiated as stilt houses and landed houses whereby people take over the traditional achitecture or local style for their houses and decorated it with Arabic ornaments. A variety of house style is found in this Kampung: limas house, gudang house and raft house. At the center of the settlement is an open space, which is used as a public space, surrounded by several houses. This pattern of settlement is quite different to the local pattern of indigenous settlement. This typology of settlement becomes a uniqueness that has been brought over by Arab community. The Chinese settlements are established around the temple Chandra Nadi. Houses are built in Chinese architecture that also differs from local architecture.

Figure 2. (a) Arabian Settlement 10 Ulu; (b) Chinese Settlement 10 Ulu.

3.2 Kampung Kapitan 7 Ulu

Kampung Kapiten is a group of 15 stilt houses which is located in the Village 7 Ulu. The village was the residence of a Chinese officer, is called Kapiten, who worked for the Dutch colonial government. At that time, Kapiten was the person in charge for collecting taxes from Chinese people and indigenous people in Seberang Ulu district. The Kapiten is also responsible for keeping the security and managing the trade system in the district.

In the core area stand three houses, considered as the largest buildings in the area, that are built facing the river Musi. The house in the middle was used for organizing parties and meetings with people in great number, while both houses at the east and west side are used as dwellings. The middle house applied a mix of culture in its architecture; Malay, Chinese and colonial style are combined into a hybrid architecture. The spatial layout is following the Chinese style, in which a typical inner court is set up as core of the house which is quite different to local style of architecture. The house has a roof form of traditional limas house, yet, in front of the building pillars in european architecture are erected and decorated by Chinese ornaments. An open space at the opposite of the houses, approximately 83 m x 12.5 m and called as Plaza Kampung Kapitan, becomes the point of orientation for the houses. The plaza is used for festivals and other public activities until now.

3.3 Traditional Kampong in 3-4 Ulu

One of the original traditional settlements in Palembang is located in 3-4 Ulu area. Houses in this settlement are large and built in limas style. Unfortunately, many of them could not retained its original state, some are sold and the owner moved out from the district.

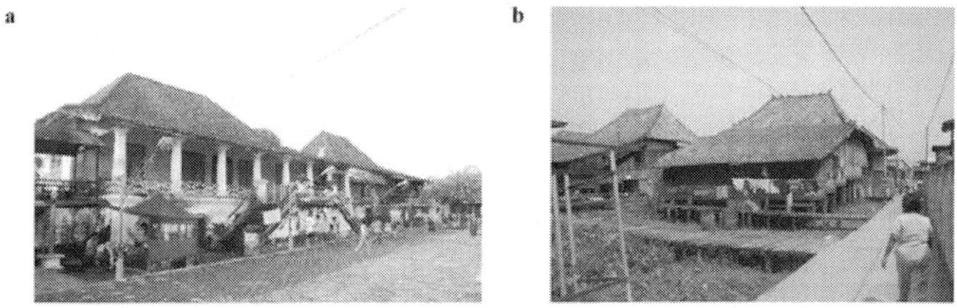

Figure 3. (a) Kampung Kapiten, 7 Ulu; (b) Limas House, 3 Ulu.

There are some religious buildings such as great mosque and temples that are worth to be preserved as cultural heritage. The Great Mosque, formerly called the Sultan Mosque, is built by Sultan Mahmud Badaruddin I (Jayo Wikramo) in 1738 AD and inaugurated in 1748M. The oldest temple Soei Goeat Kiang was built in 1733 as a replacement for the first temple in Ulu 7 which is burned down in 1732. The temple is used by the Buddhist, Taoist and Confucian. The other famous temple in Palembang is located at Pulau Kemaro in Musi River, which is nowadays being used as ceremonial and festival place for the Chinese people.

Figure 4. (a) The Great Mosque; (b) Soei Goeat Kiang Temple; (c) Kemaro Island.

Palembang has many interesting and attractive urban heritage but not enough initiatives from the government and other stakeholders in preserving and promoting them for cultural tourism. Major problems in facing this issue is the lack of people's awareness and attention in maintaining heritage buildings and the buildings are surrounded by dwellings of poor people that discourage tourists to come by. The planning for tourism development has been done, yet, the implementation process had to face funding resources.

3.4 Tourism as a tool for preservation and conservation of urban heritage

Tourism and culture have been developed worldwide as a way to revitalize obsolete buildings and open spaces for contemporary purposes, in that it provides opportunity to sustain traditional culture and contemporary values. Abandoned buildings and spaces are always at risk to be decayed or redeveloped, and not considering them as in critical situation leads to a loss of opportunity in building up the identity of a community and a place.

Cultural heritage tourism has a number of objectives that must be met within the context of sustainable development, such as conservation of cultural resources, accurate interpretation of resources, authentic visitor experience, and stimulation for revenues of cultural resources. The development of cultural heritage tourism deals not only with identification, management and protection of heritage values, but also with its impact on communities the regions, economic and social benefit, provision of financial resources, as well as marketing and promotion. (J. M. Fladmark, 1994). It is also important to think about the impact on indigenous people whose land and culture is visited by tourist. In the case of Palembang, the local people living in the neighbourhood of Kampung Kapiten and Kampung Arab are low-income people who need to be actively participated in tourism activities, so that they may get some benefit from it.

Preserving and protecting the heritage is essential for cultural heritage tourism because the commodity to be offered has to be secured for its sustainability. For this reason, the major challenge in tourism programs is ensuring that increased tourism does not destroy the very qualities that attract visitors in the first place. The increasing number of visitor and additional pressure to infrastructure and heritage sites are some problems on tourism. As a service industry, tourism programs depend on the competence of people in different jobs and location. Tourism is a highly sophisticated and fast-changing industry, therefore, it presents challenges that are not only from visitor impact, but also from visitor expectations on quality of products and services.

3.5 Strategy for cultural heritage tourism development in Palembang

Acoording to Partners and Tourism (2011) in Getting Started: How to Succeed in Heritage, there are some strategies to make cultural heritage tourism successful:

- *Collaborate*

Collaborate means that it is better to work together than to work alone. Much more can be accomplished by working together. Building partnerships is essential, not just because they help to develop local support, but also because tourism demands resources that no single organization can supply. Its success depends on the active participation of political leaders, business leaders, operators, artists, craftspeople, hotel operators, and many other people and groups. Regional partnerships are also useful to cultural heritage tourism. Cooperating in a regional arrangement will develop regional themes, pool resources, save money and expand marketing potential. Those resources include not only money for marketing campaigns, for example, but also facilities (accommodations for travelers) or expertise in tourism, preservation, and the arts.

- *Find the fit*

A few things to make cultural heritage tourism successful has to be considered, such as balancing the needs of residents and visitors, and the fit between residents and visitor expectation. It is important to ensure that cultural heritage tourism benefits everyone.

- *Focus on quality and authenticity*

Quality is an essential ingredient for all cultural heritage tourism, and authenticity is critical whenever heritage or history is involved. Existing heritage buildings must be preserved in order to maintain the authenticity in Palembang heritage area. The authenticity that is unique and charming will add real value and appeal to visitor. The true story of Kampung Kapiten, Kampung Arab, and Pulo Kemaro is worth to be explored for tourism. The story of previous generations that have made to the history and culture is interesting for visitor, because that is what distinguishes the place from every other place on earth.

- *Preserve and protect*

The cultural, historic, and natural resources of a community are valuable and often irreplaceable. The historic and cultural assets are important for tourism, and it is essential to protect them for the long term. The preservation and perpetuation of traditions is important to telling the story of the people who settled the land.

4. CONCLUSION

Palembang has many heritage buildings, but unfortunately, some are not in a good condition and some tend to decay although they are potential to be objects for cultural heritage tourism. Many urban heritage in Palembang are interesting and attractive, yet, supporting activities are not available to the preservation and promotion of cultural heritage tourism.

Linking tourism with heritage and culture will be a benefit for the local people and the local economy. The main idea in cultural heritage tourism is to save urban heritage and culture, to share it with visitors, and to get economic benefits from tourism. Some strategic steps to develop cultural heritage tourism in Palembang are to collaborate, to find the fit, to focus on quality and authenticity, to preserve and to protect.

REFERENCES

1. Fladmark, J.M., Habgood, J. (1994). Cultural Tourism: Papers Presented at the Robert Gordon University Heritage Convention, 1994. London: Donhead Publishing.
2. Hanafiah, D. (1989) Citra Palembang Tempo Doeloe. Jakarta: PT Karya.
3. Richards, G. (1996) Production and consumption of European Cultural Tourism. Annals of Tourism Research 23(2), 261-283.
4. National Trust for Historic Preservation in The United States. (1993). Getting Started: How To Succeed in Heritage Tourism. Washington,DC: National Trust for Historic Preservation.

Chapter 3

Sustainable-Responsible Tourism Discourse – Towards 'Responsustable' Tourism

Tanja Mihalic

University of Ljubljana, Faculty of Economics, Slovenia

ABSTRACT

Despite several decades of academic and practical debate on tourism sustainability, its application in practice remains difficult. The dominant tourism discourse on *sustainability*(theory, seen as a concept) and *responsibility* (practice, understood as appropriate action) calls for a solid understanding of the process of how a responsible destination actually implements a sustainability agenda, which this paper aims to provide. In this context, we explore theoretical perspectives from political economics and behavioural economics to offer a well-reasoned integrated sustainability–responsibility model comprising three stages: Awareness, Agenda and Action. This Triple-A Model complements the sustainability indicators debate and provides advice on how to continuously implement the sustainability concept and move from market-value-led and environmentally laissez-faire tourism towards more environmental- and social-value-driven responsible tourism. In addition, this paper discusses the existing sustainability and responsibility nomenclatures and their use and contributes relevant conclusions on the current understanding of sustainability and responsibility in European and UNWTO practices. The term *responsustable tourism* is suggested to join two existing terms and demonstrate that the current understanding of responsible tourism behaviour is based on the concept of sustainable tourism.

Keywords: Sustainable tourism; Responsible tourism; Responsustable tourism; CSR; Behavioural economics; Triple-A Model

1. INTRODUCTION

Concern over the natural and social environments has generated research debate on tourism–environment relationship. This debate gained momentum in the early 1970s when George Young argued that the impacts of tourism are both a blessing and a blight (Young, 1973) and Claude Kaspar, a Swiss-based tourism researcher, called for a new "dimension of [the] tourism debate" (Kaspar, 1973, p. 139), which he termed*environmental ecology*. Later, Swiss ecologist Jost Krippendorf (1984) challenged the sense of mass tourism in his book entitled "Vacation People" ("Ferienmenshen" in German, translated as "Holiday Makers" in English (Krippendorf, 1987)) and began the search for alternatives. Tourism critics in the 1980s called for "... more responsibility for the effects of travel and behaviour on host environments, both physical and human" (Butler, 1995, p. 5). This interest in "more responsibility" led to so-called alternative tourism forms and concepts which have been given many names, such as alternative, soft, quality, eco, responsible, minimum impact tourism, green and ethical tourism, with all of them representing an alternative to the mainstream mass tourism that has been becoming environmentally, socially, ethically and politically intolerable (Mihalič, 2006 and Swarbrooke, 1999). In general, global environmental concern culminated in "Our Common Future", as defined in the Brundtland report (WCED, 1987), which brought the global environmental debate and the notion of sustainability to the forefront of global and local social and political thinking and agendas. The tourism sustainability debate, e.g., the debate regarding use of the term *sustainability*, following the "Our Common Future" legacy began in the early 1990s with Edward Inskeep (1991) who defined five main criteria for *sustainable tourism*, which addressed the economic, environmental and social responsibility of tourism as well as its responsibility towards tourists (visitor satisfaction) and global justice and equity. Some of his criteria received little recognition in the following debate, which also originated in the Brundtland legacy and culminated in the next decade. The United Nations' organisations, including the UNWTO, primarily supported the three-pillar (environmental, socio-cultural and economic) concept of sustainable tourism. This concept became the focus of mainstream academic tourism literature and programmes and the input of many tourism strategies and policies, resulting in a recognised global trend towards sustainable tourism development. However, sustainable tourism research, documents and actions have recently been increasingly accompanied by the notion of responsible tourism. Examples include the new European document entitled the "Charter for Sustainable and Responsible Tourism" (TSG, 2012), and two recent books called "Responsible Tourism" (Leslie, 2012a) and "Taking Responsibility for Tourism" (Goodwin, 2011).

The trend towards sustainability has been studied and accepted by many researchers. On one hand, the sustainability concept has served for some as a magic wand pointing towards more sustainable, environmentally and socially friendlier tourism developmental models and forms (Ritchie and Crouch, 2000 and Swarbrooke, 1999). On the other hand, the concept has been persistently criticised for being flawed and inadequate (Higgins-Desbiolles, 2010). In other words, the sustainability discussion has helped draw attention to the need for a balance between economic and environmental interests in tourism. Its actual penetration into strategies and policies has resulted in many good practices and improvements such as energy savings, recycling, a reduction of waste and emissions and attempts to improve the livelihood of the local population. However, there is also significant evidence of the opposite effect. Wheeler argues that the "intellectually appealing" concept of sustainable tourism has little practical application because it has turned into a public relations tool for addressing the criticism of the impact of tourism while allowing essentially the same behaviour as before (Wheeller, 1993, p.121). Indeed, a consensus on the efficacy of sustainable tourism development remains elusive (Chettiparamb and Kokkranikal, 2012), and its implementation in practice remains difficult, leaving much of the tourism industry "... alarmingly unsustainable" (Higgins-Desbiolles, 2010, p. 117).

Tourism stakeholders are applying sustainability practices at a slow pace. Further, some destinations might inaccurately promote themselves as sustainable and increase the expectations of new tourists (Poon, 1989), who are then confronted with the gap between the actual and promised (advertised) sustainability. Extensive lists of sustainable tourism indicators have been created to measure this gap and actual sustainability (EC., 2014,ECETAT and ECOTRANS, 2004, TSG., 2007 and UNWTO., 2004). However, although these lists enable actual progress in sustainability performance to be monitored while some (aside from the three-pillar approach) measure political sustainability and customer satisfaction (EC., 2014 and ECETAT and ECOTRANS, 2004), they do not provide a measurement tool that can help destinations understand the overall transition process regarding sustainability and responsibility. For this reason, a tool to understand, measure and monitor the process of implementing sustainability is still needed. This is the primary purpose of this paper.

The existence of this paper has been provoked by two inter-related facets of the above debate. The first relates to the never-ending search for new tourism terms on the assumption that a new term will bring *more responsible tourism*. In this context, interest has been triggered by the recently growing popularity of the term *responsible tourism* in the tourism literature (Goodwin, 2011) as well as in consulting and political and business practices (TSG, 2012). The actual and academic coexistence of these two key terms can also be well illustrated by the titles of scientific conferences discussing sustainable-responsible tourism development. For example, in early October 2013, on the same days, two conferences were held on the issue, one in Istanbul, Turkey and the other in Barcelona, Spain. The first conference tried to attract participants through the conference title "sustainability" (International conference: Sustainability issues and Challenges in Tourism, 2013), the other by discussing "responsible tourism" (RTD7 Conference: Responsible Tourism in Destinations. Barcelona – Catalunya, 2013). Questions arise as to what does responsible tourism bring to the tourism debate in terms of terminology, concept and tourism type, and how does it relate to sustainable tourism.

The second facet relates to the gap between the appealing conceptual idea of sustainable tourism and its alarmingly slow penetration of action and practice, which is obviously connected to *tourism irresponsibility* or irresponsible tourism behaviour. In the context of both facets, this paper is interested in the penetration of sustainability and responsibility in the tourism industry and destination practices. Being a conceptual paper, its aim is to integrate existing theoretical and practical understandings and uses of the notions of sustainable and responsible tourism and provide new theoretical perspectives for their well-reasoned and coherent understanding. It examines the theories on the causes of environmental damage and the transition of society to a state in which environmental issues can no longer be ignored, as outlined by the Swiss welfare economist Bruno Frey (1985). His theory is applied to tourism to increase the understanding of the sustainable-responsible tourism discourse and current developments in use of the terms sustainable and responsible tourism. The increased use of the term *responsible tourism* is discussed, and an attempt to connect it with*sustainable tourism* is made. Paradoxically, the criticism of the above-mentioned never-ending renaming of tourism has led to a new tourism term. The term *responsustable tourism* has been suggested, not to offer a new tourism type or concept but as an attempt to join two existing terms to properly articulate the current *responsible tourism* debate, which is based on the concept of *sustainable tourism*.

In order to follow the primary purpose of this study, there are three specific aims: first, to understand the notions of responsible and sustainable tourism in a historical, theoretical and practical context; second, to provide a logical model to accommodate the above understandings; and third, to develop a tool to allow actual understanding and implementation of the sustainable tourism concept in a (responsible) destination.

Accordingly, the section "Introduction" explains the relevance and importance of the study. The next section explains the nature of the paper and the methodology applied. The state of the art on the sustainable-responsible tourism discourse is then presented and theoretical and practical evidence of its existence is given. Based on this discourse, the next section provides a model for responsustable tourism. Results and a discussion of the current and a proposed understanding of the sustainable-responsible tourism discourse follows, while the paper finishes with a section outlining conclusions and further research.

2. METHODOLOGY

In line with the standards for conceptual or review articles (Watts, 2011), this paper attempts to further expand and refine the understanding of the sustainable-responsible tourism discourse and suggests how to close the sustainable-responsible gap by using clear definition of each term, derived from theories on environmental damage and behavioural economics.

However, although primarily a conceptual paper, the paper's construction is informed by some explorative methods in theorising and conducting research. In this context, the paper applies engaged scholarship and action research methodology.

Engaged scholarship research emphasises advances in scientific and practical knowledge (Van de Ven, 2007) which fits the theoretical and practical nature of the present research. Being based on ongoing research and academic thinking on sustainable and responsible tourism development, this research engages academic knowledge. Further, by deriving from the current practices in strategy and policy formulation and the implementation of sustainable and/or responsible tourism, it engages with reality. Both are iterated to fit the proposed model.

In addition, the paper employs an action research methodology. Zuber-Skerritt (1996)suggests that the aim of any action research project is to bring about the practical improvement or development of social practice, and practitioners' better understanding of their practices. A key feature of this action research approach is that it involves the investigation being undertaken by the person directly concerned with the social situation under consideration (McGugan, 2002), here derived from the researcher's experience with academic research and the creation and implementation of national and European sustainable tourism strategies and policies. It means that this methodology not only focuses on the transfer of knowledge and dissemination for the purpose of adding to the existing body of tourism knowledge but, most importantly, critical thinking about this knowledge and the co-creation of new knowledge are enhanced by learning from both sides (academic and practical), whether through academic papers, conferences or work experience with the tourism industry and public and private sector in designing sustainable tourism development strategies, agendas and monitoring the implementation process.

The sustainable-responsible tourism discourse is brought to a close by the proposed Triple-A Model which joins sustainable and responsible tourism to form responsustable tourism. Some metaphoric examples of climbing a mountain are used by way of inspiration to interpret, visualise and concretise the theoretical model, its phases and penetration processes. The result of this conceptual thinking is jointly summarised in the section entitled *Framework for understanding the sustainable-responsible discourse*which fully illustrates how to shape that discourse.

3. THE SUSTAINABLE-RESPONSIBLE TOURISM DISCOURSE

In one study, Higgins Desbiolles (2010) claimed that a dominant issue in the current tourism discourse is the elusiveness of achieving sustainability. Although she argues that the sustainability concept is not adequate, the discourse she refers to actually relates to the low application of sustainability principles in the tourism industry, which "... remains alarmingly unsustainable" (Higgins-Desbiolles, 2010, p. 117). For much of the tourism industry, overall sustainability practices seem as distant as the mighty Mount Everest seems to mountain climbers. Nevertheless, Everest is on almost every professional climber's dream agenda and, although many alpinists have conquered it so far, the majority of mountain climbers have not yet – and never will – reach even its base camp for a wide variety of reasons. In this context, this paper further expands the tourism discourse by distinguishing between dream and reality agendas and actions, e.g., between sustainable and responsible tourism, which is observed as a discourse between the concept and its actual application.

Thus, the first theme of this section is sustainability as a concept. Tourism researchers (Bramwell and Lane, 1993) have traced the theoretical origins of sustainable tourism back to a publication in 1973 (Dasmann et al., 1973) and its political penetration in 1980 in the World Conservation Strategy (IUCN, 1980). As already pointed out in the "Introduction" section of this paper, the drive towards sustainable tourism has been prompted by concerns about the tourism–environment relationship as well as interest in developing concepts that might help bring this relationship more in line with the ideology of "Our Common Future" (WCED, 1987). The debate is focused on the new developmental principles based on the concept termed *sustainability*. It is beyond dispute that sustainable tourism development has been widely recognised as a *concept*, meaning "... tourism that is based on the principles of sustainable development" (UNEP & WTO, 2005, p. 11). These principles, also known as the three pillars, demand consistency with economic as well as socio-cultural and environmental objectives and enable tourism operations to remain competitive in the long run. In this context, some academics refer to economic, socio-cultural and environmental sustainability. However, feeling that it is potentially critical for the acceptability of any solution, some critics tried to integrate a fourth dimension – political sustainability (Ritchie and Crouch, 2000). Yet since it was impossible to define this political dimension as the fourth pillar, because – unlike the economic, socio-cultural and environmental pillars – it does not relate directly to the impacts of tourism, the fourth pillar of the sustainability concept has never materialised. However, parallel to this debate, some addressed the political dimension through a debate on three requirements that must be satisfied to make tourism sustainable (Fig. 1). First, sustainable tourism is based on an awareness of sustainability and ethics, supported by environmental education and information for all stakeholders on both the demand and supply sides. Second, it also relates to the above-mentioned political dimension as it requires the informed participation of all relevant destination stakeholders, a consensus, a critical mass and strong political leadership to enable its implementation. Third, sustainable tourism should maintain a high level of tourist satisfaction, thereby meeting market needs (Mihalič, 2013 and UNWTO., 2004). More specifically, in existing market economies the latter also assumes a willingness to pay for sustainability.

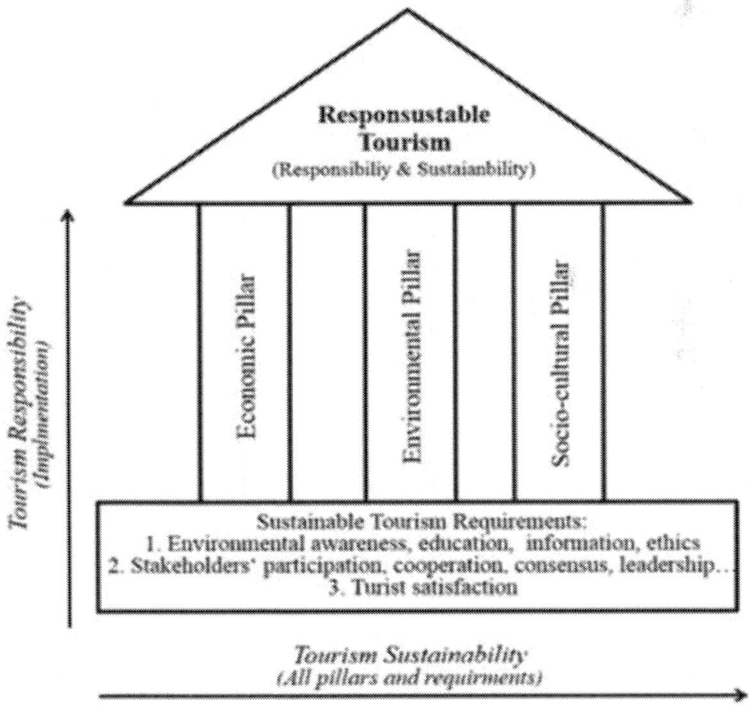

Figure 1. Proposed understanding of responsustable tourism (based on the three pillars and three requirements).

The three-pillar concept has inspired most sustainability research and penetrated numerous strategy and policy documents in the real tourism sector. However, the question remains: to what extent do sustainability principles improve the sustainability of the real tourism sector?

Notably, the tourism sustainability debate has focused on the destination level rather than the micro-level of the firm (Dwyer, 2005). Empirical evidence from the hospitality industry has shown that economic performance is the highest priority, while environmental performance is the lowest priority (Blackstock et al., 2008, Bohdanowicz et al., 2005, Bramwell et al., 2008 and Mihalič et al., 2012). It appears that tolerating and accepting sustainability agendas has been easier for public stakeholders than for private stakeholders, especially corporate businesses since they traditionally focus on economic goals. More specifically, firms have been much slower to discuss or address the environmental and socio-cultural aspects of their business practices and reluctant to talk about *sustainability*. However, other sustainability-coloured concepts such as the *triple bottom line* (TBL) or *corporate social responsibility* (CRS) seem to be closer to the business world. Reluctance to talk about sustainability has also lead to the use of notion*green tourism*, as a synonym.

Therefore, the second theme of this section involves identifying why companies have accepted CSR models more widely. To understand how firms perceive responsibility, two alternative lines of thought known as the customer and broader stakeholder orientations of the firm are relevant (Heikkurinen and Bonnedahl, 2013). According to first line of thought, the firm is market-oriented and "its social responsibility ... is to increase its profits" (Friedman, 1962, citied in van Marrewijk, 2003, p. 97). Profit is the focal point of the company, while social responsibility and actions lie in the domain of the government. In this regard, companies are interested in CSR behaviour only to the extent that it contributes to the business goal, which is ultimately driven by profit motives (Chettiparamb and Kokkranikal, 2012). This market-oriented view can be demonstrated by the argument frequently used by the tourism industry to counter larger environmental pro-activeness. Firms often claim that low consumer demand and low willingness to pay extra for environmentally friendlier products and responsible tourism practices is an insufficient incentive to improve environmental performance (Budeanu, 2005). Market orientation explicitly focuses on the creation of superior value for buyers and thus superior performance for the business and distinguishes between economic responsibilities and other responsibilities such as legal and ethical responsibilities. It prioritises economic utility, not sustainable development (Heikkurinen and Bonnedahl, 2013).

The alternative line of argument regarding a company's behaviour is its modern form which targets all three sustainability pillars, more often called business lines: the economic, environmental (e.g., natural) and social lines. Based on this understanding, CSR is now understood as a business' contribution to corporate sustainability (van Marrewijk, 2003). This corporate understanding corresponds to the societal approach to CSR which recognises that companies are an integral part of society and thus responsible to society as a whole. It assumes that the company considers a broader set of stakeholders (not only its customers) in its behaviour. The ethical behaviour of a firm reflects the concerns of its consumers, employees, shareholders and the community (Caroll, 1991). The stakeholder orientation focuses on the broader responsibility of a firm through its stakeholders who affect the business, thus incorporating all three sustainability pillars. It does not draw a strict line between the firm's economic responsibilities and other responsibilities like ethical responsibilities.

Further, the firm's reluctance to talk about sustainability is illustrated by another practice example of John Elkington who started his UK-based private consulting company, SustAinable. He invented a new terminology to sell sustainability business practices to corporate and private business. He named this new approach the *triple bottom line* (TBL) (Elkington, 1997). The TBL is just another way of wording the three pillars of sustainability, yet it is more strongly accepted by corporate stakeholders. The TBL concept has penetrated some of the tourism debate, but much less so than the sustainability concept.

A similar argument can be made for the notion of *green tourism* which organisations have in many cases used as a synonym for sustainable tourism. More specifically, the British Green Tourism Business Scheme claims "Green tourism is sustainable tourism ..." (GT, 2014). The same sentiment can be found on the web pages of other national tourism organisations as the term green tourism is often more accepted by different stakeholder groups than that of sustainable tourism (Forquan et al., 2010 and STO., 2013). For example, after trying to promote sustainable tourism for many years with a very limited response from tourism business sector, Slovenian national tourism organisation strategically introduced *green tourism* as a synonym for sustainable tourism in 2009 and succeed to mobilise diverse tourism stakeholders to join the efforts under the slogan "Slovenia is green, Slovenia acts green, Slovenia promotes green" (Mihalič, 2013 and STO., 2013), targeting all three sustainability pillars.

From a modern perspective, the debate on the differences between sustainable, CSR, TBL or green tourism seems unproductive because all converge and relate to the same pillars. However, from the aspect of business and global market reality, which delineates the roles of business, governments and civil society, the separated classical shareholders business approach still needs attention. The CSR or TBL notion seems to be more strongly rooted in the world of business corporations, and the notion of sustainability in the public sector and sustainable tourism destination development. This fact alone might be evidence of the gap between the differences presented above between an orientation towards shareholder profit and one towards wider societal responsibility and is supported in the context of the above separation.

Although there is some evidence that responsibility concepts in from of CSR models are applied to private and public sector tourism practice, the notion of responsible tourism has evolved separately. It has been present in tourism in some form since the early 1980s and is connected to many studies on environmental problems in tourism (Bramwell et al., 2008, Butler, 1995, Chettiparamb and Kokkranikal, 2012, Goodwin, 2011, Leslie, 2012b and Leslie, 2012c). Contrary to the relative consensus on the three-pillar meaning of sustainable tourism, observed as a concept, the notion of responsible tourism (or CSR) has not been as strongly established or as frequently applied.

According to some authors, responsibility and responsible tourism can mean "anything" (Chettiparamb and Kokkranikal, 2012, p. 302; van Marrewijk, 2003), and its use adds nothing to the conceptual understanding of tourism. The existing literature and documents on responsible tourism reveal many understandings that: vary regarding whether the subject is the concept, theory or practice (Leslie, 2012a, Leslie, 2012b and Leslie, 2012c); involve different stakeholders and different aims; refer to social or polar tourism (ICRT, 2011); or create environmentally friendlier tourism or better places for people to live and visit (RTP, 2002). Further, the existing literature focuses on responsible consumption or production (Budeanu, 2005 and Stanford, 2008), explores actor relations such as the tourist-host population, and questions individual versus social responsibility as well as the political assumptions and governance models behind the responsibility (Bramwell et al., 2008 and Hall, 2012).

In addition to these quite diverse research interests, it seems that the use of *responsible tourism* often occurs when Krippendorf's legacy (1987) of a new understanding of tourism is emphasised (Bramwell et al., 2008 and Goodwin, 2011) and when questions of ethical and moral responsibility are brought to the fore (Blackstock et al., 2008, Bramwell et al., 2008 and Fennell, 2006). Moreover, it occurs when the *responsibility* of tourism needs to be stressed as some implementation difficulties have been identified and a sustainability agenda is being created. In this regard, we claim that the notion of *responsibility* relates to responsible *behaviour and action*.

This behaviour- or action-based view of responsible tourism has been supported by the work of many researchers. The guru of the responsible tourism movement, HaroldGoodwin (2011), claims that:

"The idea of Responsible Tourism has at its core the imperative to take responsibility, to take action; consumers, suppliers and governments all have responsibilities. The ambition of Responsible tourism is to address the impacts of mainstream tourism, to enhance the positive and to reduce the negative ... (p. 2) .... *Responsible Tourism is about everyone involved taking responsibility for making tourism more sustainable"* (p. 31).

Goodwin clearly links responsible tourism to action in favour of making tourism more sustainable. His understanding of the concept of responsibility assumes three aspects: accountability, capacity to act, and the capacity to respond. It is the third aspect, response-ability that directly addresses tourism behaviour and involves entering into a dialogue, creating solutions and acting to make tourism more sustainable.

This behaviour-based view has also been supported by David Leslie (2012b). In his meaningfully titled book chapter, "The Responsible Tourism Debate", he argues that responsible tourism is *"a behavioural trait ... based on the basic principles of respect for others and their environment..."* (p. 20).

This view of responsible tourism enabled Leslie to go further and note that responsible tourism assumes "acting responsibly in terms of one's own actions, and moreover, in the management and operation of business" (p. 20). When faced with the dilemma of a conceptual basis for judging responsibility, he refers to the ethical and environmental criticisms of tourism and suggests behaving "... environmentally and/or ethically responsible" (p. 20).

A step further has been taken in the research by Chettiparamb and Kokkranikal (2012), meaningfully entitled "Responsible Tourism and Sustainability". The authors also conceive of responsible tourism as behaviour-based and argue that the notion is linked to sustainability initiatives, which gives a solid conceptual substance to their understanding of responsible tourism. In their research, they progress to the CSR concept and discuss a tourism-specific manifestation of the private sector CSR agenda, which focuses on *"... facilitating responsible behaviour ... through CSR-related practices targeting wider socio-economic issues"* (Chettiparamb and Kokkranikal, 2012, p. 307).

Facilitating responsible behaviour creates many questions. There are questions of whether the environmental concern of the tourism industry results in more responsible behaviour and which practical measures and policies encourage and enable greater responsibility (Bramwell et al., 2008 and Goodwin and Francis, 2003). Another important question is who the relevant actors are, who should make the relevant choices and what type of philosophy and motivation underpins the responsible tourism paradigm (Fennell and Malloy, 2007). They are all relevant to our tourism discourse as each adds a piece of evidence from a different perspective, demonstrating the discourse's complexity and differentiation, demanding that fragmented issues be placed within a common framework.

Further, many action-based tourism documents can be listed to illustrate the more practice-based use of the notion of responsible tourism. Although the term responsible tourism appeared in many older documents, the first declaration on responsible tourism was published in 2002 ("Cape Town Declaration on Responsible Tourism," 2002). In 2012, the UNWTO Ethics Committee decided to introduce and apply the term*responsible tourism* as part of the new logo for the "Global Code of Ethics for Tourism" (Diotallevi, 2013). Strongly focusing on sustainable development and adopted at the WTO General Assembly in 1999, the Code was given a new cover page and the subtitle "For Responsible Tourism". It is also expected that the UNWTO

Secretariat and Committee will work on proposing an official definition of responsible tourism (Diotallevi, 2013).

In addition, the new "European Charter for Sustainable and Responsible Tourism" (TSG, 2012) supports our argument. The charter is built on an old Tourism Sustainability Group (TSG) document (TSG, 2007) which dealt exclusively with the notion of sustainable tourism ("Action for More Sustainable European Tourism"). However, five years later, the new document uses the terms sustainable and responsible tourism but omits the word action. "Action" is now incorporated in the broader understanding of the notion of responsible tourism, as the latter is defined as:

awareness, decisions and actions of all those involved in the planning, delivery and consumption of tourism, so that it is sustainable over time (TSG, 2012: 2).

Bearing this definition and the above discussion in mind, responsible tourism is not a synonym for sustainable tourism. Responsible tourism addresses the aforementioned sustainable tourism discourse in implementation and is more of an expression to describe tourism that is sustainable because it acts sustainably. It is nothing new, neither a new alternative to other tourism forms nor a new way of "doing" sustainable tourism, as some authors claim (Blackstock et al., 2008, p. 279). Thus, responsible tourism builds on appropriate sustainability-based strategies and policies and adds appropriate behaviour, meaning sustainable (re)actions or, according to Goodwin's concept, response-ability, underpinned by tourism environmental awareness and ethics (Fennell, 2006). This sustainability–responsibility connection is illustrated in Fig. 1 in which the pillars of sustainable development stand upon the sustainable tourism requirements previously presented in the section of the paper on the "Sustainable-responsible Tourism Discourse". Indeed, the roof of the Fig. 1 structure should be termed *responsustable tourism*, demonstrating both responsibility and sustainability in tourism behaviour.

Of course, to answer the question regarding to what extent sustainability principles have improved the real tourism sector's responsustability, tourism sustainability indicators might be of assistance. In tourism research and applied practice, many models can be found, and their use is quite popular. Indeed, sustainable tourism indicators can be observed as tourism responsibility measures and monitoring that support environmentally friendlier tourism management and practice (Blackstock et al., 2008). However, the sustainability indicator schemes are very heavily measurement- and thereby result-oriented, and the majority fail to understand that achieving sustainability in a sector should be understood as a social transformation process and that old sustainability indicator vocabulary might not be sufficient at this point in time. This paper attempts to add to this perspective.

4. MODELLING RESPONSUSTABLE TOURISM

4.1. Environmental damage theories

Returning to our initial discussion, which addressed our concern about environmental damage and tourism, the question of the occurrence of environmental damage becomes relevant. Many researchers have tried to explain why "Our Common Future" awareness is not incorporated into our current behaviour. According to these scholars, the answer is threefold (Frey, 1985 and Mihalič and Kaspar, 1996). Environmentally improper behaviour is first observed as a result of the non-functionality of the global laissez-fair capitalism market system and as related to the problems of public goods, property rights and externalities. Second, the same problem can be explained based on economic and population growth, which results in overproduction and the overconsumption of resources. The final explanation blames improper behaviour based on human ignorance and/or the absence of environmental awareness and ethics.

All of the above theories are relevant to our discourse. Regarding the first theory, the environmental goods are wrongly treated as a public goods, as they don't have an owner, are assumed to be plenty and are non-excludable in their consumption. Consequently, these goods are used (and overused and polluted) freely regardless of a user's participation in the related costs. Theory sees the problem in environmental goods, being external to the market allocation mechanism which has the ability to determine the price, through the confrontation of demand (willingness to pay) and supply (willingness to sell). Thus environmental, e.g. external goods must be internalised to our current neo-liberal market allocation system in order to be properly allocated. In theory, such a system would guarantee the socially optimal consumption and use of environmental goods. However, a lead tourism researcher in the criticism of market-led tourism systems, FreyaHiggins-Desbiolles, 2006, Higgins-Desbiolles, 2008 and Higgins-Desbiolles, 2010, argues that the current system of market-based neoliberalism and its culture and ideology of consumerism are inherently unsustainable and calls for a less damaging system that will enable tourism to be a positive developmental social force. Instead of a private-sector commercial profit-driven corporatized tourism paradigm, she suggests an alternative paradigm based on equity, justice and sustainability that is civil-society- and government-driven.

The second theory on environmental damage, which addresses the questions of population and economic growth, is relevant to tourism due to tourism demand and supply growth. It is ever present in the tourism literature and is referred to in almost every work that relates to tourism development and business. This is the debate on sea, sun and sand mass tourism and its seasonal concentration (Aguiló et al., 2005, Figini and Vici, 2012, Fink, 1970, Mihalič, 2006 and Poon, 1993), which has often been observed as a dominant cause of the current sustainability tourism discourse. In this regard, non-mass-tourism developmental models, e.g., alternative tourism, are called for. At the same time, it is argued that because sustainable tourism is a set of principles, not a type of tourism, it can be applied to any tourism type or destination type, including concentrated mass tourism destinations (UNEP & WTO, 2005).

However, it is the third theory on the occurrence of environmental damage that is the most relevant to this paper. Why does tourism not behave responsibly towards Our Common Future, and why does it not develop and function in a sustainable way? On one hand, tourism businesses that are only driven by profit motives and ignore environmental and social awareness, ethics and policy in terms of strategy-based actions will necessarily face a sustainability discourse. However, on the other hand, economic business thinking is based on rationality and economic values and perceives ethical demands for ecology and justice as irrational. Based on this argument, the theory claims that the absence of so-called social environmental ethics has caused the current negative attitude to our environment (Frey, 1985). In this regard, the existence of environmental damage is explained through the absence of environmental social ethics (and awareness) and as a product of human ignorance. Thus, application of this theory calls for increased environmental awareness and ethics. It should, however, be noted that the theory assumes that people are good by nature and that a proper course of action is followed once they have enough information and knowledge on proper and just behaviour, regardless of the current economic allocation system in place.

4.2. Social stages of responsible behaviour

The information, knowledge, awareness and ethical aspects have been further addressed by the model of the penetration of environmental awareness into society and its policy. To explain how these aspects are addressed, we can turn to political economy. The Swiss-born welfare economist Bruno Frey (1985) proposed the famous four-stage scheme for new environmental trends: *Stage 1: pre-environmental-protection thinking*(society is simply unaware of environmental problems); *Stage 2: awareness of environmental harm* (which explodes after a society can no longer deny the facts); *Stage 3: discussion about environmental policy* (with the intention to mobilise for proper action); and *Stage 4: implementation of environmental policy* (which involves carrying out the appropriate action and assumes proper results).

In the continuation, Frey discussed the above scheme in relation to different stakeholders such as the voters, government, bureaucracy and producers (Frey, 1992). Although this is also highly relevant to our tourism-related debate, this paper only focuses on the application of Frey's model to the tourism sustainability discourse.

4.3. The Triple-A Model on sustainable-responsible tourism: towards responsustable tourism

One can easily apply the same four phases to the way our tourism stakeholders address environmental concern and sustainability. The idea of Frey's social stages was first applied to the tourism field by Mihalič and Kaspar (1996). Here the original four stages were reinvented and renamed. The first stage is the so-called *ignorance* stage that occurs before the destination considers environmental problems. Tourism stakeholders are unaware of such problems and thus express no concern at all; other, non-environmental sets of values determine their behaviour. As this phase acts somewhat as the pre-history of our sustainability discourse, the next stage is more relevant. This is a stage of environmental *awareness*, as exemplified by exposure to information on tourism environmental impacts and relationships that have made destinations sufficiently aware of tourism impacts that they can no longer be denied. Since tourism environmental concern, our starting point, is deeply connected with the current sustainable tourism ideology, sustainability issues inevitably enter the destination and create and shape so-called tourism sustainability awareness. The next stage involves the conceptual inclusion of sustainability issues, as defined above, because these become a matter of debate and are transferred into goals, coded into the destination's strategy and placed on its *agenda*. To shape and form the willingness to act consistently with the sustainability concept, the discussion of relevant policy instruments begins. Finally, the social-sustainability ethics and the need to act to achieve the strategic goals drive this process toward and into the last stage, which is the successful implementation of a sustainability policy and, hence, the stage of implementing sustainable *action*, thereby manifesting environmental responsibility. The destination, e.g., its stakeholders, behave in a sustainable manner, thus meeting the objectives and standards for sustainable development, business, products, etc.

The crucial words here are *Awareness–Agenda–Action*, all starting with the letter "A" and thus creating our Triple-A Model. Henceforth, we refer to this model as our Triple-A Model, which is presented and explained in Fig. 2.

Fig. 2 presents the top three stages of Frey's model after environmental concern has entered society. The pyramidal form associates the sustainability–responsibility process with mountain climbing. To reach the top of a mountain, a mountain climber needs to start at the bottom. He knows that the stepwise passage through the lower mountain station is a precondition for reaching the higher levels and that each climbing expedition up needs to be supported by a base station below. Thus, the narrowing shape of the pyramid suggests that each previous stage is a base upon which the next stage is dependent. The vertical and horizontal arrows also demonstrate that the penetration of sustainability into a destination is a twofold process. First, the process moves stepwise because the destination has to pass through different phases – from the starting phase of the destination's awareness, through the agenda stage to achieve the implementation action phase at the top of the pyramid. This process is demonstrated by the vertical progression line. Second, the process is continuous because the destination may proceed to the next phase before the previous phase has fully matured. More specifically, the top stage might be reached by only a few of the mountain expedition's climbers; the rest might stay in lower base stations and climb up later or not at all. This characteristic is presented by a horizontal arrow line that measures the penetration into each stage. Theoretically, a destination and its tourism stakeholders might find themselves at any of the stages on the sustainability–responsibility progression and penetration scale. At the same time, the destination might reach the top phase without having fully developed the characteristics of the previous phases, thus simultaneously positioning its stakeholders in all phases.

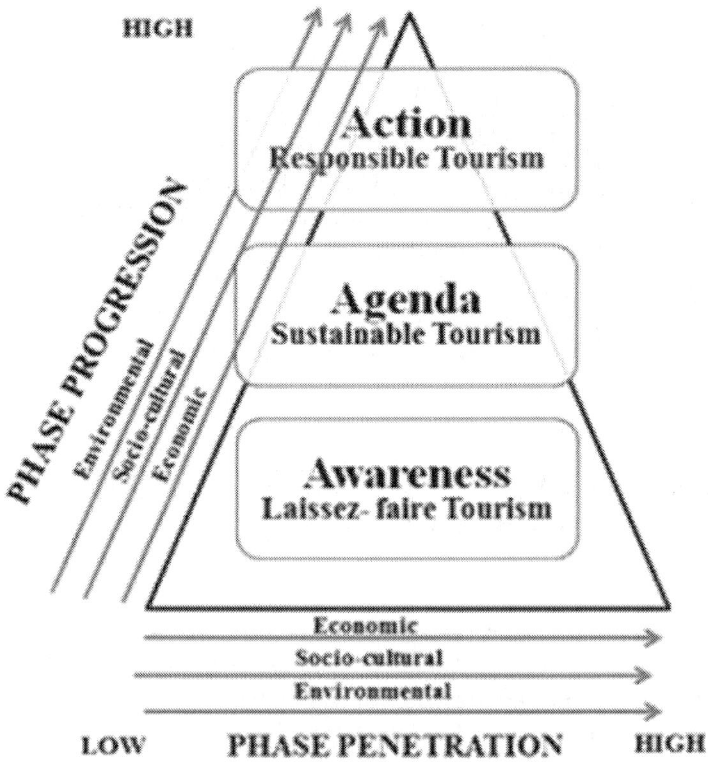

Figure 2. The Triple-A Model – for responsustable tourism.

The challenge of our pyramid is that it is based on a three-pillar sustainability understanding and thus reflects awareness, agenda and action in three different areas: economic, environmental and socio-cultural. This is illustrated by the three penetration and progression arrows in Fig. 2. In fact, the pyramid in Fig. 2 is the sum of three different pyramids and can be broken down into those three, one for each pillar. Tourism stakeholders might be at different positions in terms of both progression and penetration in each of the three pyramids. The existing global market system prioritises progression in the economic pillar pyramid; for some stakeholders, the other two pyramids (environmental and social) might become less important, thus affecting the triple sustainability of the complete pyramid. For example, one hotel study that measured firm behaviour with regard to the presence of different strategies showed that most of the surveyed hotels have an economic strategy in place (reported for 80% of firms) and only 35% of firms confirmed they had an all-pillars sustainable development strategy (Mihalič et al., 2012).

In reality, many destinations and tourism businesses do not know exactly where on this pyramidal scale they are and how sufficiently or insufficiently they have developed each of the phases. For example, a destination or firm might be well informed about the occurrence of environmental damage and be aware of environmental problems but might lack the appropriate information and knowledge on how to respond to those problems. The greater the information gap, the further to the left it will be positioned on the phase penetration scale. Some destinations or firms might have recently discussed sustainability agendas and created partnerships to adopt one, while others might have already included sustainability in a strategy and are confronted with policy-efficacy or implementation problems because a destination's stakeholder behaviour might lag behind the declared sustainability goals, meaning that the critical mass has not conquered the apex of the sustainability mountain.

5. RESULTS AND DISCUSSION: A FRAMEWORK FOR UNDERSTANDING THE SUSTAINABLE-RESPONSIBLE TOURISM DISCOURSE

Table 1 links our understanding of responsustable tourism presented in Fig. 1 with our Triple-A Model presented in Fig. 2. Further, it demonstrates how the whole process is driven by visitor satisfaction and their will to pay for sustainability, an ultimate measure and driver of a destination's success.

Table 1. A framework for understanding the sustainable-responsible tourism discourse.

Frey's model	Responsustable tourism pillars and requirements (Fig. 1)	The Triple-A Model (Fig. 2)	Stage-related tourism concept	Description
1	2	3	4	5
4. Implementation of environmental policy	Stakeholders' participation, cooperation, consensus, leadership, action & Visitor satisfaction	**3. Action**	Responsible tourism behaviour	The process of engaging in actual behaviour to achieve a full set of sustainability aims (responsible behaviour), taking actions in line with all three pillars of the sustainability concept, implementation of the agenda.
3. Discussion of environmental policy	Economic and/or Environmental and/or socio-cultural pillar & Visitor satisfaction	**2. Agenda**	Sustainable tourism concept	A blueprint for action for more sustainable tourism. The agenda is based on a sustainable tourism strategy and informed by sustainability value sets that demand action to implement the sustainable tourism concept. Economic considerations are extended to environmental and socio-cultural aspects. All are observed as important pillars.
2. Awareness of environmental harm	Tourism environmental awareness, education, information, ethics & Visitor satisfaction	**1. Awareness**	Environmentally laissez-faire tourism	Despite growing concern about tourism and its effects on the social, cultural and natural environment, the tourism sector is fully led by market rules, transactions among private parties are free from all non-economic considerations.
1. Pre-environmental-protection thinking		Ignorance(*Not presented in our Tripple-A Model in* Fig. 1)	Environmentally laissez-faire tourism	A pre-historic stage of our sustainable-responsible tourism discourse, the tourism sector is unaware of socio-cultural and environmental tourism issues.

Column 4 of Table 1 employs the term *environmental laissez-faire tourism* to describe a tourism realm that is free of any environmentally based interference, including values, enabling its system to operate according to economic rules and only one (economic) business line. Tourism is free of any environmental demands or standards, such as legal interventions, public policy, voluntary environmental actions or social mobilisation and other non-economic instruments and demands. As long as tourism is only based on the sustainable tourism concept, the sector still behaves in much the same way as before. Practically speaking, responsible tourism behaviour only begins in the third stage of our tourism model, when a sustainability-based agenda for responsible tourism starts to be implemented.

One may argue that this paper is dwelling on old findings because the concept-action tourism discourse has been present in the sustainability debate for quite some time. For example, the original Brundtland Report offered "... the recommendations for a sustainable course of development" (WCED, 1987: 16). Indeed, the Agenda 21 approach used to be quite popular, referring to a blueprint for action for more sustainable real tourism. However, this approach has not reduced our sustainability discourse because*agenda* is our final Alpine base camp and thus serves as the second A in our Triple-A Model. *Agenda* is a precondition for our mountain climber to start their journey to the peak of the mountain, to start the implementation action phase. Agenda extends a theoretical sustainability concept into a concept with specifically defined actions, meaning that it codifies the sustainability principles into a strategy and policy and offers policy instruments. Thus, tourism that enters the implementation action stage needs a new name. Based on the proper action or responsible actual behaviour, the term*responsible tourism* would be appropriate. This interpretation argues that responsible tourism is not a new tourism concept and especially not an alternative concept to sustainable tourism. It simply refers to acting responsibly and in line with sustainability values. Thus, responsible tourism is a result of the transition towards actual tourism based on more sustainable values.

The Triple-A Model pyramidal view of the progression from environmentally laissez-faire tourism towards responsible and thus more sustainable tourism integrates new parts that have developed within the tourism environmental and sustainability debate. It demonstrates an increase in the interconnectedness and complexity of the debate, integrating its stages, which include and transcend the previous ones. However, vertical and horizontal progression up our Awareness–Agenda–Action pyramid is not guaranteed or self-evident; however, as time progresses, it cannot be stopped. For example, the awareness- and ethics-driven transition towards responsible tourism might be slowed down for many reasons. One study found that European tour operators have a high level of awareness about tourism's negative impacts and recognised their responsibility to reduce them. Nevertheless, facing low customer demand for responsible tourism practices and having little regulatory pressure to behave in a more environmentally friendly manner, tour operators engage in few concrete activities and these remain superficial relative to their potential due to their central distribution role and capability to direct customers to suppliers and destinations (Budeanu, 2005 and Sigala, 2008). However, tour operators are financing sustainability and environmental research and customer education activities, which will straighten the vertical position of tourism in our awareness stage and allow more stakeholders to progress to the next stage in the pyramid where more activities will be demanded and designed in both the private and public sectors. This case clearly demonstrates the shareholder market-led approach to corporate sustainability. However, ignoring the progression in time might be dangerous as standards for environmental behaviour are nevertheless changing due to the above-mentioned processes. In other words, behaviour is changing because "... the way the world is going, the companies are having to become more responsible ... it ... helps them to remain competitive, so they are going to have to do it" (Miller, 2001, p. 595).

Our Awareness–Agenda–Action understanding of tourism responsibility is in full harmony with the European Commission's latest standing on sustainable versus responsible tourism (Mihalič, 2011), which demands "*... awareness, decisions and actions ..., so that it is sustainable over time*" (TSG, 2012, p. 2). However, this explains little about the corporate and economic tensions underlying it.

These tensions might be best illustrated by the confrontation of two popular concepts: sustainability versus CSR. In this context, the question of the similarities and differences between sustainability and CSR may arise. This discussion can take place at the conceptual level where both sustainability and CSR are observed as two concepts that converged over time and are now both linked to the same three pillars: economic, social and environmental. However, it is evident that CRS is more popular in the corporate world and sustainability among tourism destinations and public bodies or organisations. One may even claim that the term sustainability has been avoided by corporate business practice and consultants.

We argue here that CSR, which is relevant to the corporate world, operates from the standpoint of corporate values and may add the other two sets of sustainability values, e.g., social and environmental, in a way that makes real business or economic sense. At the profit-driven level of sustainability motivation (market-oriented view), sustainability is narrowly defined inside the business fence and is only promoted if profitable. Quite differently, debate based on sustainable tourism (broader stakeholder orientation) claims that all pillars are equally important, with no priorities given to any pillar and certainly not to the economic one. In this case of a holistic level of sustainability, the motivation for a sustainability orientation is derived from the awareness that all beings and phenomena are mutually interdependent, and universal responsibility towards all other beings is fully recognised (van Marrewijk, 2003). As research has shown, in the laissez-faire environmental tourism phase, tourism managers publicly admit economic but not ecological responsibility, believing that they are a priori contradictory. After society enters the awareness stage, where the connections between the economy, society and ecology become obvious and cannot be ignored, such statements become rare and many managers participate in the discussion on the sustainability agenda. However, the question of how to convince competitive corporations to follow the full set of sustainability values remains unsettled. Ultimately, the idea of sustainability has emerged completely isolated from a thorough analysis of the principles governing market-based economic systems, and the corporate world seems to have a recognised right to acknowledge and protect the importance of economic values and even challenge the other two pillars. In this regard, sustainability seems to live in an isolated world and CSR (if applied) in a so-called green capitalism characterised by "green capitalists" and "green consumers" (Elkington, 1997, pp. ix, x). Quite the contrary, tourism destinations are expected to behave in a sustainable-responsible fashion and a destination's responsibility is broadly defined inside and outside the destination's fence. The question here is whether the all-pillars sustainability position is observed as a threat to businesses and profit-driven sustainability as a legitimate mundus operandi in line with corporate values and practice. The obvious conclusion is that a CRS orientation assumes that tourism can happily live in "sustainable capitalism", underpinned by green customers who are demanding more green products and thus more responsible tourism, and are willing to pay for it.

To summarise the above discussion, there are two reasons for the rise of *responsible tourism*. The first is the need to overcome what has become evident, the actual sustainability discourse, due to the inability of sustainable tourism itself to be more strongly implemented in real tourism. The second reason is more ideology-based and refers to an attempt to merge the tourism-based environmental debate with the language used in other sectors, most importantly the corporate or business world, with the aim of increasing the acceptance and attractiveness of the relevant tourism concept among all tourism stakeholders.

6. CONCLUSIONS AND FURTHER RESEARCH

The low efficacy of *sustainability* in tourism coupled with the increasing attention to the notion of *responsible tourism*, as demonstrated by current theoretical and practical developments, call for critical reflection. The changes in our understanding occur in stages that follow both a logical and a chronological order. More holistic understanding develops later because it has to wait for the emergence of the parts that will then make up a complete theory and interconnections with the new stages including and transcending the previous ones (van Marrewijk, 2003).

In this context, the present conceptual paper integrates and challenges different existing theoretical and practical private and public stakeholder understandings and uses of the notions of responsible and sustainable tourism. It offers a well-reasoned integrated understanding of responsible tourism and the process that leads towards *responsustable tourism*. The proposed Triple-A Model helps to analyse and diagnose the main sustainability tourism discourse on enduring unsustainable tourism practices. The creation of the model was also provoked by the current inconsistency in use of the terms sustainable and responsible tourism and theories on environmental damage and behavioural reaction to tourism damage. As the two developments, theoretical and practical, are relevant to this paper, we focussed our discussion on both.

By putting the sustainability–responsibility discourse into a relevant context of political economics and behavioural economics, this paper adds to tourism concepts and nomenclatures. Although the proposed Triple-A Model and its phases have existed in some form in the tourism debate and while one might argue that they are common sense, they contribute to understanding the process and the reality of the sustainable tourism discourse. This understanding, placed in a real context, brings us to the tourism-responsibility paradox. Although the Awareness-Agenda-Action process theoretically leads towards more socially and environmentally responsible tourism, it is the economic set of interests that is attempting to react and adapt to the growing environmental awareness and demand and is leading the process and promoting the notion of responsible tourism. Here, we refer to our observation that its use is pushed by the 'responsibility' (e.g., in the form of CRS) debate, recognising the need to go green but while strongly protecting the values of the economic pillar. Thus, more research is needed to assess whether the above-mentioned green movement determined by the capitalistic system is in fact leading towards a sufficiently sustainable future.

Further, our model can help increase tourism responsibility by analysing the implementation chances of agendas and environmental policies. To avoid the expectation that the model leads to the implementation (e.g., responsible tourism) phase itself, the model must recognise that potential conflicts among the stakeholder groups are an important element in the political process, enabling progression towards the higher phases. In that regard, the roles of tourists, governments, the tourism industry, NGOs and local populations will also need to be further discussed and researched.

Our model can be applied to tourism destinations of any level, tourism businesses, governments, non-governmental organisations, institutions and tourists. All of these relevant stakeholders have a responsibility to seek to reduce the negative and enhance the positive environmental, socio-cultural and economic impacts of tourism. In this regard, more research is needed from the aspect of each group, focusing on awareness (and ethics) creation, instruments for achieving the goals and behavioural context.

One may extend the understanding of responsible tourism behaviour to all stages of our pyramid. That is, if one understands that the destination is responsible for being aware of tourism damage and minimising it, that responsibility also requires the developing of an action plan. Thus, responsible tourism might refer to all of the stages, not only the last one, as presented in Fig. 2. Indeed, we used the term responsible tourism in a narrower sense, limited to actual behaviour that makes real tourism more sustainable. In our model, we have connected responsible behaviour with the sustainable tourism concept and thus added the sustainability value to the behavioural concept. In this context, another, joint nomination would better describe our top pyramidal tourism phase. Merging the words responsible (behaviour-based) and sustainable (concept and values-based) produces the new term responsustable tourism. It is argued that this new term fully reflects the academic and practical debate and action that is increasingly labelled "responsible" tourism, yet de facto based on sustainability. In this context, paper proposes a Triple-A Model as a tool that helps to understand the process of how a responsible tourism destination or firm actually implements sustainability agenda.

REFERENCES

1. Aguilo, E., Alegre, J.n., Sard, M., 2005. The persistence of the sun and sand tourism model. Tour. Manag. 26 (2), 219–231.
2. Blackstock, K.L., White, V., McCrum, G., Scott, A., Hunter, C., 2008. Measuring responsibility: an appraisal of a Scottish National Park's sustainable tourism indicators. J. Sustain. Tour. 16 (3), 276–297.
3. Bohdanowicz, P., Simanic, B., Martinac, I.V.O., 2005. Environmental training and measures at scandic hotels, Sweden. Tour. Rev. Int. 9 (1), 7–19.
4. Bramwell, B., Lane, B., 1993. Sustainable tourism: an evolving global approach. J. Sustain. Tour. 1 (1), 1–5.
5. Bramwell, B., Lane, B., McCabe, S., Mosedale, J., Scarles, C., 2008. Research perspectives on responsible tourism. J. Sustain. Tour. 16 (3), 253–257.
6. Budeanu, A., 2005. Impacts and responsibilities for sustainable tourism: a tour operator's perspective. J. Clean. Prod. 13 (2), 89–97.
7. Butler, R., 1995. Introduction. In: Butler, R., Pearce, D. (Eds.), Change in Tourism: People, Places and Processes. Routledge, New York, pp. 1–11.
8. Caroll, A.B., 1991. The Pyramid of Corporate Social Responsibility: toward the Moral Management of Organizational Stakeholders Business Horizons. July – August.
9. Chettiparamb, A., Kokkranikal, J., 2012. Responsible tourism and sustainability: the case of Kumarakom in Kerala, India. J. Policy Res. Tour. Leis. Events 4 (3), 302–326.
10. RTD7 Conference: Responsible Tourism in Destinations, 1–4 October 2013. Catalunya, Barcelona. Retrieved 1.12.2013, from: http://www.accessibletourism.org/?i=enat.en.events.1473.
11. Dasmann, R.F., Freeman, P.H., Milton, J.P., 1973. Ecological Principles for Economic Development. John Wiley & Sons, Washington.
12. Diotallevi, M., 2013. Personal Correspondence, 14.3.2013, Chief of Programme Ethics $ Social Dimensions of Tourism. UNW, Madrid.
13. Dwyer, L., 2005. Relevance of triple bottom line reporting to achievement of sustainable tourism: a scoping study. Tour. Rev. Int. 9 (1), 79–938.
14. EC, 2014. European Tourism Indicators System Toolkit for Sustainable Destinations. European Commission, Brussels.

15. ECETAT, ECOTRANS, 2004. The Visit Initiative, Tourism Eco Labelling in Europe e Moving the Markets towards Sustainability. European Centre for Eco Agro Tourism, Amsterdam.
16. Elkington, J., 1997. Cannibals with Forks : the Triple Bottom Line of 21st Century Business. Capstone, Oxford.
17. Fennell, D.A., 2006. Tourism Ethics. Channel View, Clevedon.
18. Fennell, D., Malloy, D., 2007. Codes of Ethics in Tourism. Practice, Theory, Synthesis. Channel View Publications, Clevedon.
19. Figini, P., Vici, L., 2012. Off-season tourists and the cultural offer of a mass-tourism destination: the case of Rimini. Tour. Manag. 33 (4), 825–839.
20. Fink, C., 1970. Die Massentourismus (Mass Tourism). Verlag Paul Haupt und Stuttgart.
21. Forquan, A.M., A., P., Hussin, R., 2010. Promoting Green tourism for future sustainability. Theor. Empir. Res. Urban Manag. 5 (8), 64–74
22. Frey, B.S., 1985. Umweltoekonomie (Engl.: Environmental Economics). Vandenhoeck & Ruprecht in Goettingen, Goettingen.
23. Frey, B.S., 1992. Umweltokonomie (Engl.: Environmental Economics). Vandenhoeck € & Ruprecht, Gottingen.
24. Goodwin, H., 2011. Taking Responsibility for Tourism. Goodfellow Publishers Limited, Oxford.
25. Goodwin, H., Francis, J., 2003. Ethical and responsible tourism: consumer trends in the UK. J. Vacat. Mark. 9 (3), 271–284
26. GT, 2014. Why Green Tourism? Green Tourism. Retrieved 15.4.2014 from: http:// www.green-tourism.com/-
27. Hall, M., 2012. Governance and responsible tourism. In: Leslie, D. (Ed.), Responsible Tourism. Concepts, Theory and Practice. CABI, Wallingford, pp. 107–118.
28. Heikkurinen, P., Bonnedahl, K.J., 2013. Corporate responsibility for sustainable development: a review and conceptual comparison of market- and stakeholderoriented strategies. J. Clean. Prod. 43 (0), 191–198.
29. Higgins-Desbiolles, F., 2006. More than an "industry": the forgotten power of tourism as a social force. Tour. Manag. 27 (6), 1192–1208.
30. Higgins-Desbiolles, F., 2008. Justice tourism and alternative globalisation. J. Sustain. Tour. 16 (3), 345–364.
31. Higgins-Desbiolles, F., 2010. The elusiveness of sustainability in tourism: the culture-ideology of consumerism and its implications. Tour. Hosp. Res. 10 (2), 116–115.

32. ICRT, 2011. Alberta declaration on responsible tourism in destinations. In: 5th International Conference on Responsible Tourism in Destinations. Canada.

33. Inskeep, E., 1991. Tourism Planning: an Integrated and Sustainable Development Approach. John Wiley & Sons, New York.

34. International Conference: Sustainability Issues and Challenges in Tourism, 3e5 October 2013. Retrieved 1.12.2013 from. http://www.butrmconferences.org/.

35. IUCN, 1980. World Conservation Strategy: Living Resource Conservation for Sustainable Development. International Union for Conservation of Nature and Natural Resources, Gland.

36. Kaspar, C., 1973. Fremderverkehrsoekologie – eine neue Dimension der Fremdenverkehslehre. (Engl: tourism ecology – a new dimension of tourism sience). In: Walter, E. (Ed.), Festscrift zur Vollendung des 65. Lebensjahres von P. Bernecker, pp. 139–143. Wien

37. Krippendorf, J., 1984. Die Ferienmenschen. (Engl.: Holiday People). Orel Fuessli Verlag, Zuerich.

38. Krippendorf, J., 1987. The Holiday Makers – Understanding the Impact of Leisure and Travel. Heinemann, Oxford.

39. Leslie, D., 2012a. Responsible Tourism. Concepts, Theory and Practice. CABI, Wallingford.

40. Leslie, D., 2012b. Introduction. In: Leslie, D. (Ed.), Responsible Tourism. Concepts, Theory and Practice. CABI, Wallington, pp. 1–16.

41. Leslie, D., 2012c. The responsible tourism debate. In: Leslie, D. (Ed.), Responsible Tourism. Concepts, Theory and Practice. CABI, Wallingford, pp. 17–42.

42. McGugan, S., 2002. Asynchronous computer mediated conferencing to support learning and teaching: an action research approach. J. Hosp. Leis. Sport Tour. 1 (1), 29–42.

43. Mihalič, T., 2006. Tourism and its Environments: Ecological, Economic and Political Sustainability Issues. Ekonomska fakulteta. Univerza v Ljubljani, Ljubljana.

44. Mihalič, T., 2011. Discussion on Sustainability, Responsibility and CSR, Based on the TSG Working Document. TSG meeting 20.6.2011. European Commission, Brussels.

45. Mihalič, T., 2013. A Green tourism Barometer in the time of economic crisis e the concept and case of Slovenia. In: Paper presented at the conference Tourism in Southern and Eastern Europe. Retrieved 1.12.,

2013, from. http://www.fthm.uniri.hr/files/Kongresi/ToSEE/2013/papers/1%20Mihalic.pdf

46. Mihalič, T., Kaspar, C., 1996. Umweltokonomie im Tourismus. (Engl.: Environmental € Economics in Tourism.). Paul Haupt, Bern.
47. Mihalič, T., Žabkar, V., Knežević Cvelbar, L., 2012. Hotel sustainability business model: evidence from Slovenia. J. Sustain. Tour. 20 (5), 701e719.
48. Miller, G., 2001. Corporate responsibility in the UK tourism industry. Tour. Manag. 22 (6), 589–598.
49. Poon, A., 1989. Consumer strategies for a new tourism. In: Cooper, C. (Ed.), Progress in Tourism, Recreation and Hospitality Management, vol. 1. Bellhaven Press, London, pp. 91–102.
50. Poon, A., 1993. Tourism, Technology and Competitive Strategies. CAB International, Wallingford.
51. Ritchie, J.R.B., Crouch, G.I., 2000. The competitive destination: a sustainable perspective. Tour. Manag. 21 (1), 1–7.
52. RTP, 2002. Cape Town Declaration on Responsible Tourism. Responsible Tourism Partnership. Retrieved 15.6.2013 from: http://www. responsibletourismpartnership.org/RTWorld.html
53. Sigala, M., 2008. A supply chain management approach for investigating the role of tour operators on sustainable tourism: the case of TUI. J. Clean. Prod. 16 (15), 1589–1599.
54. Stanford, D., 2008. Exceptional visitors: dimensions of tourist responsibility in the context of New Zealand. J. Sustain. Tour. 16 (3), 258–275.
55. STO, 2013. Green Tourism e Sustainable Development Framework in Slovenia. Retrieved 16.3.2013 from: http://www.slovenia.info/?ps_trajnost_ slovenija=0&lng=2.
56. Swarbrooke, J., 1999. Sustainable Tourism Management. CABI, Wallingford.
57. TSG, 2007. Action for More Sustainable European Tourism. European Commission, Brussels. Retrieved 5.12.2013 from. http://www.tolearn.uni-kiel.de/tolearn/ index.php?id=128.
58. TSG, 2012. Charter for Sustainable and Responsible Tourism. European Commission, Brussels.
59. UNEP, WTO, 2005. Making Tourism More Sustainable. Paris: United Nations Environment Programme. World Tourism Organization, Madrid.

60. UNWTO, 2004. Indicators of Sustainable Development for Tourism Destinations. A Guidebook. United Nations World Tourism Organisation, Madrid.
61. Van de Ven, A.H., 2007. Engaged Scholarship: a Guide for Organizational and Social Research. Oxford University Press, Oxford.
62. van Marrewijk, M., 2003. Concepts and definitions of CSR and corporate sustainability: between agency and communion. J. Bus. Ethics 44 (2e3), 95–105.
63. Watts, R.E., 2011. Developing a conceptual article for publication in Counseling Journals. J. Couns. Dev. 89 (3), 308–312.
64. WCED, 1987. Our Common Future. Oxford University Press, Oxford.
65. Wheeller, B., 1993. Sustaining the ego. J. Sustain. Tour. 1 (2), 121–129.
66. Young, G., 1973. Tourism Blessing or Blight? Penguin Books, Inc., Baltimore
67. Zuber-Skerritt, O., 1996. New Directions in Action Research Taylor & Francis. Retrieved 12.4.2014 from: http://proxy.library.brocku.ca/login?url=http://lib. myilibrary.com?id=2055.

Chapter 4

Ecological Conservation, Ecotourism, and Sustainable Management: The Case of Penang National Park

Sara Kaffashi [1,*], Alias Radam [2], Mad Nasir Shamsudin [1], Mohd Rusli Yacob [1] and Nor Hanim Nordin [3]

[1]*Faculty of Environmental Studies, Universiti Putra Malaysia, 43400 UPM Serdang, Selangor, Malaysia*

[2]*Department of Management and Marketing, Faculty of Economics and Management, Universiti Putra Malaysia, 43400 UPM Serdang, Selangor, Malaysia*

[3]*Faculty of Economics and Management, Universiti Putra Malaysia, 43400 UPM Serdang, Selangor, Malaysia*

ABSTRACT

Penang National Park (PNP), as Malaysia's smallest national park, is one of the few naturally forested areas left on Penang Island, in Peninsular Malaysia. The main objective was to analyse users' preferences and willingness to pay to enhance improved management of PNP for the dual aim of conservation and recreation. Structural equation modelling (SEM) was used to analyse the formation of attitudes towards different aspects of PNP. Results showed that implementing enforcements with rules and regulations and imposing permits and charges on certain activities were the most influential variables of PNPs' perceptions. The results of a random parameter logit model (RPL) demonstrated that visitors placed the highest value on having adequate information about PNP, and the second-highest value on improvements in the park's ecological management. The welfare measure for improvement in management of PNP against status quo is estimated at about MYR 9. Results also showed that demand for better conservation and management of PNP is relatively price-inelastic.

Simulations of the results showed, under a MYR10 admission fee, that improvement in management would have 96% of market share compared with status quo. This study concluded that visitor entrance fees can and ought to be introduced as a means of financing conservation initiatives and possibly preventing congestion.

Keywords: conservation charge; nature based tourism; park visitor preferences; structural equation modeling; choice experiment; provision of information; willingness to pay

1. INTRODUCTION

National parks were established primarily to conserve and enhance natural scenery, wildlife, and cultural heritage. However, over the years this aim has slightly broadened to include education, recreation, and local poverty reduction. Nowadays, national parks all over the world are important tourist destinations. Ecotourism is a perfect example of direct monetary values provided by nature [1,2]. Increasing interests in nature-based tourism made it one of the largest and fastest growing segments of tourism industry in the last decade. However, a key consideration for managers is to be aware that achieving the dual goals of conservation and tourism requires regular assessment of tourism satisfaction and preferences to find out whether ecotourism needs are in accordance with conservation goals [3].

Based on annual reports from Malaysia's Department of Wildlife and National Parks (DWNP), the number of visitors to the protected areas of Malaysia has doubled in the past 10 years. This shows the great current demand for nature-based tourism and the need for facilities and infrastructures to support this increasing demand. Therefore, parks authorities in Malaysia need to develop facilities for visitors which minimize any possible conflicts with their conservation goals [4]. An effective approach in providing services that can balance both ecological conservation and income generation from parks is to understand and incorporate conservation goals, tourists' preferences for nature appreciation, infrastructure, and other attributes of protected areas [5]. In doing so, knowing the visitors' priorities between conservation of the ecological and biological future of a park and its tourism use is essential. Moreover, managers who must budget carefully and invest in conservation, resource management, and tourist accommodations can inform themselves about how these managerial decisions could impact their revenue through commissioning and carefully studying research on visitors' perceptions and preferences concerning natural environment protection and possible tradeoffs with recreational facilities, if budgets are tight [3].

Located on the northwestern tip of Penang Island, Penang National Park (PNP) is Malaysia's smallest national park. PNP is the only considerable natural forested areas left on Penang Island. Penang Island is considered one of the most developed places in Malaysia. In 2012, PNP and ten other protected areas under the management of the Department of Wildlife and National Parks (DWNP) attracted 1.3 million tourists, a 6.5% increase compared with 2011. This figure, which represents a 22.5% increase compared to 2011, is evidenced by 108,276 visits to PNP (DWNP 2012). The current practice in PNP ecotourism is concentrating on "conventional carrying capacity technique" to calculate a "magic number" as a "threshold limit" [6]. Lack of enforcement of laws and regulations, encroachment of urban and agricultural land uses into the park, and lack of zoning are sources of natural environment disturbance and visitor dissatisfaction. PNP management needs to find approaches for comprehensive improvement from current conditions, which are unsatisfactory in many respects. Calculating the extent of resource improvements from the current conditions and how those improvements affect visitor satisfaction can inform PNP managers in their decisions whether to match visitor preferences with long-term, medium-range, and short-term park goals (if applicable) [7]. Previous studies suggested that visitor satisfaction directly affects their willingness to pay to adopt conservation programs [1,8,9]. Therefore, the educational mission of the park may require extra funds that some managers might not have in-house. This is where consulting with experts can inform their decision to institute an entrance fee.

There is no doubt that better achieving management goals and improving overall visitor satisfaction requires money that most often is beyond the national government's budget for such purposes [10]. GEF-UNDP funded an extensive project of "Enhancing Effectiveness and Financial Sustainability of Protected Areas in Malaysia" (2012–2019). The aim of this project is to improve management and funding for the nation's national parks. During the project period, Department of Wildlife and National Parks Malaysia (DWNP) will try to optimize, balance, and broaden its long-term management plan and share with visitors the costs of protecting protected areas of Malaysia which they are visiting [4]. At this moment, PNP only relies on the government's budget allocation, and visitor entries into the park are free. Finding strategies for a diverse revenue steam can improve the protection, maintenance, and upgrades of natural attractions of PNP, but that is another challenge for park management. Diverse source(s) of income could reduce the park's over-reliance on any one funding source. It can also help PNP to establish its financial self-sufficiency, which is beneficial for long-term sustainability of the park. For many tourist areas, charging visitors an entry fee is believed to be the only sustainable approach to self-financing, other than relying upon government budget allocations [11]. Entry fees also can help to prevent congestion by creating an entry barrier for those below the WTP threshold. However, any realistic entry fee amount must be evidence-based and match the WTP of the respondents [12].

A choice experiment can help to disaggregate preferences and tradeoffs among different attributes of the park, including price. It can inform PNP managers about the relative value of each of the different attributes being evaluated. Estimation of visitors' marginal WTP and compensated surplus can provide valuable information on whether or not to introduce an admission fee and what such an admission fee should be. Discovering visitors' preferences and their most favored recreational uses of the park can help park managers prioritize the improvement of park features which best support the preserve financially, without which the resource could die from inadequate funding [7,13]. Several studies applied choice experiment techniques to gather information on visitors' preferences (e.g., [3,7,14,15,16,17,18,19,20,21,22,23]. These studies further concluded that respondents are ready to pay to gain maximum utility from their preferred activities.

The aim of this study was to discover how visitors might differentially value and thus be willing to trade off management and recreational aspects of PNP, including restricted access to some areas and packaged and/or layered pricing, in the form of an admission fee, leaving open the possibility of additional fees for other premium uses of the park. Specifically, this study will present the use of a structural equation modelling (SEM) and choice experiment (CE) as mechanisms to analyze preferences and WTP of tourists in relation to the future development of PNP. Results of this study could be useful on obtaining values of different aspects of the park (from the viewpoint of either conservation or tourism) from respondents with different perspectives and backgrounds. Results of this study could also be useful for decision makers of other natural recreational sites of Malaysia as they try to discern visitors' general preferences and the way they trade them off with money.

2. PENANG NATIONAL PARK

Penang National Park (PNP) is located in the northwest corner of Penang Island, a metropolis in Northern peninsular Malaysia (Figure 1). PNP is the first national park established under Malaysia's National Park Act of 1980, Act No. 226 enacted by the federal and state governments as a measure to protect the rich biodiversity endangered by the nearby metropolis [4]. The park, formerly a forest reserve which until 1996 had been used for logging, is known for its coastal ecosystem of approximately 1266 hectares, of which 70 hectares are still regarded as virgin forest [24,25]. PNP, which has both forest and marine features, had attracted a wide range of visitors. Penang National Park was set up to preserve and protect "flora and fauna, geological, archaeological, historical, ethnological features and other scientific value and scenic interest" [26].

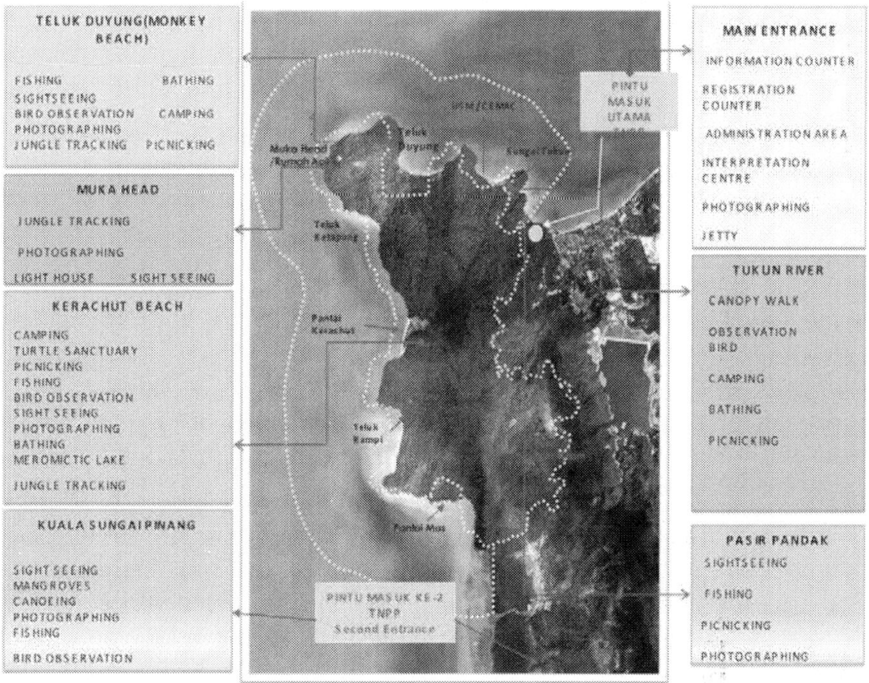

Figure 1. Penang National Park (Locations and Activities) [4]. Source: Modified fromhttp://www.journeymalaysia.com/islandpenangtrekking_map.htm.

PNP is the only national park in Peninsular Malaysia with a coastal habitat which provides diverse range of habitats from sandy and rocky shores to mangroves, for wide range of flora and fauna species [4].

Teluk Bahang, Pantai Kerachut, Pantai Mas and Teluk Duyung (Monkey Beach) are the famous four of the eight pristine and sandy beaches around the forest [27]. Over 1000 plant species and one unique species, namely Stag Horn's Fern (*Platycerium coronarium*), have been reported in PNP [28]. The park is also home to 90 species of medicinal plants. In addition, almost 200 species of birds, 25 species of mammals, 53 species of butterflies, and a sizable diversity of marine species have been recorded from PNP [25,28]. The park's inimitable features and habitats such as the meromictic lake and turtle nesting sites are unique to PNP.

Some of the coastal areas in PNP are abodes to mangrove forests. They are found on the northern, western and southern coasts. Kuala Sungai Pinang consists of mangrove areas, making it a good bird-watching site for avid bird watchers. Bird species that can be found there include Little Heron, Haliaeetus Leucogaster (*Helang Siput*), Haliastur Indus (*Lang Merah*), and some other small colorful birds, such as sunbirds (Kelicap) and sandpipers (Kedidi).

The first main entrance to PNP is located in Teluk Bahang, and visitors need to register for the entry permit at the National Park's illumination counter. The second entrance is in Balik Pulau, nestled in Kuala Sungai Pinang. From there, visitors must get a visitors' boat to get to PNP. While entrance to PNP is free, tourists need to register by name at the office before entering the park and sign out before leaving. Information signs, mostly near the registration point, point out walking trails, directions, and other useful and interesting information. Penang National Park only receives day visitors. For overnight camping, visitors must receive permission from the Park's management. Due to limited water supplies and possibility of rain, the overnight camper limit is 80 at the Pantai Kerachut campsite and 30 at the Sungai Tukun campsite.

From the main entrance of PNP, three trails diverge, heading to different natural habitats. The park has provisions for varied ecotourism activities: climbing, trekking, swimming, fishing, wildlife observation, or simply enjoying the natural environment of sea and forest scenery. Camping and picnic facilities are provided in Pantai Kerachut (a white sandy beach). This beach also hosts a Green Turtle (*Chelonia mydas*) nesting ground between April to August, and one for the Olive-Ridley turtle (*Lepidochelys olivacea*) from September and February [24]. Nevertheless, most visitors do not have enough information about these two species, and there is no guarantee that turtles' habitat will not be disturbed by visitors.

Although visitors are reminded to deposit their litter in the "monkey-proof" trash in monkey beach, discarded, mostly plastic, containers used by visitors are scattered across the beach and along the trails. Hence, environmentalist groups have launched an effective awareness and environmental education system for all stakeholders, involving the development of a sustainable management plan and environmentally-friendly infrastructure, and focusing in PNP on ecotourism rather than mass tourism [25].

The park's relatively aggressive and unbalanced infrastructure development has greatly disrupted its natural tranquility and other ecological harmonies, such as species balance. An example is the construction of roads and strings of camping facilities within and just outside the park. Further, the Wildlife Department is the sole decision-maker in park management; its failure to consult relevant governmental and non-governmental organizations in key decisions, whether deliberate or through negligence, has resulted in noncompliance and a number of government policies which contradict one another. However, ecological threats to PNP are not merely a result of PNP's currently excessive attention to mass tourism development and the unsustainably large numbers of tourists who enter the park. Inadequate park management, including lack of proper zoning and failures to protect park borders, has resulted in destruction of critical habitats and of the park's overall ecosystem. This is beyond the encroachment of urban and agricultural land use into the park. Poor enforcement, lack of trained guides, and lack information about sensitive areas or allowed practices in different ecological zones are other important challenges to the park's natural tranquility and sustainable tourism [25].

3. METHODOLOGY

Valuation methods can be divided into stated and revealed preference techniques. While revealed preference techniques can only capture use values, the stated preference methods are the only way to estimate non-market and non-use values. These methods include contingent valuation method (CVM) and choice experiment (CE) analysis. National parks managers need to consider both protection priorities and visitor satisfaction, so methods must be able to study the multi-dimensionality of parks [7]. However, CVM and CE are similar in their theoretical basis, survey design, and the capability to estimate welfare for moving from a current condition (status quo) to an alternative level. The choice experiment (CE) method can also provide multidimensional information for a certain park and calculation of trade-offs between goods in the choice set or attributes and price helps to highlight any prioritized park attributes. Accordingly, a choice experiment method was used in this study to weigh the complex dimensions of both park preservation and the visitors' experience of enjoyment.

3.1. Survey Design

The choice experiment (CE) questionnaire used for data collection highlights problem statements, choice sets, and socioeconomic and attitudinal questions. A clear definition of these sections can assure more reliable responses [29,30].

The first step in designing the questionnaire required determining the most important attributes of management with the aim of saving both ecological integrity and tourism development. The selection of alternative attributes can be done through expert surveys, scientific research, and focus group discussions. Hence, the most policy-relevant attributes which have future practical influence in PNP management were selected. Present short-term practice in PNP is geared towards attracting mass tourism, while we are more concerned about sustainable tourism medium-range and long-term ecological integrity. Current practices involve medium- and long-term risks to the operation. Therefore, the attributes chosen for this study were ecological management (with emphasis on the protection of ecological integrity), recreational facilities (with emphasis on considering park capacities and being as nature friendly as possible), provision of information and price. We defined minimum and maximum attribute levels, where the minimum level reflects the current situation. These selected attributes ensued from a synthesis of opinions based on the outcome of focus group discussions with stakeholders including park managers, tour guides, DWNP experts, and researchers who are familiar with PNP. Nonetheless, the final decision on attributes and levels was also based on our personal observations at the park. The final priorities selected, therefore, were those directly affecting decision making and park management in ways which would optimize tourist comfort with minimal ecological impact. In conformity to the study's objective, it was ensured that the selected attributes are those related to recreation, tourism and conservation.

3.1.1. Attributes and Levels

The attributes considered for the assessment of visitors' preferences were ecological management, recreational facilities, information boards, and entrance fee. The inclusion of the monetary attribute might help park managers consider the introduction of an entrance fee for PNP, as admission is currently free of charge. Imposing an admission fee creates an entry barrier and could help to reduce congestion by imposing a marketplace decision on spending on potential visitors, who now must sort their preferences for visiting the site with their other possible uses for the money. However, on the positive side, the possibly greater overall revenue it generates could be spent on programs which would enhance public education, guards' training, research, and conservation projects, which could help mitigate environmental problems inside the PNP.

The four attributes used in the final design, as discussed earlier, are listed in Table 1. One of the attributes is treated as a cost, while the other three attributes represent improvement in management of the park, although they also impose costs, which may or may not be charged back to visitors as fees.

Table 1. Attributes and Levels Selected for PNP.

Attribute	Attribute Levels	Current Management Practices
Ecological management (EV)	1. *Current condition* 2. Satisfied 3. Very good	Not satisfied
Recreational facilities (RF)	1. *Current condition* 2. Satisfied 3. Very satisfied	Not satisfied
Provision of information (PIS)	1. *Current condition* 2. Medium 3. High	Low
Conservation Charge (CC)	1. *Current condition* 2. RM 5 3. RM 10 4. RM 15	RM 0

- Ecological Management: The term "ecological management" refers to the proper management of the park with minimum negative effect on flora, fauna, and the quality of the physical environment, such as water purity and natural scenery. From our focus group discussion with stakeholders, we learned that waste disposal, urban encroachment into the park, lack of zoning, and poor infrastructure were major threats to the ecosystem. Thus, the effects of encroachment, improper waste disposal and declining numbers of animal and plant species, either due to visitor or management disturbance, were highlighted and presented to the respondents. The three levels considered for this attribute include:

    1. Current condition (not satisfied): This reflects the status quo, where encroachments are ignored, waste disposal is uncontrolled, zoning of the park is not clear, and infrastructure construction is not environmentally friendly enough. This status quo level leads to natural environment disturbance.

    2. Satisfied: This denotes a relatively more desirable level where encroachments are addressed and controlled, and waste disposal is controlled through the provision of enough trash collection bins along the trails and the coast. In addition, the park management distinguishes restricted areas for tourism, while infrastructure construction inside the park is minimized.

    3. Very good: This implies the best conditions with prevention of encroachment, a clean environment and scheduled waste management, provision of enough trash collection bins along the trails and coast, and the provision of education and enforcement programs through park guides. These measures

are implemented in addition to park zoning, allowing visitors patronage only to recreational and economic activity zones. Moreover, strict access controls are enforced in and around construction projects, even in areas adjacent to the park.

- Recreational Facilities: This study contemplated the facilities' quality and numbers of washrooms; availability of showers, shelter, and picnic areas; safety nets or guard rails; and the cleanliness of beach areas. The levels of this attribute include:

  1. Current condition (Unsatisfactory): This denotes the status quo, where only limited numbers of washrooms and showers are available at the entrance: Sungai Tukun and turtle beach. Guard rails and safety nets are rare, and waste has been allowed to litter surroundings, most visibly along the beach.
  2. Satisfactory: This envisions relatively better facilities including the provision of washroom and camping facilities for 150 campers at a time in allowed zones, providing safety ropes in dangerous area especially for children and elderly people. This is besides keeping the beach areas fairly clean.
  3. Very satisfactory: This level describes the best provision of facilities including washrooms, water supply, and camping areas for 200 campers at a time, providing safety ropes wherever needed along trails, and keeping beach areas very clean.

- Provision of Information: This includes the provision of information about flora, fauna, and park attractiveness. This could be useful to educate visitors and communities living near the ecotourism sites. This enhanced and relevant awareness might later encourage the nearby residents to support nature protection in the nearby park and to reduce the degradation of the park's resources. Providing information can be achieved through availability of interpretive programs, deftly positioning information boards, signboards, and posters in the right places, where relevant brochures and pamphlets are also made available. The effectiveness of this program depends on how well the park's visitors can get useful information inside the park. This means the design of the materials and their distribution and availability at appropriates place will determine how effectively even well-designed, well-written information can be used. Inside PNP at present, the tour guides are neither licensed nor trained; training them would be very helpful in providing information for tourists.

  1. Current condition (Low): This denotes limited information, direction, and signs only for management and safety purposes;

informative paper factsheets, brochures, and pamphlets are only available at the information counter; guides are not educated about the park's mission, features, history, and ecological utility.

2. Medium: This describes a better condition, where directions and signs for management and user safety are provided in key intersections. Signs are easy to read and identifiable, park guards and educated guides provide information on turtle beach and for some critical habitats and species of the park.

3. High: This proposes the best level of providing information services, where directions and signs for management and safety are provided in intersections and near camping area and restrooms, and where park guards and guides act as one-on-one guides to visitors in the park, enlightening them on how to contribute to the sustainability of the park and environmental friendly use of the park.

- Cost: Although entrance to the PNP is free at this time, PNP managers are being asked to reflect on the acceptability of introducing an entrance fee. The payment vehicle attribute must be defined carefully so that all the respondents are able to act as they make the real choices and, therefore, any hypothetical bias in the study can be reduced [3,15]. Suggestions for the entrance fee include possible payments of RM 0, RM 5, RM 10, and RM 15.

3.1.2. Choice Sets Design

Applying the full factorial design from an original set of attributes and levels allowed 108 (3 × 3 × 3 × 4) possible combinations. Implementing the full factorial design is difficult because it contains so many combinations. Hence, we applied an orthogonal main effect design (using SPSS) to generate a smaller number of combinations. The final design included 8 alternatives in four choice sets, each choice set including two proposed alternatives plus the status quo. Alternatives A and B entailed various combinations of suggested management policy-attributes, along with cost, while alternative C always represented the current situation, at no additional monetary cost. The choice of alternative A, B, or C to each question yielded information about the selected scenario for a given respondent (Table 2).

Table 2. Sample of choice sets.

ATTRIBUTE	MANAGEMENT OPTION A	MANAGEMENT OPTION B	CURRENT MANAGEMENT PRACTICE
Ecological management Proper management of the park with minimum negative effect on flora, fauna and physical environment quality such as water quality and natural scenery	Very Good Enough trash bins everywhere, provision of information by educated guards, tourism activity only in recreational and economic zone, no construction is allowed	Very Good Enough trash bins everywhere, provision of information by educated guards, tourism activity only in recreational and economic zone, no construction is allowed	Not Satisfied Uncontrolled waste disposal, no zoning, environmental disturbance
Recreational facilities Quality and numbers of washrooms, availability of shower room, shelter and picnic areas, availability of interpretive programs, safety nets or guard rails and cleanliness of beach areas	Very Satisfied Provision of facilities including washroom, water supply and camping area for 200 campers at one time, providing safety ropes wherever needed along trails, very clean beach areas	Satisfied Provision of enough washrooms and facilities for 150 campers at one time, providing safety ropes in needed area considering kids and elderly people, fairly clean beach area	Not Satisfied Limited numbers of washrooms and showers, Guard rails and safety nets are rare and waste is allowed to litter surroundings most especially along the beach

ATTRIBUTE	MANAGEMENT OPTION A	MANAGEMENT OPTION B	CURRENT MANAGEMENT PRACTICE
Information Provision of information about flora, fauna and park attractiveness	High Directions and signs are provided in intersections, near camping area and restrooms, Educated Park guards and guides enlighten visitors with environmentally friendly use of the Park	Low Limited information, brochures, and pamphlets are only allocated at the information counter, guides are not educated	Low Limited information, brochures, and pamphlets are only allocated at the information counter, guides are not educated
			RM 0
Your Choice			

While the sets of alternatives change between questions, an analysis of numerous responses allows a statistical estimation of each attribute level's perceived value. It is assumed that responses from a given respondent are independent [31]. As such, in regression analysis, the total number of valid responses to the choice questions indicates the sample size and not the number of sampled respondents [31]. Once the set of attributes and levels were defined and choice sets were developed, the questionnaire was prepared in Bahasa Malay and English versions, to present to the surveyed tourists.

3.2. Sampling and Data Collection

Using outcomes from an initial pilot study on visitors at the park, the final version of the questionnaire was prepared, consisting of four parts. The first part was introductory script and information about PNP, highlighting the values of the park and the problems it is facing. This part was followed by information on PNP choice sets and definitions on attributes of the choice experiment and related levels such as ecological management, tourist guides, recreational facilities, provision of information, and entrance fees.

Follow-up questions were included to capture the respondents' reason for selecting current condition or "status quo." "The questions are ambiguous", or "I do not accept paying an entrance fee to enter the national parks" or "I agree with current management policies, and it should continue operating with only tax revenues, as in past years" were some of the options presented to the respondents.

The second part contained questions to gauge visitors' attitudes towards the park. This part included questions related to respondents' information about the park, their reasons for interest in visiting this park, and their number of previous visits to this park. The third part contained some questions to elicit the respondents' preferences on "improvement in management strategies", "ecotourism development", and their views on "implementing an entrance fee for PNP" with a five-point (Likert-like) scale: extremely important, important, of average importance, of little importance, or not important at all. The final section included questions to obtain the socioeconomic status of the respondents. This section included questions about their age, gender, occupation, education level, income level, residential status, and nationality.

The survey was conducted at PNP from February through April 2012. Based on the National Oceanic and Atmospheric Administration (NOAA) panel guidelines [32], in-person interviews with park visitors were considered appropriate. The target population included both domestic and foreign visitors. A systematic random sampling method was used for sampling purpose. It means every nth visitor aged over 18 years old was intercepted when passing by the interviewer, then asked about their interest and time to fill out the survey. The study location was chosen to ensure that the tourists had visited the PNP and that they were walking toward an exit from the park. This approach was taken because the experience of the respondents from the park and its existing conditions could help them to respond more accurately. The surveys were carried out simultaneously at specified locations (the park's two main exits are Monkey Beach and the main gate). This chosen method was considered fair and representative, as it ensures that all populations would have an equal chance to be selected as respondents. The data collection was carried out continuously from 9:00 am to 6.00 pm. The Mitchell and Carson CV method was applied to estimate the sample size [33]. Mitchell and Carson suggested considering a coefficient of variation of 2.0 when determining sample size [33]. Therefore, using their formula and tolerating the disparity of 20% between the true population mean and sample estimate, the mean sample size of 286 was estimated [33]. Considering this and our available budget, a sample size of 300 respondents was determined for this study. Overall, 12 responses were eliminated on account of either continuous selection of status quo—without considering it as their best choice—or incomplete answers to choice questions. Therefore, their answers were considered a mimic of protest vote and thus were eliminated from the analysis. However, those respondents with genuine zero-bid were retained. These are those respondents who "However, agreed with improvements but could not afford to pay" and "preferred to spend that money on other things". Statistical analysis and model estimation was carried out using SPSS (Ver.17) and Nlogit 5 software.

3.3. Econometric Models

A random parameter logit model (RPL) which is a generalization of the standard multinomial logit model (MNL) is applied in this study for two main reasons. First, it explicitly accounts for unobserved preference heterogeneity across respondents; and second, it assumes that the alternatives are not independent [34] A.C. Cameron and P.K. Trivedi, Microeconometrics: Methods and Applications, Cambridge University Press, New York (2005). Thus, it has an advantage of relaxing the assumption of Independence of Irrelevant Alternatives (IIA) [35].

The RPL relies on random utility theory [36,37] and the economic theory of value [38] to provide a utility theoretic interpretation of the discrete responses derived from respondents. The individual utility function (U) for each respondent (i) could be disaggregated into two parts. One such part is the deterministic component (V), which would typically be specified as a linear index of the attributes (X) of the *j*th alternative in the choice set. The second component is a stochastic element (ε) which represents the error term:

$$U_{ij} = V_{ij}\,(X_{ij}) + \varepsilon_{ij} = \beta_i X_{ij} + \varepsilon_{ij} = \beta' X_j + bX_j + \varepsilon_{ij} \tag{1}$$

where U_{ij} is the utility individual *i* obtains from alternative choice set *j*, V_{ij} is a non-stochastic (determinist) utility function, ε_{ij} is a random component. In the right side of the model the decomposition of utility functions, β refers to sum of population mean (β′) and individual deviation from the mean (b).

For instance, consider a case where an individual (*i*) is asked to choose between alternative goods (V_j and V_k) that are assumed to be differentiated by their attributes and levels. In choosing between them, the respondent is assumed to compare the utility that he or she could get with either choice before selecting the alternative or the higher utility:

$$P_{ij} = (V_{ij} + \varepsilon_{ij}) > (V_{ik} + \varepsilon_{ik}) = P[(V_{ij} - V_{ik}) > (\varepsilon_{ij} - \varepsilon_{ik})] \tag{2}$$

Each attribute, except for the cost in terms of monetary value (CC), was divided into three levels. The status quo, or Level 1, was set as the baseline, and Levels 2 and 3 were agreed to imply proposed management options, respectively. Attribute levels were "dummy coded," meaning they were set equal to 1 if the corresponding level was present, and set to 0 otherwise. In all models, the base level was the status quo.

For an estimation of willingness to pay (WTP), the following formula can be applied:

$$MRS = \beta_{X1}/\beta_{C} \qquad (3)$$

This value for any attributes (other than price) is called implicit price or marginal rate of substitution (MRS). For marginal change in an attribute, WTP value is typically derived by dividing the β value of each non-monetary attribute by the β value of the price attribute. The specification of RPL can be found in Hensher and Greene [34] and Hensher *et al.* [39,40].

3.4. Structural Equation Modelling

3.4.1. Exploratory Factor Analysis

Factor analysis has been widely used in social science literature to identify and categorize the latent structure of a set of variables. Therefore, a large set of variables which measure similar things will get into a smaller set of unobserved variables called factors. In the present study, 18 attitudinal items were collected from respondents. These variables were measured on a five-point (Likert) scale ranging from 1 (not important at all) to 5 (extremely important). Nevertheless, including all 18 items in the WTP model seems rather impractical. Hence, in the present study, exploratory factor analysis was applied in order to reduce 18 attitudinal items into a smaller number of identified factors [41]. In order to examine the appropriateness of items to use in factor analysis, the reliability test using Cronbach's alpha was conducted. Reliability of factors in this research was guaranteed with Cronbach's alpha value higher than 0.7 threshold [42]. Using a reliability test, one item was eliminated. Conducting correlation estimation between variables showed that multicollinearity is not an issue. To measure the validity of the latent variables, exploratory factor analysis with Promax rotation was applied in SPSS software.

3.4.2. Confirmatory Factor Analysis

Structural equation modelling (SEM) was employed in order to analyse factors influencing each person's choice using observed and latent variables (Confirmatory Factor Analysis). SEM as a statistical tool can be used to analyse and estimate the casual relationships between different observed and latent variables. Latent variables (or hidden variables) are those variables which are not observed directly but which are inferred from other observed variables. It is worth noting that SEM can take both direct and indirect relationships. This means that some relation between variables might be improved by interconnecting them. SEM results can be decomposed to the measurement model and the structural model [43]. While the measurement model defines the relationships between latent variables and their observed indicators, the structural model explains the relationships between and among exogenous and latent endogenous variables. Endogenous variables are those that are dependent at least in one of the models equations. In the present study, as suggested by Hess and Beharry-Borg [44], it was assumed that answers to attitudinal questions are only a function of underlying attitudes rather than a direct measure of them. In contrast, exogenous or independent variables are observed variables, such as the socio-demographic information of respondents.

If we assume $y = (y_1, y_2, y_3,\ldots y_p)$ is vector of observable endogenous variables and $x = (x_1, x_2, x_3, \ldots, x_q)$ is vector of observable exogenous variables, the measurement model can be written as follow [43]:

$$y = \Lambda_y \eta + \varepsilon \tag{4}$$

$$x = \Lambda_x \zeta + \delta \tag{5}$$

where Λ_y and Λ_x are (p × m) and (q × n) matrices of regression coefficients. The *y* variables are function of some latent endogenous variables so called η's, while *x* variables are vector of latent exogenous variables so called ζ's.

The structural equation model which specifies the relations between these latent variables can be expressed as follows [43]:

$$\eta = B\eta + \Gamma\zeta + \zeta \tag{6}$$

where *B* is an (m × m) coefficient matrix with the regression coefficient measuring the effect of the η's endogenous variable on the other η's endogenous variables; Γ is a (m x n) coefficient matrix representing the effect of the ζ's exogenous on the η's endogenous variables; ζ is the vector of the error term.

The conceptual model based on equations (4–6) reads as:

$$y = \begin{pmatrix} y_1 \\ y_2 \\ y_3 \\ . \\ . \\ . \\ y_{17} \end{pmatrix} = \begin{pmatrix} 1 & 0\,0 & \dots & 0 \\ \gamma_{2,1} & 0\,0 & \dots & 0 \\ \gamma_{3,1} & 0\,0 & \dots & 0 \\ & . & & \\ & . & & \\ & . & & \\ 0\,0\,0 & & \dots & \gamma_{17,n} \end{pmatrix} \begin{pmatrix} \eta_1 \\ \eta_2 \\ \eta_3 \\ . \\ . \\ . \\ \eta_n \end{pmatrix} + \begin{pmatrix} \varepsilon_1 \\ \varepsilon_2 \\ \varepsilon_3 \\ . \\ . \\ . \\ \varepsilon_{17} \end{pmatrix} \quad (7)$$

$$\mathrm{x} = \begin{pmatrix} \text{Age} \\ \text{Gender} \\ \text{Education} \\ . \\ . \\ . \\ \mathrm{x}_8 \end{pmatrix} = \begin{pmatrix} 1\,0\,0 \dots 0 \\ 0\,1\,0 \dots 0 \\ 0\,0\,1 \dots 0 \\ . \\ . \\ . \\ 0\,0\,0 \dots 1 \end{pmatrix} \begin{pmatrix} \text{Age} \\ \text{Gender} \\ \text{Education} \\ . \\ . \\ . \\ \mathrm{x}_8 \end{pmatrix} \quad (8)$$

The structural model as then can be expressed as:

$$\begin{pmatrix} \eta_1 \\ \eta_2 \\ \eta_3 \\ . \\ . \\ . \\ \eta_n \end{pmatrix} = \begin{pmatrix} 0 & \beta_{1,2} & \beta_{1,3} & \cdots & \beta_{1,n} \\ \beta_{2,1} & 0 & \beta_{2,3} & \cdots & \beta_{2,n} \\ \beta_{3,1} & \beta_{3,2} & 0 & \cdots & \beta_{3,n} \\ & & . & & \\ & & . & & \\ & & . & & \\ \beta_{n,1} & \beta_{n,2} & \cdots & \beta_{n,n-1}, & 0 \end{pmatrix} \begin{pmatrix} \zeta_1 \\ \zeta_2 \\ \zeta_3 \\ . \\ . \\ . \\ \zeta_n \end{pmatrix} \quad (9)$$

It is important to note that in the measurement model for identification purposes, one of the coefficients of each latent variable set as 1. In structural model, however, absence of impact between variables results in setting that coefficient at zero.

4. STATISTICAL RESULTS

4.1. Sample Characteristics

The characteristics of respondents are shown in Table 3. Large percentages of visitors at the PNP were locals (84.7% *vs.* 15.3%). The respondents were between 18 to 76 years old and majority of them were men (64% *vs.* 36%).

The number of married respondents (58%) was slightly larger than unmarried ones (42%). In terms of occupation, a majority of the respondents were working in the private sector (59%), 28% in governmental organizations, 4.8% were students, and 7.6% were retired. About 34% of the respondents were from a rural area, and the rest (60%) from an urban area. In terms of education, 30% had a university degree, and the rest had some college education (13%), secondary school (54%), or only primary school (3%). The average income of visitors was RM 3600 per month. With respect to the local demographics, the Malaysian household average monthly income is MYR4025 per month, while about 60% of households earned around MYR3500 or less [45]. The income levels as gauged by this study were therefore representative of the Malaysian people broadly.

Table 3. Socio demographic characteristics of the respondents.

Variable	Frequency		Mean	St. Deviation	Type of Varible
	Number	(%)			
Gender			0.64	0.48	dummy
Male	159	63.9			
Female	90	36.1			
Age			36	13.03	continues
Citizenship			0.18	0.39	dummy
Malaysian	211	84.7			
Other	38	15.3			
Strata					
Urban	163	65.5			dummy
Rural	86	34.5			
Education Level			14.77	3.04	categorical
Primary school	7	2.8			
Secondary school	135	54.2			
College	33	13.3			
University	74	29.7			
Marital Status			0.58	0.49	dummy
Single	105	42.2			
Married	114	57.8			
Variable	**Frequency Number**	**(%)**	**Mean**	**St. Deviation**	**Type of Varible**
Employment			2.23	1.14	categorical
Student	12	4.8			
Government Sector	72	28.9			
Private sector	146	58.7			
Retired	19	7.6			continues
Income (MYR)			3649.2	6323.8	
Sample size	288				

4.2. Results of Exploratory and Confirmatory Factor Analysis

As Table 4 shows, the results of exploratory factor analysis revealed three interpretable factors with Eigen values greater than 1. The first latent variable was formed from 6 attitudinal variables related to respondents' concern on PNP's importance as a national park (NP). The second latent construct was established from four attitudinal variables related to PNP management concerns (MNG). The third latent variable was constructed from three attitudinal variables related to an ecotourism perspective on PNP (TOUR).

Table 4. Factor Structure of Visitor's Perceptions on PNP

Variables	Factors			Cronbach's
	NP	MNG	TOUR	alpha = 0.797
Visiting Penang National Park has encouraged me towards conservation efforts	0.530			0.787
Activities in PNP promote Environmental Education programs	0.608			0.781
PNP provides an important view point on interaction between human and nature	0.551			0.785
Conservation efforts at PNP have increased my awareness of the importance of national parks	0.434			0.784
PNP has an important role in provision of environmental information	0.678			0.789
PNP has an important role as ecotourism destination	0.710			0.778
Interactive programs like fishing and canopy walking and camping increased my knowledge about the value of protected areas	0.688			0.779
PNP has a role in wildlife conservation	0.732			0.780
PNP is of great value for conducting research	0.619			0.779
Provision of a better integrated management plan in PNP		0.569		0.781
Better enforcement of rules and regulation		0.752		0.788
Implementing entrance fees for conservation charge		0.728		0.784
PNP authorities and managers find additional sources of revenue for efficient management purposes		0.615		0.785
Implement licenses or permits and charges, on certain activities such as fishing, using cameras and camping		0.462		0.786
Develop and implement the interpretation programme			0.612	0.787
Providing licensed guides			0.665	0.783
Improvement of recreational facilities			0.546	0.787
Implementing tax on tour operators or tourist guide				0.868
Eigenvalue	5.63	1.93	1.18	

Extraction Method: Maximum Likelihood. Rotation Method: Promax with Kaiser Normalization.

The results of SEM with underlying path diagrams for estimated models are presented in Figure 2, where observed components are shown in rectangles and unobserved (latent) variables in ellipses. In order to reach or determine the best model fit, several models were computed in AMOS. The final model resulted in eliminating highly insignificant variables and consisting of a covariance path to link error terms of latent or observed constructs. Hence, the final result was achieved with 13 variables out of the primary 18. The final model shows all factors' loading are highly significant, with coefficients considerably higher than minimum recommended 0.2 [46]. The standardized total effect on PNP, MNG, and TOUR corresponding to each variable in the model was estimated as well. Results show that interaction with nature and awareness role of PNP have the largest effect on NP with total effect of 0.71

and 0.70, respectively. Implementing enforcements with rules and regulations and permits and charges on some activities are the two most influential variables of MNG with total effects of 0.75 and 0.73, respectively. Finally, providing licensed guides, with total effect of 0.73, and improving recreational facilities, with total effect of 0.58, are the variables with the largest effect on TOUR. From exogenous variables, only education was found to have a direct and positive effect on latent endogenous variables.

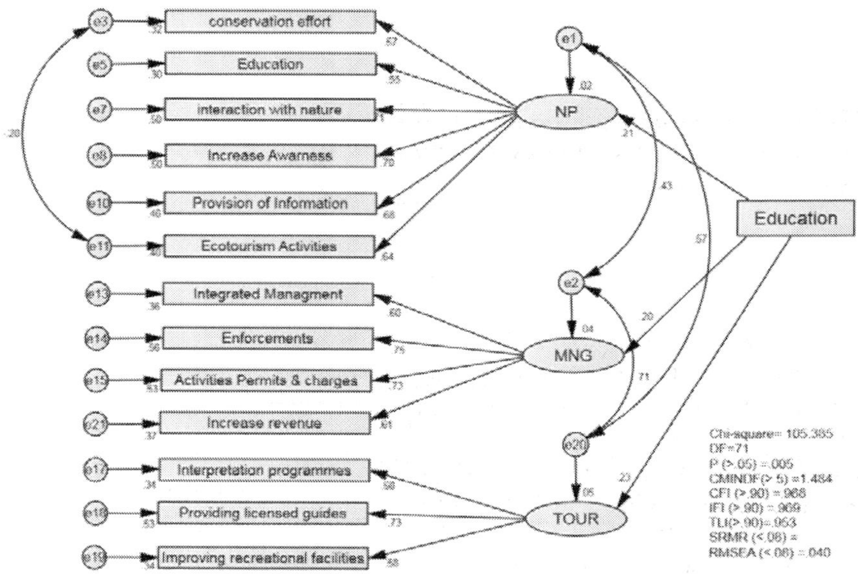

Figure 2. Results of confirmatory factor analysis. All path coefficients shown are completely standardized.

The estimated model has an adequate fit according to the measures of absolute, incremental, and parsimonious fit. The final model shows a normed chi-square value of 1.48 (within the recommended interval of 1 to 3), root mean square error of approximation (RMSEA) values of 0.04 is below the threshold maximum value of 0.08. Comparative fit index (CFI) value and incremental fit index (IFI) value of 0.96 also indicate that the model has a good fit [47].

4.3. Results of Choice Experiment

In the simple multinomial logit model (MLM) (Table 5), all variables except ecological management (level 2), recreational facilities (levels 2), and cost were significant and exhibited the expected positive sign. This evidences that improvement in ecological management along with recreational facilities and information about PNP is supported by visitors.

Table 5. Results of Simple MNL and RPL Models.

	Multinomial Logit [1]		Random Parameter Logit Model [2]	
Variable	Coefficient	S.E	Coefficient	S.E
EM2	1.12	1.06	0.07	0.40
EM3	3.36 ***	0.65	2.09 ***	0.49
RF2	0.12	0.15	0.12	0.30
RF3	2.50 **	1.15	1.07 ***	0.36
PIS2	9.26 ***	3.24	5.77 ***	2.02
PIS3	4.87 ***	1.15	3.00 ***	0.49
CC	−0.10	0.11	−0.39 ***	0.11

Notes: ***, **, * ==> Significance at 1%, 5%, 10% level; [1] Log likelihood function = −532.71, Number of observations = 996; [2] Log likelihood function = −524.23, Chi squared = 1167.96, Significance level = 0.00000, McFadden Pseudo R^2 = 0.52, Number of observations = 996.

Since the use of an RPL model has advantage of counting for heterogeneity in preferences and paves the way for relaxation of the IIA assumption, we tried to apply an RPL model. Several approaches can be employed to improve model fit and to increase the estimating models' accuracy. Each attribute, except for the conservation value in terms of monetary value (CV), was divided into three levels and recoded as dummy (0, 1): The status quo was set as the baseline, and Levels 2 and 3 were agreed to imply medium and high levels of each attribute, respectively. In all models, the base level was the status quo, or first level of each attribute. Results of simple RPL model are shown in Table 5. Using the likelihood ratio test the estimated value of 16.96 with 6 degrees of freedom was higher than critical value of 16.812 at the 1% significance level. This implies that the RPL model fits data significantly better than the MNL model. However, the RPL model fit data better than the MNL but neither parameters of EM2 nor RF2 were significant.

In view of the fact that socio-demographic variables remain the same for any given respondent regardless of selecting any given alternative, these variables enter the model through interactions with the attribute variables. The interaction variables comprised 8 socio-demographic and three attitudinal factors interacting with each levels of main attribute. These variables included age, gender, education level, employment situation, income, region, marital status, and visitor's origin. Age, education, and income are continuous variables; others are 1–0 dummies. Only significant variables were retained and reported in the final model. The interaction model is shown in Table 6. The results demonstrated that the pooled RPL model was superior to the simple RPL due to the model fit and expected variables signs. Both simple and extended RPL models were estimated assuming bounded normal distribution [34] based on 150 Halton draws under panel data assumption. We assumed that all variables other than conservation value or cost variable (CC) are random. The cost coefficient was kept fixed to avoid random variations in the cost coefficient. A variable was treated as random if its coefficient of standard deviation was statistically significant at the 10% level. Results presented in Table 6 are from the final model where the variables EM2, EM3, RF2, RF3, PIS2, and PIS3 were found significant and comprising expected signs.

Table 6. Results of pooled RPL Model and IP.

Variable	Coefficient	S.E	Coeff. std.	Marginal WTP (MYR)
		Random parameters		
EM2	1.82 ***	0.52	1.82 ***	5.22 * (−0.11 to 10.55)
EM3	2.17 ***	0.62	2.10 ***	6.23 ** (1.07 to 11.39)
RF2	0.84 **	0.42	0.84 **	2.42 ** (0.0078 to 4.84)
RF3	2.10 ***	0.56	2.17 ***	6.03 *** (1.90 to 10.16)
PI2	1.11 **	0.45	1.11 **	3.17 * (−0.029 to 6.65)
PI3	2.97 ***	0.65	2.97 ***	8.51 *** (2.16 to 14.86)
		Non-random parameters		
CC	−0.34 **	−0.13		
RF3_JOB	0.31 *	0.18		
RF3_INC	0.89 ***	0.14		
PIS2_EDU	0.67 ***	2.91		
EM3_EDU	0.79 ***	0.26		
EM3_MNG	1.96 ***	0.52		

Notes: ***, **, * ==> Significance at 1%, 5%, 10% level. Log likelihood function = −512.52; Chi squared = 1163.39; Significance level 0.00000; McFadden Pseudo $R^2 = 0.53$; Number of observation = 996.

The significant standard deviation of all random variables indicates variation of these parameters in the population. This confirms our earlier assumption on presence of preference heterogeneity in the sample.

Since visitors' WTP may be influenced by their demographic characteristics or their attitudes, it is therefore important to consider these factors when implementing a potential policy. In our study, visitors' income, education, employment conditions, and management concerns were the most important variables in the CE model.

4.4. Willingness to Pay Estimation

Marginal values were estimated by dividing the coefficient of each attribute by the coefficient of price. The results of marginal WTP is presented in the last (rightmost) column of Table 6. Estimated marginal values for various attributes were estimated for hypothetical improvement from current condition to level two and three of each attribute. Results showed that PI3 and EM3 received highest value.

Once these parameters are estimated, welfare measures for a desired change in national park management can be calculated. Welfare measures are particularly beneficial for policy purposes with estimating changes in welfare associated with a particular policy [48]. Using Hanemann's method, welfare measure is represented by following formula:

$$CS = 1/\alpha\,[\ln \Sigma \exp (V_i 1) - \ln \Sigma \exp (V_i 0)] = -1/\beta cv\,(V_i 1 - V_i 0) \quad (10)$$

In estimating the related utility following Hensher *et al.*, 2005 [39], distributional assumptions related to random parameters were considered. The welfare measure for improvement in management of PNP against status quo is estimated at about MYR 9. The optimal rate in this study was determined by PNP management, based on perspectives of the majority of visitors to prevent any disproportionate effect on low-income citizens or other social groups. Therefore, the study suggests that the optimal entrance fee of MYR 10 (for year 2014) would be accepted by most visitors and would maximize PNP overall revenue, as well. Aggregate results were obtained by multiplying the estimated WTP with the number of annual visitors. The findings show that the net economic value of the PNP would have been about MYR 1,082,760 (US$ 338,362) in 2012. This estimated amount can be increased by growth in the number of visitors and by imposing a larger admission fee.

5. DISCUSSION OF THE RESULTS

According to SEM results, the importance of national parks and approaches to improve management and tourism activities in the PNP are strongly related. This indicates that people's appreciation of the national park and their awareness about value of the park is a prerequisite for growing concerns about management and tourism improvement and, conversely, concerns about management improvement and meeting tourists' expectations promote and facilitate people's appreciation and awareness. The results have special importance for PNP management since they suggest that park visitor satisfaction is, on average, increased through the provision of informational resources that inform their experiences. The PNP managers also need to specify necessary programmes that engage visitors in contemplative experiences and ensure the long-term efficiency of sustainable environmental behaviors. As people get more opportunities to be involved in park-related issues, they change from passive visitors to active visitors that care about management of the park and other visitors' behavior. The significant effect of education on latent variables indicated that persons with higher education have more positive attitudes about the value of PNP and are more concerned about seeing improvements in management and the tourism facilities in PNP.

The results of the choice experiment showed that respondents also valued the PNP proposed management scenarios highly. This emphasizes respondents' expectations for park managers to move away from the current condition of the park by providing better management alternatives. Accordingly, variables EM2 and EM3 were positive in accordance to prior expectations and significant at the 1% level. This means improvement in ecological management of the PNP to the satisfied and very good level is preferred to the current condition level. This result is in agreement with earlier findings (e.g., [3,49,50]), showing that better biodiversity conservation and environmental protection contributed to increases in overall utility for the respondents.

Level 2 and level 3 of recreational facilities were significant at the 5% and 1% levels, respectively, with an expected positive sign. The lack of facilities and their quality (such as cleanliness) results in complaints from visitors. Waste disposal especially affects the beauty of beaches in PNP and makes it undesirable to visitors. Therefore, the positive sign of this attribute implies that visitors are willing to pay to improve the basic recreational facilities of PNP. Similarly, previous studies showed that the quality and quantity of facilities such as washrooms, camping and picnic areas, and suitable beach areas can affect experiences and satisfaction among visitors [13].

Providing information directly to the visitors may be their only opportunity to learn about the ecological importance of these protected areas in harmoniously balancing relationships between plant and animal species in nature. Hence, an effective and enlightening approach explaining and interpreting the park to the visitors can contribute to sustaining waves of tourists by educating them about what they are seeing and experiencing, assisting them in minimizing their potentially negative impacts, and encouraging those activities which ensure ongoing maintenance of the quality of the environment. Accordingly, variables PIS2 and PIS3 (provision of information) were significant at the 5% and 1% levels, respectively, with an expected positive sign. The positive sign means that visitors are concerned about information and priorities education. The results are consistent with [3,5,17,51], in that they confirmed that interpretation and information provision is favored by visitors who had been interviewed. The cost variable (CC) which was offered as possible entrance fee was significant at the 5% level with an expected negative sign. Visitors prefer those programs that do not require any additional cost. Thus, the sign of the payment coefficient was negative, which indicated the effect on utility of choosing a choice set with a high payment level, as expected.

The interactions between income and employment with level 3 recreational facilities, education with improvement in ecological management of PNP, and provision of information were significant and positive. This implies that persons with higher income and higher education who are currently employed were more concerned about improvement in management and conservation of PNP.

This result seems reasonable since a higher level of education and income could be related to a better understanding of the problem and greater ability to pay. The positive relationship between these variables is generally supported by the literature. For example [16,50,52,53,54,55] indicated that highly educated people and those with higher income are more supportive of nature preservation and have higher WTP.

The positive relationship between perception of improvement in PNP management and EM3 (level 3 of Ecological Management) was significant at the 1% level with an expected positive sign. This means that people with high appreciation and knowledge about PNP and more concern about improvement in park management have higher WTP.

Marginal willingness to pay estimation indicated the high value of provision of information between respondents, as most of them visit the park with their families or with educational groups. Hence, they appreciate the provision of information about PNP. Results also showed that visitors preferred more adequate management of the park's ecological and biological aspects. This quantitative information about visitors' preferences can help park managers and policy makers decide about future park programs [56] and also possibly introduce an entrance fee as a revenue stream to fund park upkeep and improvement. Introducing and collecting an entrance fee can provide the park with additional income beyond what the government budgets. The revenue generated from tourists who value the ecosystem for its inherent qualities can be used to support the values of selected tourists, to protect biodiversity and ecological integrity, mitigate environmental problems, employ educated guards and tour guides, and enhance public education programs. Also, benefits from strengthened park management could strengthen local communities whose members could participate in park activities, and as a result, this helps to enhance quality of life for local residents.

The estimated WTP amount in this study is generally equal to other Malaysian protected areas. For example, in addition to PNP, entry is currently free for Malaysia's National Elephant Conservation Center and the Tasek Bera Ramsar Site. While the entrance fee for foreign visitors to some national parks such as Taman Negara Kelantan, Taman Negara Pahang, and Taman Negara Terengganu is $3.33 per visitor, the estimated entrance fee is lower than other developing countries such as Thailand, Nepal, Costa Rica, or Ecuador. Taking into account the fact that the number of visitors to the national parks (with low entrance fee) is relatively low compared to total visitors, other policy instruments such as a departure green tax could be considered in countries such as Malaysia,.

In our study, the point and cross elasticity measurement between alternatives gave us the opportunity to predict how likely tourists would react to any future change in the entrance fee. In simple terms, elasticity measures the relationship between percentage change of variable (or program) and percentage change in quantity demanded, *ceteris paribus* [39].

Based on our findings, a 1% increase in the entrance fee will decrease the probability of selecting option 1 and option 2 management alternatives each by 0.6%. However, the price elasticity for the option 1 alternative is relatively inelastic. Therefore, overall results suggest that improving management along with imposing any admission fee above zero up to MYR 10 would increase PNP revenue. Since aggregated revenue can be estimated from price multiplied by the number of likely visitors, a 10% increase in entrance fee should increase revenue by about 4.5% (we assumed MYR 1 as the baseline price).

A simulation method in Nlogit5 was used to estimate a "what-if" scenario. The selected scenario assumed imposing an admission fee of MYR10 (compared to zero, or no entrance fee at the moment). Imposing an RM10 entrance fee would produce an estimated market share of 47.43% (down from 48.9%) and 49.05% (down from 50.17%) for option 1 and option 2 alternatives, respectively. These two options propose better management of PNP. The same change would produce a market share of 3.51% (up from 0.93%) for the status quo. This means that increasing the entrance fee from zero (reflecting the status quo) to RM10, decreases the share of options which contain or involve better park management (namely option 1 and option 2) by 1.45% and 1.12% respectively. Since these two improvement options are unlabeled, it implies that 2.57% (26) of the original choices for these alternatives will prefer the status quo option of "no management improvement" with zero entrance fees. Findings also indicated that for any improvement in the management of PNP, imposing an admission fee of MYR10 will result in almost 96% market share of these alternatives compared to the status quo. Based on Malaysian GDP growth, the income increment is around 4.5% per year. Therefore, we calculated a 5% annual price and simulated the results for 5 years. Our results demonstrated that under the 5% price increment scenarios, the decline in original choices will be 0.8% each year. Considering this and improvement in management along with awareness could be a useful tool to determine possible future revenue amounts based on the distribution of incomes in likely tourists.

6. CONCLUSIONS

Malaysia's tourism vision involves making the industry a primary source of national revenue and a major contributor to the socio-economic development of the nation. The tourism sector has great potential for contributing numerous financial benefits, directly or indirectly. It helps generate growth and development for a nation, job opportunities, foreign exchange earnings, and government revenue.

The present study utilized a choice experiment as one feasible approach to analyze users' preferences for improving the management of PNP for the dual purposes of conservation and recreation. The results of the study revealed that visitors prefer improvement in PNP management. These results demonstrated that the expectations of tourists are more likely to be fulfilled when there is improvement in the ecological conditions of the park. This is in addition to their desire for readily-available information on primary recreational facilities. In general, more than 86% of respondents selected one condition above the current situation of PNP. The survey respondents expressed the highest value and preference for very good ecological management of park, with efforts to provide more information about its valuable flora and fauna, sensitive areas, and the characteristics of its habitats.

The results of this study can contribute to better-informed decisions for two levels of managers and regional planners. First, park managers need to more robustly understand how nature-based tourism can educate a broader and more socioeconomically diverse base of prospective tourists. Second, the longer-term ecological goals of conservation require implementing sustainability principles, and national park managers must integrate all of these understandings with their current projects and the continuous improvement of their management practices. Important conservation and development decisions can be made by considering users' preferences and perceptions. Third, estimated marginal value, which is expressed as an entrance fee and used in trade-off between improved attributes of the park, can reflect the discovery that the park can consider the introduction of an entrance fee. In light of limited budget allocations from the national government, this additional revenue could help fund improved management of the park. It is worth noting that, since demand for ecotourism is usually unpredictable and unstable, it is better to consider entry fees as a "supplementary" source of revenue to government funding and not a "replacement" for it [11]. The successful management of PNP and its development are highly dependent on visitors' satisfaction with its resources. From the visitors' perspective, better park conservation will improve their aesthetic appreciation and enhance their experience and satisfaction from the park [13]. Application of a choice experiment which enables decision-makers to estimate visitor WTP for each particular level of improvement could help park managers focus and schedule their development and management efforts. Since the main aim of establishing national parks is the preservation of sensitive habitats and biodiversity, introducing an admission fee could help the managers concentrate on effectively managing the park rather than on building mass tourism, from which increased congestion would increase stresses on the ecosystems they are trying to protect.

This study results suggests an admission fee of MYR 10 (USD 3.3) for PNP (based on 2013 prices). However, in the sample composition, the proportion of foreign visitors was small. This did not permit the estimation of stratum-specific models. Therefore, in future studies, a sufficient number of domestic and foreign visitors could be targeted to produce better information on the willingness to pay of this group of visitors. Hence, a two-tier charging system would be more efficient. Applying more efficient designs rather than an orthogonal main effect design and more advanced methods to model heterogeneity in preference of visitors could also lead to more realistic and useful results, which our study was unable to consider because of funding constraints.

ACKNOWLEDGMENTS

The authors thank the Department of Wildlife and National Parks Malaysia, Penang National Park staff, and all the respondents who kindly participated and responded to the questions. The authors also would like to thank the editors and two anonymous reviewers for their valuable comments on the manuscript.

AUTHOR CONTRIBUTIONS

Alias Radam, Mohd Rusli Yacob and Nor Hanim Nordin contributed to research design, questionnaire design and administration and data collection. Sara Kaffashi contributed to data analysis and writing the paper.

REFERENCES

1. Liu, C.; Xiao, W.; Li, J.; Pechacek, P. Attitude of tourists visiting nature reserves in China. *Tour. Manag. Perspect.* **2013**,*5*, 1–4.
2. Hein, L. Economic benefits generated by protected areas: The case of the Hoge Veluwe forest, the Netherlands. *Ecol. Soc.* **2011**, *16*, 13.
3. Juutinen, A.; Mitani, Y.; Mäntymaa, E.; Shoji, Y.; Siikamäki, P.; Sventom, R. Combining ecological and recreational aspects in

national park management: A choice experiment application. *Ecol. Econ.* **2011**, *70*, 1231–1239.

4. Journey Malaysia. http://www.journeymalaysia.com/islandpenangtrekking_map.htm (accessed on 10 November 2014).
5. Hearne, R.R.; Salinas, Z.M. The use of choice experiments in the analysis of tourist preferences for ecotourism development in Costa Rica. *J. Environ. Manag.* **2002**, *65*, 153–163.
6. Hamzah, A. Managing the Impact of Visitors to Protected Areas in Malaysia: Success Stories & New Challenges. In Conference Proceedings of the International Workshop on Governance in Asian Protected Areas: Follow-up on the Aichi Target (Post 2010 Target) and Protected Areas Decision from the CBD/COP 10, Akita, Japan, 8–11 January 2010.
7. Turner, R.W. Using Contingent Choice Surveys to inform national park management. *J. Environ. Stud. Sci.* **2013**, *3*, 120–138.
8. Theodori, G.L.; Luloff, A.E.; Willits, F.K. The Association of Outdoor Recreation and Environmental Concern: Reexamining the Dunlap-Heffernan Thesis1. *Rural Sociol.* **1998**, *63*, 94–108.
9. Zaradic, P.A.; Pergams, O.R.; Kareiva, P. The impact of nature experience on willingness to support conservation.*PLoS ONE* **2009**, *4*, e7367.
10. Mansourian, S.; Dudley, N. *Public Funds to Protected Areas*; WWF International: Gland, Switzerland, 2008.
11. Reynisdottir, M.; Song, H.; Agrusa, J. Willingness to pay entrance fee to natural attractions: An Icelandic case study.*Tour. Manag.* **2008**, *29*, 1076–1083.
12. Shahabuddin, Gh. Dynamics between protected areas and economic use. 2009. Available online: http://www.fao.org/docrep/011/i0627e/I0627E10.htm (accessed on 22 July 2014).
13. Loomis, J.; Santiago, L. Economic Valuation of Beach Quality Improvements: Comparing Incremental Attribute Values Estimated from Two Stated Preference Valuation Methods. *Coast. Manag.* **2013**, *41*, 75–86.
14. Yacob, M.R.; Radam, A.; Shuib, A. A contingent valuation study of marine parks ecotourism: The case of Pulau Payar and Pulau Redang in Malaysia. *J. Sustain. Dev.* **2009**, *2*, 95.
15. Ahmad, A.S. Visitors' Willingness to Pay for an Entrance Fee: A Case Study of Marine Parks in Malaysia. PhD Thesis, Department of

Economics, Faculty of Law, Business and Social Science, University of Glasgow, 2009.

16. Han, S.Y.; Lee, C.K.; Mjelde, J.W.; Kim, T.K. Choice-experiment valuation of management alternatives for reintroduction of the endangered mountain goral in Woraksan National Park, South Korea. *Scand. J. For. Res.* **2010**, *25*, 534–543.
17. Lindberg, K.; Veisten, K. Local and non-local preferences for nature tourism facility development. *Tour. Manag. Perspect.* **2012**, *4*, 215–222.
18. Tratalos, J.A.; Sugden, R.; Bateman, I.J.; Gill, J.A.; Jones, A.P.; Showler, D.A.; Watkinson, A.R. The conflict between conservation and recreation when visitors dislike crowding: A theoretical and empirical analysis of the spatial distribution of recreational beach users. *Environ. Resour. Econ.* **2013**, *55*, 447–465.
19. Cerda, C.; Losada, T. Assessing the value of species: A case study on the willingness to pay for species protection in Chile. *Environ. Monit. Assess.* **2013**, *185*, 10479–10493.
20. Wuepper, D. A Recreational Demand Model for Jasmund National Park: Taking into Account Anchoring and Averaging Bias. In Proceedings of the 20th Annual Conference of the European Association of Environmental and Resource Economists (EAERE), Toulouse, France, 26–29 June 2013.
21. Vincent, J.R.; Carson, R.T.; DeShazo, J.R.; Schwabe, K.A.; Ahmad, I.; Chong, S.K.; Potts, M.D. Tropical countries may be willing to pay more to protect their forests. *Proc. Natl. Acad. Sci. USA* **2014**, *111*, 10113–10118.
22. Czajkowski, M.; Budziński, W.; Campbell, D.; Giergiczny, M.; Hanley, N. *Spatial Heterogeneity of Willingness to Pay for Forest Management*, Available online: ftp://188.128.204.246/ spatial_WP .pdf (accessed on 10 March 2015).
23. Mayer, M. Can nature-based tourism benefits compensate for the costs of national parks? A study of the Bavarian Forest National Park, Germany. *J. Sustain. Tour.* **2014**, *22*, 561–583.
24. Chan, N.W. Issues and challenges of managing natural heritage in Penang National Park, Malaysia. Badaruddin Mohamed & Morshidi Sirat. In Proceedings of the 2nd National Symposium on Tourism Research–Theories & applications, Penang, Malaysia, 18 July 2009; pp. 111–123.

25. Hong, C.W.; Chan, N.W. The potentials, threats and challenges in sustainable development of Penang National Park.*Malays. J. Environ. Manag.* **2010**, *11*, 95–109.
26. Laws Of Malaysia Act 226 National Parks Act 1980. Available online: http://www.agc.gov.my/Akta/Vol.%205/Act%20226.pdf (acc essed on 20 February 2014).
27. Samdin, Z.; Aziz, Y.A. The Role of Socio-demographic and Economics on Ecotourism Activities at the Penang National Park, Malaysia: A Services Marketing Perspective. In Proceedings of The 2nd International Research Symposium in Service Management, Yogyakarta, Indonesia, 26–30 July 2011.
28. Addo-Fordjour, P.; Rahmad, Z.B.; Shahrul, A.M.S. Effects of human disturbance on liana community diversity and structure in a tropical rainforest, Malaysia: Implication for conservation. *J. Plant Ecol.* **2012**, *5*, 391–399.
29. Morrison, M.; Bennett, J. Valuing New South Wales Rivers for use in benefit transfer. *Aust. J. Agric. Resour. Econ.* **2004**,*48*, 591–611.
30. Amalia, M. Designing a Choice Modelling Survey to Value the Health and Environmental Impacts of Air Pollution from the Transport Sector in the Jakarta Metropolitan Area (No. rr2010111). In *Economy and Environment Program for Southeast Asia (EEPSEA)*; 2010.
31. Weber, M.A.; Stewart, S. Public values for river restoration options on the Middle Rio Grande. *Restor. Ecol.* **2009**, *17*, 762–771.
32. Arrow, K.; Solow, R.; Portney, P.R.; Leamer, E.E.; Radner, R.; Schuman, H. Report of the NOAA Panel on Contingent Valuation. *Fed. Regist.* **1993**, *58*, 4601–4614.
33. Mitchell, R.C.; Carson, R.T. *Using Surveys to Value Public Goods: The Contingent Valuation Method*; Resource for the Future: Washington, DC, USA, 1989; p. 225.
34. Greene, W.H.; Hensher, D.A. A latent class model for discrete choice analysis: Contrasts with mixed logit. *Transp. Res. Part B: Methodol.* **2003**, *37*, 681–698.
35. Birol, E.; Cox, V. Using choice experiments to design wetland management programmes: The case of Severn Estuary Wetland, UK. *J. Environ. Plan. Manag.* **2007**, *50*, 363–380.
36. Thurstone, L.L. The method of paired comparisons for social values. *J. Abnorm. Soc. Psychol.* **1927**, *21*, 384.

37. McFadden, D. The measurement of urban travel demand. *J. Public Econ.* **1974**, *3*, 303–328.
38. Lancaster, K.J. A new approach to consumer theory. *J. Polit. Econ.* **1966**, *74*, 132–157.
39. Hensher, D.A.; Rose, J.M.; Greene, W.H. *Applied Choice Analysis: A Primer*; Cambridge University Press: New York, NY, USA, 2005.
40. Hensher, D.A.; Rose, J.; Greene, W.H. The implications on willingness to pay of respondents ignoring specific attributes. *Transportation* **2005**, *32*, 203–222.
41. Maier, R. *Knowledge Management Systems: Information and Communication Technologies for Knowledge Management*; Springer: Berlin-Heidelberg, Germany, 2002.
42. Voon, B.H. Service environment of restaurants: Findings from the youth customers. *J. ASIAN Behavioural. Stud.* **2011**,*1*, 45–56.
43. Tang, J.; Folmer, H.; Xue, J. Estimation of awareness and perception of water scarcity among farmers in the Guanzhong Plain, China, by means of a structural equation model. *J. Environ. Manag.* **2013**, *126*, 55–62.
44. Hess, S.; Beharry-Borg, N. Accounting for latent attitudes in willingness-to-pay studies: The case of coastal water quality improvements in Tobago. *Environ. Resour. Econ.* **2012**, *52*, 109–131.
45. Muhammad-Sukki, F.; Munir, A.B.; Ramirez-Iniguez, R.; Abu-Bakar, S.H.; Yasin, S.H.M.; McMeekin, S.G.; Stewart, B.G. Solar photovoltaic in Malaysia: The way forward. *Renew. Sustain. Energy Rev.* **2012**, *16*, 5232–5244.
46. Jöreskog, K.G.; Sörbom, D.; Du Toit, S.H.C. *LISREL 8: New Statistical Features*; Scientific Software International: Lincolnwood, IL, USA, 2000.
47. Toma, L.; Mathijs, E. Environmental risk perception, environmental concern and propensity to participate in organic farming programmes. *J. Environ. Manag.* **2007**, *83*, 147–157.
48. Kjær, T. A review of the discrete choice experiment-with emphasis on its application in health care. Syddansk Universitet: Odense, Denmark, 2005.
49. Kaffashi, S.; Shamsudin, M.N.; Radam, A.; Yacob, M.R.; Rahim, K.A.; Yazid, M. Economic valuation and conservation: Do people vote for better preservation of Shadegan International Wetland? *Biol. Conserv.* **2012**, *150*, 150–158.

50. Wang, P.W.; Jia, J.B. Tourists' willingness to pay for biodiversity conservation and environment protection, Dalai Lake protected area: Implications for entrance fee and sustainable management. *Ocean Coast. Manag.* **2012**, *62*, 24–33.
51. Moscardo, G. Interpretation and Sustainable Tourism. *J. Tour. Stud.* **1998**, *9*, 2–13.
52. Naidoo, R.; Adamowicz, W.L. Biodiversity and nature-based tourism at forest reserves in Uganda. *Environ. Dev. Econ.* **2005**, *10*, 159–178.
53. Westerberg, V.H.; Lifran, R.; Olsen, S.B. To restore or not? A valuation of social and ecological functions of the Marais des Baux wetland in Southern France. *Ecol. Econ.* **2010**, *69*, 2383–2393.
54. Ghosh, P.K.; Mondal, M.S. Economic valuation of the non-use attributes of a south-western coastal wetland in Bangladesh. *J. Environ. Plan. Manag.* **2012**, *56*, 1403–1418.
55. Kreye, M.M.; Adams, D.C.; Escobedo, F.J. The Value of Forest Conservation for Water Quality Protection. *Forests* **2014**,*5*, 862–884.
56. Adams, D.C.; Bwenge, F.; Lee, D.J.; Larkin, S.; Alavalapati, J.R.R. Public Preferences for Controlling Upland Invasive Plants in State Parks: Application of Choice Model. *For. Policy Econ.* **2011**, *13*, 465–472.

Chapter 5

Towards Sustainable Tourism Development in Urban Areas: Case Study on Bucharest as Tourist Destination

Andreea Zamfir and Razvan-Andrei Corbos

Faculty of Management, The Bucharest University of Economic Studies, Piata Romana 6, Bucharest 030173, Romania

ABSTRACT

The issues of tourism sustainability and urban development have become major priorities for public policy makers across the globe. Today, maybe more than ever, there is a need for managing sustainable tourism development, and this cannot be attained without taking into account environmental problems and their global dimension. Various problems and requirements of society and of the development of urban areas may be solved by transforming the cities into attractive tourist destinations. Therefore, this study explores how sustainable tourism development in urban areas can be basically achieved and managed. The paper discloses some success factors for managing sustainable tourism development in urban areas and emphasizes a case study regarding Bucharest, the capital of Romania, as a tourist destination. The originality and value of this study consist of identifying the main ways of developing sustainable tourism in Bucharest based upon empirical research conducted with the aid of a survey. The findings of this study may be helpful for upcoming research in the area of managing sustainable urban tourism development.

Keywords: sustainable tourism development; urban areas; urban development; tourist destination; tourism management

1. INTRODUCTION

Nowadays, maybe more than ever, there is a need for managing sustainable tourism development, and this cannot be attained without taking into account environmental problems and their global dimension. There are many problems and needs of society and of the development of urban areas that may be solved by transforming the cities into attractive tourist destinations. Therefore, the aim of this study is to explore how sustainable tourism development in urban areas can practically be achieved and managed.

Sustainable development is a multidimensional concept, which is related to the environment and resources as well as to the population and industrial and agricultural production [1]. A sustainable behavior of consumption is the key element of long-term development [2] (p. 143). Therefore, achieving sustainable development implies solutions aiming at the same time at the social, political, economic, demographic, and technical fields [1]. Sustainable development may be achieved if it relies on the harmony between economic progress and the boundaries of nature, particularly between the quantity and the regeneration time of natural sources, human-made emissions, and the neutralization capabilities of nature [3].

Therefore, to be more specific, the focus and purpose of this study is to identify the main success factors for managing sustainable tourism development in urban areas, based on a literature review and on a case study regarding Bucharest as a tourist destination. The research was conducted using evidence provided by conceptual papers, research papers, and literature reviews, as well as books and internet sources. In order to answer the research question, we have first revealed some modern approaches in defining sustainable tourism. Secondly, the study disclosed some success factors for managing sustainable tourism development in urban areas. Thirdly, we have emphasized a case study regarding Bucharest, the capital of Romania, as a tourist destination with an eye to identify the main ways of developing sustainable tourism in Bucharest based upon empirical research conducted with the aid of a survey. In order to answer the research question, we have analyzed the published sources, and we have evaluated and interpreted the empirical data collected through the questionnaire created.

The contribution this paper will make to the literature and practice derives from the logical connections of the arguments and counter-arguments for identifying the success factors for managing sustainable tourism development in urban areas, as well as from the possibility of using the findings in comparative urban studies or in order to conceive a strategy for developing Bucharest as a sustainable tourist destination.

2. MODERN APPROACHES IN DEFINING SUSTAINABLE TOURISM

There are many approaches and ways of defining the concept of sustainable tourism that can be found in the literature, and many of them start the inquiry by making a transition from the concepts of sustainability or sustainable development. The abundance of definitions, meanings, and implications will be further reviewed in order to reveal the essence of the sustainable tourism concept.

Sustainable development may be defined as an all-embracing concept which involves all aspects of human activity and in which all the nations of the world should be involved in Zaharia *et al.* [4] (p. 159). Moreover, it is seen as one of the major challenges of modern evolution. In order to achieve sustainable development, there is a need to identify the main causes of environmental degradation firstly, to assess its size secondly, to find reasonable solutions to counteract it thirdly, and to reduce its negative impacts on economic and social life fourthly [5] (p. 99).

As Rifkin [6] asserts, a significant impact into the 21st century will be the Third Industrial Revolution, which will radically change all facets of working and living. The social, economic, and political life of the previous industrial revolutions based on fossil fuels was characterized by a conventional top-down organization of society. However, this type of organization is rapidly changing, and new distributed and collaborative relations emerge as a foundation of the green industrial era [6].

The concept of sustainable tourism emerged in the early 1990s, by connecting sustainable development ideas and principles with tourism. This new concept, at that time, was quickly accepted and promoted by many international and national organizations. For instance, the International Union for Nature Conservation, World Federation for Nature Protection, and the European Federation of National and Natural Parks defined sustainable tourism in 1991 as development, marketing, and management of all forms/types of tourism taking into account the natural, social, and economic environment and preserving the natural and cultural resources for future generations [7,8]. From this definition it follows that tourism (ecotourism, green tourism, rural tourism, business tourism, urban tourism, *etc.*) should be based on the principles of sustainable development [9] (p. 250). Currently, there is a strengthening of political concerns for nature conservation due to the aggravation of environmental problems and due to the acknowledgment of the importance of maintaining the quality of the environment for future generations [5] (p. 100).

According to the World Tourism Organization, the development of sustainable tourism responds to the present needs of tourists and to those of the host regions, and also protects and increases future chances and opportunities. Through sustainable tourism, all resources can be managed so as to satisfy needs and to maintain cultural integrity, biological diversity, ecological dimensions, and the life system [10]. Accordingly, sustainable tourism can be defined as a form of tourism that takes into account present and future impacts (economic impact, social impact, and environmental impact), and responds to the various needs of tourists, industry, and local communities, as well as the environment [11] (p. 85).

Tourism sustainability is a complex concept because it has a latent, relative, and multidimensional nature [12] (p. 277). The multidimensional nature of tourism sustainability is based on three distinct sustainability dimensions: environmental, social, and economic sustainability [13] (pp. 5–6). As a result, sustainable tourism must undertake the environmental, social, and economic principles of sustainability (Figure 1) [14] (p. 3). This way the synergy effect is produced at the level of tourism sustainability. As a general rule, synergy means that certain features occur within the whole system without occurring in any of the subsystems [15] (p. 68).

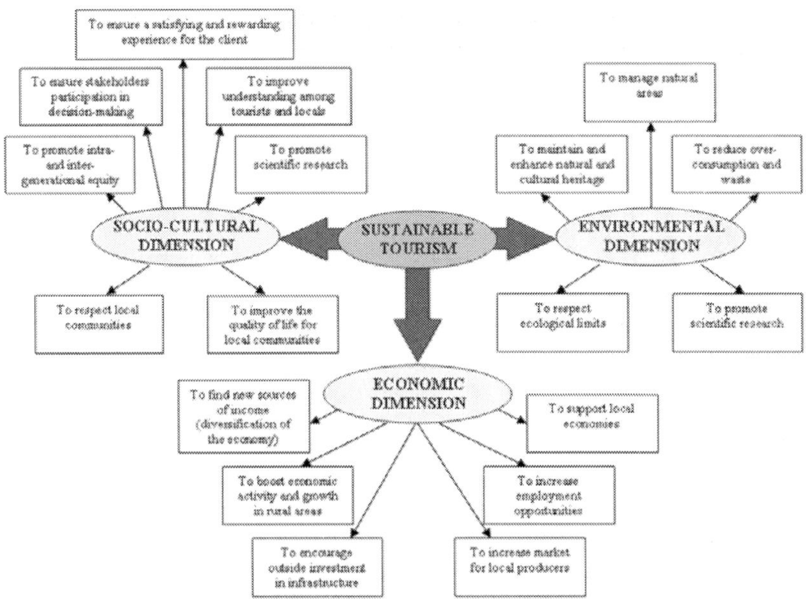

Figure 1. Principles of sustainable tourism. Source: Adapted from White *et al.* [14] (p. 4).

Sustainable tourism covers all types/forms of tourism: conventional mass tourism, cultural tourism, mountain tourism, seaside tourism, spa tourism, business tourism, medical tourism, rural tourism, urban tourism, *etc.* Regardless of the type or tourism activities, they should take on the principles of sustainable tourism: (1) the local community should initiate the tourism activity by its own means, maintaining control over tourism development in the area; (2) tourism should provide jobs to the residents, thus improving the quality of life; (3) the accepted international standards in tourism should be used when the guidelines for tour operators, for monitoring the impact of various tourism activities, and for setting the acceptability limits for different areas are established; (4) educational and training programs to improve management in the area of protecting natural and cultural resources must be implemented [16].

3. SUCCESS FACTORS FOR MANAGING SUSTAINABLE TOURISM DEVELOPMENT IN URBAN AREAS

Nowadays, cities must assume an important role in addressing environmental problems. They are innovation, efficiency, investment, and productivity centers, but they are also an important cause of climate change. The solutions to sustainable development and global climate change are to be found in the ecological urbanization of cities. Therefore, the campaigns and projects carried out by urban managers in order to reach the sustainable development of regions, cities, and rural areas around are becoming extremely important these days [9] (p. 254).

The role of urban areas in sustainable development is more and more recognized at a global level. Providing services to residents and businesses, creating jobs, stimulating research, and development within and outside the economic sector are only some of the functions of the cities. However, these functions are not only limited to administrative boundaries of cities, but they also create benefits for the whole region in which cities are included. A success condition for a modern city is to demonstrate that it meets the environmental requirements [17] (p. 379).

Even if local socio-economic development is achievable and accelerated in urban areas, these are, at the same time, places of environmental problems [18] (p. 79). Cities are special local environments, and the manner in which they should be governed and managed is determined by their features [19] (p. 31). A feasible urban development strategy should include elements relating to the image and brand of the city, elements which will ensure a competitive advantage for the local community [20] (p. 734). Today, business organizations and not-for-profit organizations, among which is the government, need to create and maintain permanent and effective

the Web, the museum "breaks" the traditional value chain in the sense that it enhances its role from being just a "commodity" with value added to the local tourism services to becoming the catalyst of an innovative "niche" supply developed around the cultural event. Moreover, if effectively integrated in the service process, new technologies could actively support the development of strategies related to innovation, collaboration, and value co-creation, playing a key role in providing competitive services [27] (p. 85).

4. CASE STUDY REGARDING BUCHAREST AS A TOURIST DESTINATION

4.1. Data and Methodology

Bucharest, the capital of Romania, was included in the online contest "The Best City to Visit Travel Tournament 2013 Championship" initiated by the site Foxnomad to designate the city visited in 2013 by the blogger Anil Polat [30]. Even though Bucharest did not win the final online contest, it still remained the second destination to visit according to the voters. This is not surprising if we are taking into account the variety of places of interest for tourists in Bucharest such as museums, parks, administrative buildings, restaurants, clubs, *etc.* The Romanian capital fought hard against and defeated Aswan, Rome, Cape Town, Edinburgh, and Granada and it was finally eliminated by another Romanian city, Sibiu [30].

The urban essence that makes Bucharest a unique city, with its strengths, weaknesses, opportunities, and threats, must be found in business urban life, as well as in resident and tourist life [31] (p. 277). Therefore, the approach of the sustainable urban tourism development should start from the people, from the residents and tourists. They should be partners of the government, of the public administration, of the policy-makers. Consequently, we assert that the opinion of residents and tourists matters because they are the beneficiaries of the public and private services provided. When choosing a destination, the tourists are searching for those services that best suit their needs.

Furthermore, we argue that any policy or strategy in the area of sustainable urban tourism development should start from the needs and desires of the people and should be connected to the principles of sustainable development. Thus, we assert that the method we used in our research was the best approach in this context.

In order to identify people's opinions regarding the main ways of developing sustainable tourism in Bucharest, we have created a questionnaire which was applied between 14–17 March 2013 at Romania's Tourism Fair. The questionnaire used was constructed based on a five-level Likert scale of increasing intensity where 1 indicates strongly disagree (very badly) and 5 designates strongly agree (very good). The sampling method was random and 236 subjects, residents aged 18 and over, participated in the research.

Most of the respondents are aged between 18 and 25 years (34% of the respondents), followed by respondents aged between 36 and 45 years (26% of the respondents), respondents aged between 26 and 35 years (25% of the respondents), respondents aged between 46 and 55 years (11% of the respondents), and respondents aged over 55 years (4% of the respondents). Regarding gender, 56% of the respondents are female, while 44% of them are male. The research adequately reflects gender distribution in Bucharest in statistical terms.

The information collected was processed using Microsoft Excel 2003, both for centralizing and analyzing data.

The next section of the article briefly presents the results of the survey, emphasizing people's opinions regarding the main attractions of Bucharest as tourist destinations, recreational facilities and sources of information about them, institutions involved in promoting sustainable tourism in urban areas, and the role of education in promoting sustainable urban tourism.

4.2. Results of the Survey

4.2.1. Tourist Attractions and Places to Visit

With reference to the main tourist attractions of Bucharest (Figure 2), among the responses to the question "What comes to mind when you think about the main attractions of Bucharest as a tourist destination?" the most common answer of the respondents was "Palace of Parliament (People's House)", with 21% of all persons surveyed. In second place came "Dimitie Gusti National Village Museum", with 13% of the respondents, followed by "Grigore Antipa Natural History Museum", with 12% of the respondents. Less than 5% of persons surveyed associated the attractions of Bucharest with churches or parks.

For example, Dimitrie Gusti National Village Museum has a website only in the Romanian language. Therefore, we strongly recommend the translation of museums' official websites not only in English, but also in other foreign languages, so as to increase information availability and to attract a bigger number of foreign visitors. Furthermore, museums may rely on multimedia information systems in order to redesign their traditional services and to promote new cultural experiences. Multimedia information systems allow museums to involve a worldwide network of potential visitors in the production of the cultural service [32] (p. 407).

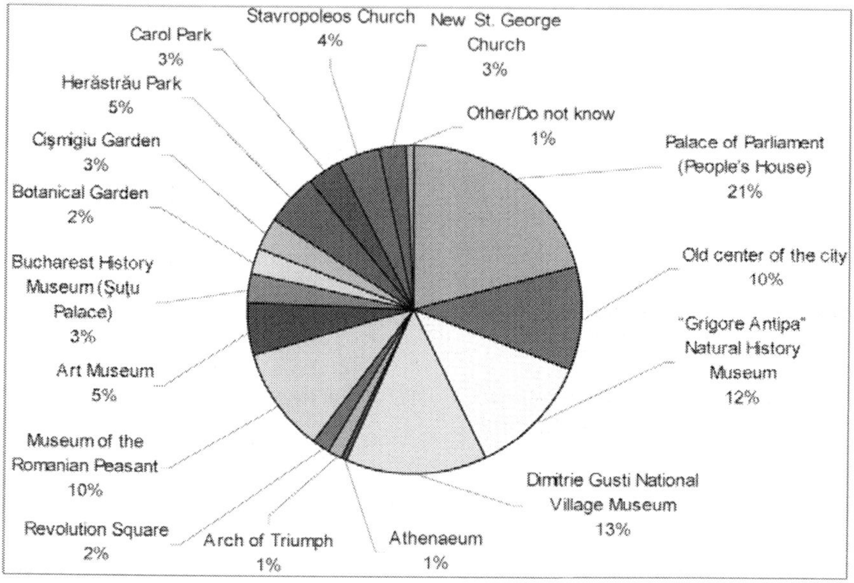

Figure 2. Main tourist attractions in Bucharest. Source: Authors, based on the survey.

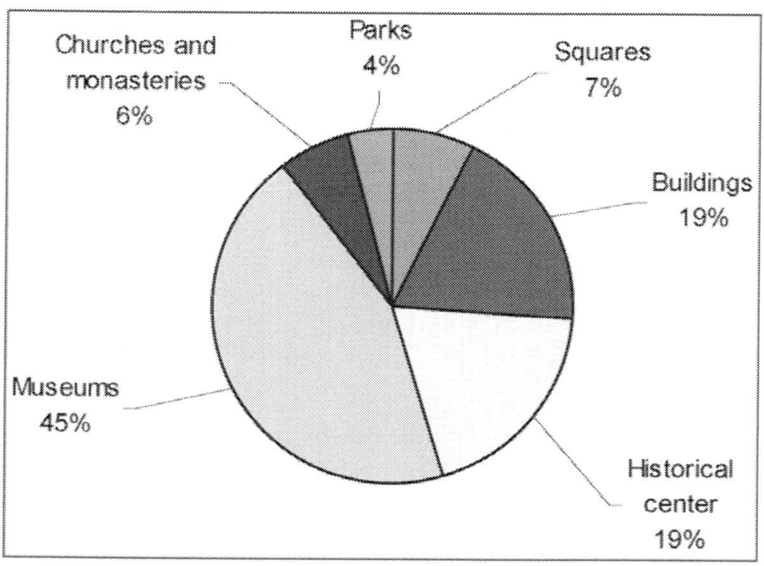

Figure 3. Types of places chosen by the respondents to visit in Bucharest. Source: Authors, based on the survey.

Organizing different festivals and other cultural events in the city may be another way of attracting tourists [33] (p. 19).

Regarding the preferences of the respondents (Figure 3), among the responses to the question "What do you like to visit in Bucharest?" the most common answer of the respondents was "museums", with 45% of the responses, followed by "buildings" together with "historical center", each accounting for 19% of respondents. At the opposite pole, the "parks" are preferred only by 4% of the respondents.

The heterogeneousness of the Romanian museums may help in playing an important role in educating visitors as long as there are appropriate management and marketing strategies implemented. Moreover, these strategies may benefit the development of information and communication technologies so as to lead to providing competitive cultural services [32] (p. 408).

On the topic of the recreational facilities (Figure 4), 10% of the respondents appreciate them as "very good", 43% as "good", 22% as "neither good nor bad", 19% as "bad", and 6% as "very bad".

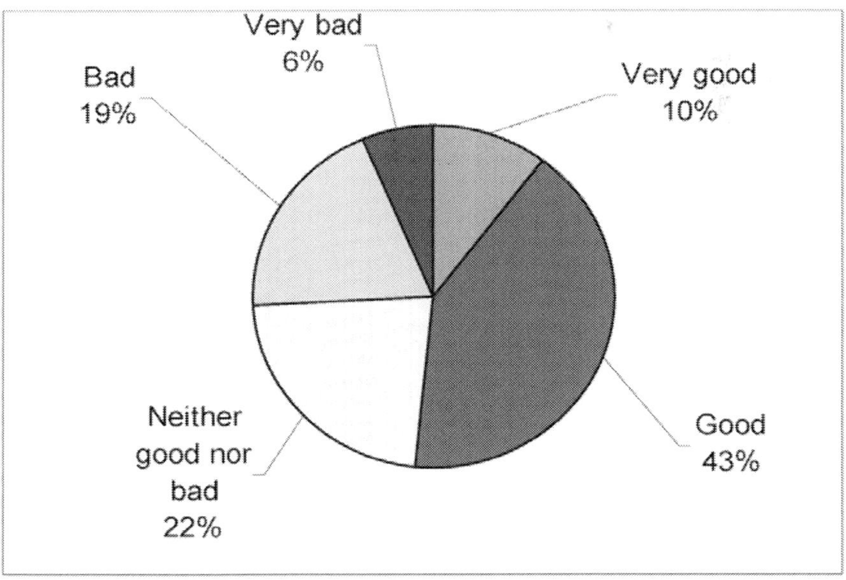

Figure 4. Recreational facilities in Bucharest. Source: Authors, based on the survey.

4.2.2. Sources of Information

The Internet is the main source of information regarding tourist destinations (Figure 5) for the respondents participating in the study, with 88% of them choosing it. Way behind as information source are television (9%), radio (2%), and newspapers (1%).

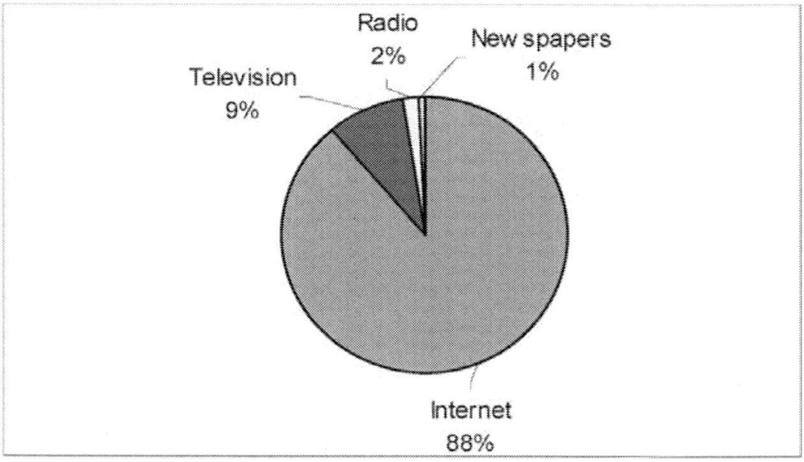

Figure 5. Sources of information regarding tourist destinations. Source: Authors, based on the survey.

4.2.3. Ways of Developing Sustainable Tourism in Bucharest

Regarding the role of institutions in promoting sustainable tourism in urban areas (Figure 6), a majority of respondents (60%) consider that the public administration (regional and local public authorities −36% and central public authorities −24%) are in the best position to implement programs to promote and develop sustainable tourism in urban areas. In second place came the non-governmental organizations (19% of the respondents), followed by citizens (16% of respondents).

Public authorities may create an urban brand in order to attract tourists, and to convince them to recommend the city to other visitors [34] (p. 132).

According to the performed study, the educational activities on environmental issues contribute to achieving sustainable urban tourism for 60% of the respondents, of which 11% strongly agree and 49% agree (Figure 7).

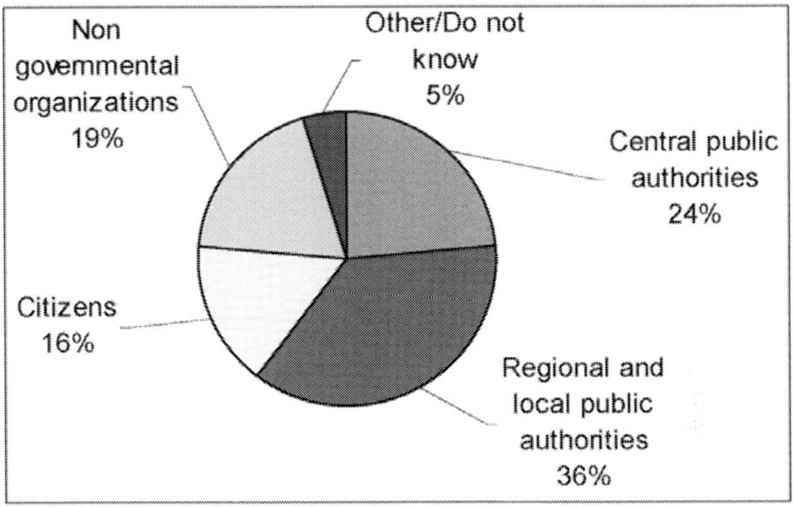

Figure 6. Institutions involved in promoting sustainable tourism in urban areas. Source: Authors, based on the survey.

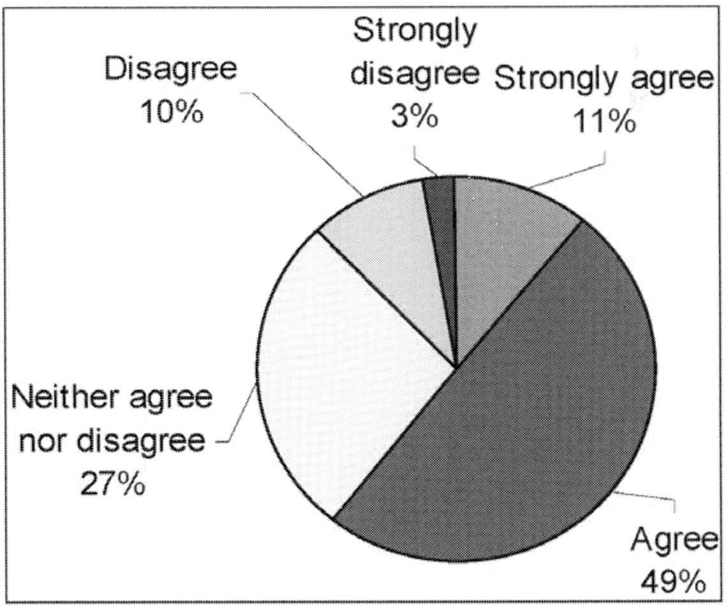

Figure 7. Important role of education in achieving sustainable urban tourism. Source: Authors, based on the survey.

The main ways of developing sustainable tourism in Bucharest (Figure 8) are the limitation of waste and environment degradation, as well as the integration of tourism into planning (both with 18% of respondents). In addition, applying better management based on sustainable urban tourism strategies is seen as a way of developing sustainable tourism in Bucharest by 15% of the respondents, followed by educating people and maintaining the natural and cultural heritage (both with 14% of the respondents).

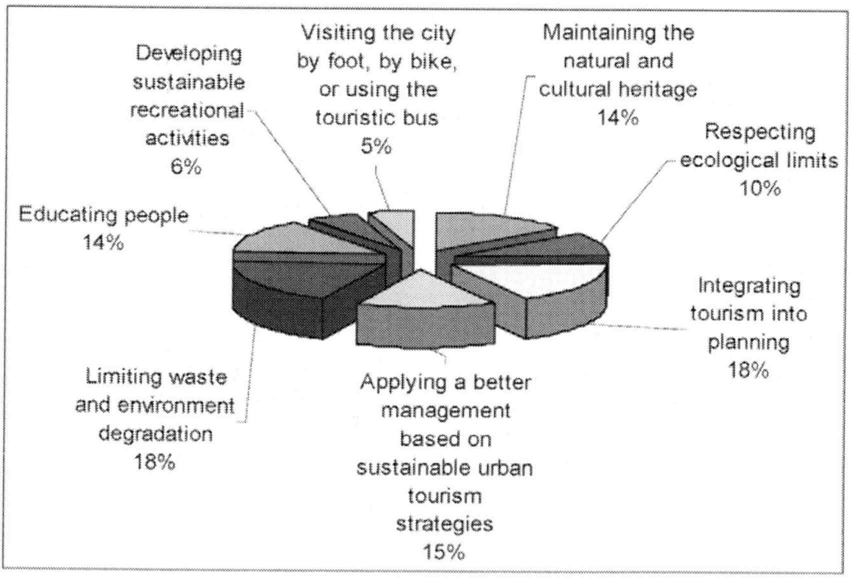

Figure 8. Ways of developing sustainable tourism in Bucharest. Source: Authors, based on the survey.

Ecological education aims to increase the awareness and understanding of the relationships between people and nature in order to maintain a balance between the present requirements and the future needs [5].

Also, sustainable tourism in Bucharest may be developed through respecting ecological limits (10% of the respondents), developing sustainable recreational activities (6% of the respondents), and visiting the city by foot, by bike, or using the touristic bus (5% of the respondents).

The public authorities may analyze the deficiencies and barriers faced in the process of strategic planning of sustainable tourism in Bucharest, and they may identify the main problems and may indicate the development priorities.

5. DISCUSSION

The results of the survey reveal that there are some points worth highlighting. On one hand, the capital city may be a good destination for cultural tourism due to the heterogeneousness of the museums and other cultural places and events (festivals, concerts, *etc.*) that can be organized. On the other hand, there are many actions that need to be done in order to promote Bucharest as a cultural tourist destination. For instance, increasing information availability, the translation of official websites into English and other foreign languages, and using multimedia information systems are only some of the recommendations to redesign the traditional services and to promote new cultural experiences for tourists.

Another idea emerging from the data analyzed is that the Internet is the main source of information regarding tourist destinations. Therefore, we strongly recommend using the advantages of the Internet to attract tourists, and to further involve them in the production of tourist services. Social networks like Twitter or Facebook may be used to promote Bucharest as a tourist destination due to rapid dissemination of information. Not unintentionally, we have presented the results of the blog online contest "The Best City to Visit Travel Tournament 2013 Championship" in the beginning of Section 4 of this article. Such online contests or ways of interaction attract people, and may be used within the strategies to promote tourist urban destinations.

The public administration is in the best position to implement programs, policies, and strategies to promote and develop sustainable tourism in urban areas. These strategies may benefit the development of information and communication technologies so as to lead to providing competitive services. Public authorities may create an urban brand in order to attract tourists and to convince them to recommend the city to other visitors.

The main ways of developing sustainable tourism in Bucharest identified by the respondents suggest that they care about the environment, as well as about the natural and cultural heritage.

On the basis of what has been revealed above, we may bring the paper to a close by saying that it may be appropriate to find a way to apply better management based on sustainable urban tourism strategies. The public administration, the residents, and the private companies, as well as other stakeholders, play an important role in this endeavor. The development of Bucharest as a sustainable tourist destination is possible only if there is a real partnership between all actors more or less involved in tourist activities.

6. CONCLUSIONS

This study has revealed some success factors for managing sustainable tourism development in urban areas and has emphasized a case study regarding Bucharest, the capital of Romania, as a tourist destination. The main conclusions of this study may be summarized as follows: (1) among the success factors for managing sustainable tourism development in urban areas, the strategic planning plays an important role; (2) sustainable tourism development in Romania may be achieved through developing urban areas as touristic destinations; and (3) the subjects comprised in this study appreciate that sustainable tourism development in Bucharest may be achieved in different manners.

The originality and value of this study consist of identifying the main ways of developing sustainable tourism in Bucharest based upon empirical research conducted with the aid of a survey. These ways of developing sustainable tourism in Bucharest are the following: limitation of waste and environment degradation, integration of tourism into planning, practice of better management based on sustainable urban tourism strategies, education, preservation of the natural and cultural heritage, avoidance of exceeding ecological limits, and development of sustainable recreational activities. Any policy or strategy in the area of sustainable urban tourism development should start from the needs and desires of the people and should be connected to the principles of sustainable development.

Our current experience advises us that there is no panacea for successful sustainable urban tourism development. The findings of this study can inform future research on urban sustainability by using them in comparative studies or as a starting point for developing policies and strategies in the field. We suggest that further research should be done in the direction of identifying the opinions of the representatives of the public administration and of tourism organizations, and to use them along with the findings of this study to conceive a strategy for developing Bucharest as sustainable tourist destination. Another direction for future research could be to replicate the survey in various cities from different regions in order to find out how local culture may influence the approach of sustainable urban tourism and what its effects are (if any) on the development preferences.

AUTHOR CONTRIBUTIONS

Andreea Zamfir and Razvan-Andrei Corbos designed the research and identified the logical connections of the arguments and counter-arguments; Andreea Zamfir collected the data; Razvan-Andrei Corbos organized the data; Andreea Zamfir and Razvan-Andrei Corbos analyzed the data; Andreea

Zamfir and Razvan-Andrei Corbos wrote the paper. All authors read and approved the final manuscript.

REFERENCES

1. Lazar, C.; Lazar, M. The Quantification of the Sustainable Development at Local Level. *WSEAS Trans. Bus. Econ.* **2008**,*5*, 310–319.
2. Marinescu, P.; Burcea, M. Information and Ecological Behaviour towards the Natural Resources Consumption of the Population of Bucharest. *Amfiteatru Econ.* **2012**, *XIV*, 142–156.
3. Pozeb, V.; Krope, T. Importance of Legal Protection and International Quality Standards for Environmental Protection. In IASME/WSEAS, Proceedings of the 2nd IASME/WSEAS International Conference on Energy & Environment, Portoroz, Slovenia, 15–17 May 2007; WSEAS Press: Athens, Greece, 2007.
4. Zaharia, R.M.; Stancu, A.; Stoian, C.; Diaconu, M. Commercial Activity's Contribution to Sustainable Development by Social Responsability Actions: A Vision of SMEs. *Amfiteatru Econ.* **2010**, *XII*, 155–167.
5. Minciu, R.; Pădurean, M.; Popescu, D.; Hornoiu, R. Demand for Vacations/Travel in Protected Areas—Dimension of Tourists' Ecological Behavior. *Amfiteatru Econ.* **2012**, *XIV*, 99–113.
6. Rifkin, J. *La Tercera Revolución Industrial. Cómo el Poder Lateral Está Transformando la Energía, la Economía y el Mundo*; Paidós: Barcelona, Spain, 2011. (In Spanish)
7. Istrate, I.; Bran, F. *Economia Turismului şi Mediul Înconjurător*; Editura Economică: Bucharest, Romania, 1996. (In Romanian)
8. Bran, F.; Simion, T.; Nistoreanu, P. *Ecoturism*; Editura Economică: Bucharest, Romania, 2000. (In Romanian)
9. Popescu, R.I.; Zamfir, A. Strategic Role of Ecotourism for Romania's Regional Development. In Proceedings of The 5th International Conference—The Scale of Globalization, Think Globally, Act Locally, Change Individually in the 21st Century, Ostrava, Czech Republic, 8–9 September 2011.
10. Stănciulescu, G. *Strategii si Politici Regionale de Dezvoltare Durabilă a Spaţiului Dunărean Provocări Pentru Turism*; Editura ASE: Bucharest, Romania, 2004. (In Romanian)

11. Minciu, R.; Popescu, D.; Pădurean, M.; Hornoiu, R.; Băltăreţu, A. Commercialization of Holidays in the Protected Natural Areas—Form of the Sustainable Development in Tourism. *Amfiteatru Econ.* **2010**, *XII*, 83–98.
12. Fernández, J.I.P.; Sánchez Rivero, M. Measuring Tourism Sustainability: Proposal for a Composite Index. *Tour. Econ.* **2009**, *15*, 277–296.
13. Kožić, I.; Mikulić, J. An Exploratory Assessment of Tourism Sustainability in Croatian Coastal Destinations. 2011. Available online: http://bib.irb.hr/datoteka/516656.Kozic_Mikulic_Tourism_Sustainability_Croatia_FEBWPS11-04.pdf(accessed on 25 May 2015).
14. White, V.; McCrum, G.; Blackstock, K.L.; Scott, A. Indicators and Sustainable Tourism: Literature Review. 2006. Available online: http://www.macaulay.ac.uk/ruralsustainability/LiteratureReview.pdf (accessed on 25 February 2015).
15. Corboş, R.A. Integration and Competition—Appropriate Approaches for Achieving Excellence in Management. *Bus. Excell. Manag.* **2011**, *1*, 67–73.
16. Jamieson, W.; Noble, A. A Manual for Sustainable Tourism Destination Management. 2000. Available online: http://www.gdrc.org/uem/eco-tour/etour-principles.html (accessed on 10 May 2015).
17. Zamfir, A.; Popescu, R.I. Promoting Renewable Energy—A Local Solution to Global Environmental Problems of Competitive Cities and Regions. In Proceedings of The 5th International Conference, the Scale of Globalization, Ostrava, Czech Republic, 8–9 September 2011.
18. Alpopi, C.; Manole, C.; Colesca, S.E. Assessment of the Sustainable Urban Development Level through the Use of Indicators of Sustainability. *Theor. Empirical Res. Urban Manag.* **2011**, *6*, 78–87.
19. Bačlija, I. Reconceptualisation of Urban Management: Evidence from EU Cities. *Theor. Empirical Res. Urban Manag.* **2013**, *8*, 30–50.
20. Ţapardel, A.-C.; Alexe, F.-A. Strategic Directions for the Bucharest Strategy and City Brand. *Amfiteatru Econ.* **2012**,*XIV*, 720–737.
21. Jimenez Figueroa, Y.H.; Muro Pico, J.F. *Desarrollo Organizacional y Humano*; El Cid Editor: Santa Fe, Argentina, 2009; Available online: https://www.clubensayos.com/Informes-de-Libros/Resumen-Libro-Desarrollo-Organizacional-Y-Humano-JIMENEZ/886501.html (accessed on 25 May 2015). (In Spanish)

22. Argüden, Y. *Keys to Governance, Strategic Leadership for Quality of Life*; Palgrave Macmillan: Hampshire, UK, 2011.
23. Pantelescu, A.M.; Ioncică, M. Researching the Mass Media Role in Promoting and Selling Tourists' Packages in Romania. *Amfiteatru Econ.* **2012**, *XIV*, 802–810.
24. Stănciulescu, G.C.; State, O. Potential Interpretations of the Ecological Behaviour in Tourism. *Amfiteatru Econ.* **2012**,*XIV*, 114–128.
25. Huibin, X.; Marzuki, A.; Razak, A.A. Conceptualizing a Sustainable Development Model for Cultural Heritage Tourism In Asia. *Theor. Empirical Res. Urban Manag.* **2013**, *8*, 51–66.
26. Roberts, S.; Tribe, J. Sustainability Indicators for Small Tourism Enterprises—An Exploratory Perspective. *J. Sustain. Tour.* **2008**, *16*, 575–594.
27. Zamfir, A. *Management of Services within the Knowledge-Based Society*; Editura ASE: Bucharest, Romania, 2010.
28. Nedelea, A.; Bălan, A. E-tourism and Tourism Services Consumer Protection. *Amfiteatru Econ.* **2010**, *XII*, 492–503.
29. Minghetti, V.; Moretti, A.; Micelli, S. Reengineering the Museum's Role in the Tourism Value Chain: Towards an IT Business Model. *Inf. Technol. Tour.* **2002**, *4*, 131–143.
30. Polat, A. The Best City to Visit Travel Tournament 2013: Championship. 2013. Available online: http://foxnomad.com /2013/03/26/the-best-city-to-visit-travel-tournament-2013-championship/ (accessed on 15 June 2015).
31. Popescu, R.I.; Corboş, R.A. Strategic Options in the Construction of the Bucharest Brand through the Application Analysis of the Measuring Instruments for the Urban Brands. *Ann. Univ. Petrosani Econ.* **2010**, *10*, 267–278.
32. Zamfir, A.; Corboş, R.A. Development of Cultural Services within the Knowledge Economy: Case Study on the Romanian Museums. *Int. J. Arts Sci.* **2012**, *5*, 399–409.
33. Popescu, R.I.; Corboş, R.A. The Role of Festivals and Cultural Events in the Strategic Development of Cities. Recommendations for Urban Areas in Romania. *Inf. Econ.* **2012**, *16*, 19–28.
34. Popescu, R.I.; Corboş, R.A. The Role of Urban Brand Measuring Tools in Grounding the Strategic Actions for Promoting Cities. *J. Adm. Public Manag.* **2009**, *13*, 116–132.

Chapter 6

Ecotourism *versus* Mass Tourism. A Comparison of Environmental Impacts Based on Ecological Footprint Analysis

Mehdi Marzouki [1,*], Géraldine Froger [2] and Jérôme Ballet [2]

[1]*Local government (Yvelines department), 2 place Pierre Bonnard, Guyancourt 78280, France*

[2]*CEMOTEV-UVSQ (EA 4457), 47 boulevard Vauban, Guyancourt 78280, France;*

ABSTRACT

Academic and policy interest in ecological footprint analysis has grown rapidly in recent years. To date, however, the application of ecological footprint analysis to tourism has been limited. This article aims to discuss the potential of ecological footprint analysis to assess sustainability in tourism. It is about a comparison of the global environmental impacts of different forms of tourism in southern countries where tourism is a major source of foreign exchange earnings. It illustrates how an ecotourism destination has a larger ecological footprint than a "mass" tourism destination.

Keywords: ecological footprint; sustainable tourism; environmental impacts; global scale; Tunisia; Seychelles

1. INTRODUCTION

The ecological footprint (EF) concept was introduced in the 1990s by William Rees and Mathis Wackernagel [1,2]. The concept is rooted in the search for indicators of sustainable development and more in particular the wish to measure how the human appropriation of the earth's resources relates to the carrying capacity of the earth. The aggregated use of land is seen as a good common denominator to express the impact of humans on the earth's natural resources.

The EF measures the amount of biologically productive land and sea area an individual, a region, a given population or a human activity requires to produce all the resources it consumes and to absorb the corresponding emissions (such as carbon dioxide from fossil use), using prevailing technology and resource management practices. It compares this measurement to how much land and sea area is available [3,4,5]. The EF is usually measured in global hectares. Biologically productive land and sea includes area that (1) supports human demand for food, fibre, timber, energy and space for infrastructure and (2) absorbs the carbon dioxide emissions from the human economy.

The total EF of an individual or a given community is broken down into a number of components. Often six components are distinguished [6,7,8,9]: use of arable land (for food, feed and other agricultural products), use of pasture land (for animal grazing), use of forest/woodland (for timber), use of built-up space (for living, *etc.*), use of productive sea space (for fish), and use of forest land to absorb CO_2 that was emitted due to human activities. Ecological footprint analysis (EFA) aims at expressing demand for all impact components as an equivalent, imaginary land area (global hectares or gha) [8,10].

According to [4] (p. 1964), "the EF deviates from other sustainability indicators in two respects: it expresses the impacts of humanity on the environment in one common unit (use of bioproductive space) and it can be related to the carrying capacity of the earth (the available bioproductive space or so-called "biocapacity")".

Many potential applications of EFA have been proposed e.g., [11]. To date, however, the application of EFA to tourism has been limited. EFA can be used in the tourism sector to calculate the biological demands of this activity [12]. This is a pertinent measure because it takes into account the impact of air travel and the areas of land used to cater for the various types of tourism-related consumption. The question of travel and transport is likely to call into question some types of tourism intended to be sustainable, such as ecotourism, particularly when they take place in destinations that are remote from the home country of the tourists [13]. For example, ecotourists generally originate from developed countries and frequently holiday in less developed countries in the South. Potential beneficial environmental impacts of ecotourism at the local level (in the sense that it puts a minimum threat on local ecosystems through the conversion of lands, trampling, collection of species, *etc.*) may be outweighed by harmful effects at the global scale (air travel).

This article then seeks to discuss EFA as a tool to assess sustainability in tourism. Our analysis reports a case study concerning the application of the EFA to Tunisia, which is often qualified as a "mass tourism" destination [14]. Our objective is to compare our findings with those reported by Gössling *et al.* [15] for the Seychelles. According to Gössling *et al.*[15] (pp. 206–207), "The development of travel mirrors a process of increasing competition between destinations, and can also be seen as an indicator of the growing inter-changeability of destinations. Particularly destinations in the tropics become replaceable because sun-sand-sea travel choices are increasingly made on the basis of two factors only: travel (flight) duration and hotel standard. Considering these trends, the Seychelles are unique in their effort to not become a mass-tourism destination and to attract the high-value segment of international tourism". Tourism is the second major source of foreign exchange earnings for the Seychelles, and a pristine environment is understood as the precondition for attracting an exclusive, wealthy clientele [16]. The Seychelles have safeguarded a wide range of ecosystems in protected areas, which are ultimately dependent for their existence on financial resources derived from tourism. They have based their marketing on the image of a pristine, exclusive eco-destination that seriously attempts to integrate environmental conservation and development [17,18].

Using EFA to assess sustainability of tourism in these two southern destinations is interesting because it focuses on the environmental impacts of tourism scenarios not on a local scale [19,20] but on a more global scale. It takes into account the global environmental aspects of travel. It illustrates that a destination branding itself as ecotourism destination has a larger EF than a "mass" tourism destination. Section 2 clarifies the method of calculating the EF applied to tourism. Section 3compares the EF of tourism between Tunisia and Seychelles. Section 4 discusses the potential and limitations of the EFA and its outcome.

2. METHODS

EF aims at expressing—using space equivalents—the appropriation of biologically productive area by individuals or nations. The idea of the concept is to compare the area required to support a certain lifestyle with the area available, thus offering an instrument to assess if consumption is ecologically sustainable [3,21]. The method used for our EFA applied to several forms of tourism is based on a reference model developed by Gössling *et al.* [15]. This method consists of several steps, each of which yields one of the components of the total EF. Generally, the method remains the same for studies that have applied the EFA to the tourism sector [5]. It relates human consumption and waste production to six major components of productive space: built-up land (that refers to spaces where the biological

Planet Report [24] and resumed by Gössling *et al.* [15], as well as those based on Ewing *et al.* [10] are shown in Table 1.

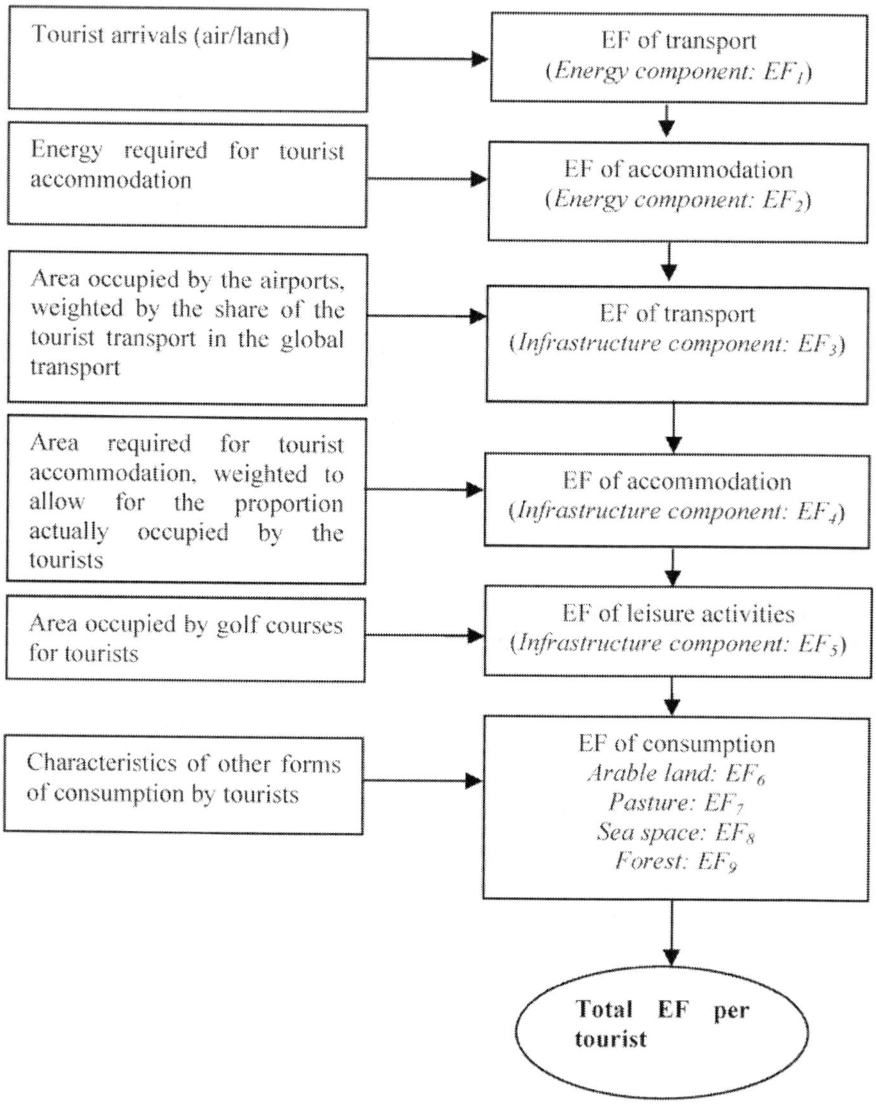

Figure 1. EF of tourism.

Table 1. Equivalence factors (based on relative biomass yield), Source: [15].

Area type	ef_i (global hectares per hectare) 2000
Fossil energy land (newly planted forest area needed to absorb emitted CO_2 (ef_1)	1.8
Built-up land (required for roads, buildings, *etc.*) (ef_2)	3.2
Arable land (for growing crops) (ef_3)	3.2
Pasture (for grazing animals) (ef_4)	0.4
Sea space (for harvesting fish and other sea food) (ef_5)	0.1
Forest area (for producing wood for furniture, paper, etc.) (ef_6)	1.8

In this way, the EF is calculated for the areas (EF_i), and the predetermined equivalence factors (ef_j) for every year on the basis of the equation below:

$$EF = \sum_{k=1}^{2} EF_k.ef_1 + \sum_{m=3}^{5} EF_m.ef_2 + \sum_{n=6}^{9} EF_n.ef_{n-3}$$

For the components associated with fossil energy, d_a and d_v designate the mean distances travelled per tourist by air and by road, and e_{ni} the mean energy consumed per tourist for accommodation. For the components associated with infrastructures and leisure activities, s_t, s_h, and s_g are reported, corresponding to the mean surface areas per tourist required for travel, accommodation, and golf courses and other leisure activities respectively. δ and η are respectively the fraction corresponding to the tourism-related use of the transportation infrastructures and occupation of accommodation. Finally, c_a, c_p, c_e, and c_f designate the mean relative consumption in hectares per tourist in terms of arable land, pasture, sea space, and forest. The footprint can then be expressed as follows:

$$EF = ef_1.\left[\beta.\left(d_a.\alpha_1 + d_v.\alpha_2\right) + e_{mi}.\mu\right] + ef_2.\left(s_t.\delta + s_h.\eta + s_g\right) + ef_3.c_a + ef_4.c_p + ef_5.c_e + ef_6.c_f$$

The EF describes to what degree tourism remains within the scope of the environment's ability to regenerate, and reveals the distribution of the pressure brought to bear on the environment as a function of the type of consumption [10].

Previously, the preferred index used to analyse the environmental impact of tourism was usually the carrying capacity of the earth [21]. Moreover, during the last ten years many sustainability indicators have been proposed and discussed [26]. Nevertheless, all of these indicators are indicators of environmental impact at the local level, and the advantage of the EFA is to provide an indicator of sustainability at the global level by taking air travel into account [27,28,29]. As a result, a tourist's footprint will include a wider range of impacts than most of the other indicators currently applied to the environment (at the local level but also at the global level with the carbon footprint which takes only carbon issue into account). Our approach will then compare tourism in Tunisia and the Seychelles with regard to their EF. Tunisia and the Seychelles have been chosen as study cases because tourism is a major source of foreign exchange earnings in both cases and each destination corresponds to a different kind of tourism: the Seychelles seek to brand themselves as an ecotourism destination, while a large part of some coastal areas of Tunisia can be considered as mass tourism.

3. RESULTS

For the Seychelles, the study carried out by Gössling *et al.* [15] provides some very detailed data about the country's EF. No such information is available for Tunisia in any paper. We therefore propose to calculate the ecological footprint of tourism in Tunisia by using the data of the ONTT report of 2004 [30] and using some parameters and coefficients introduce in the article of Gössling *et al.* [15], to compare our results with those reported in this paper for the Seychelles in 2000.

The use of data for 2004 for the Tunisian case is not problematic because of the stability of the inputs required by the EFA (origin of tourists, categories of accommodation establishments, airports and golf courses) between these two years. The quantitative data such as tourist arrivals and length of stay (5.1 days for 2004 [30]) have few consequences for the footprint calculation because the output is a mean value. Nevertheless, different hypotheses were required due to the limited availability of data.

We first formulated two hypotheses for the energy component linked to transport. The first concerns the distances to be taken into account. So far these have tended to be homogenized, and refer solely to the locations of the capital cities of the countries of origin of the tourists (see Appendix 1).We then refer to the details of kilometres on the airline website. The second hypothesis concerns road travel by tourists once they have arrived at their destination, which we have assumed to have a mean value of the order of 300 km [30]. This value is assumed as a mean of the tourism trip in Tunisia where the primary motivation of the tourists (about 80%) is sea-sand-sun with few trips, and the others 20% of tourists having more trips with desert excursions in the South for example or other forms of excursion in the North.

Secondly, for the infrastructure component linked to means of transportation, we also formulated some new hypotheses. As the two most significant means of transportation are air and road, we focussed our calculations solely on these two types of travel. Maritime transport has been left out of the equation due to the tiny fraction this means of locomotion for tourists represents for tourists visiting Tunisia (less than 3% of the tourism transport referring to ONTT data). Furthermore, since the roads also carry heavy traffic unrelated to tourism, the area occupied per tourist for this type of construction has also been ignored. Thus, the area related to the share of building for transport is in fact focussed on the areas occupied by the airports (see Appendix 2).

Thirdly, for the infrastructure and energy components related to accommodation, we have adopted several hypotheses to offset the lack of data. The first concerns the theoretical data for the different parts of the calculation. The area per capita has been taken from article [15], with a few modifications. As this study was carried out in the Seychelles, where we assumed that the space required per tourist was greater than that required per tourist in Tunisia because of the concentration of tourists in the accommodation establishments which is the major characteristic of a mass tourism destination (see Appendix 3), so we assumed that it was necessary to reduce this theoretical space considerably. We have considered 50% of the space required per tourist in the Seychelles. Energy consumption was taken to be the same for both of these tourism zones. As a result, the data provided by Gössling *et al.* [15] were also used to construct this aspect of the EFA applied to Tunisia, but we adjusted the parameters used for the Seychelles by considering that energy consumption is the same for each category of accommodation establishments (about 80 MJ per bed night) because the distinction between categories in Tunisia refers essentially to the quality of services proposed.

Finally, the data for the leisure activities and consumption were also lacking, so once again we had to adopt several hypotheses. There are only a few golf courses in Tunisia, and they are assumed to occupy a mean area of 90 ha as in the case of the Seychelles. We also adopted hypotheses related to consumption to assess footprint of food and fibres and to overcome the poor official statistical database and the unwillingness of the tourist industry to provide data. First, it was assumed that the quality and quantity of the food and fibre consumed at the destination is similar to consumption at home. Evidence suggests that there is at least some similarity [31]. We assumed that the consumption of tourists during their stay in Tunisia would be similar to consumption in their home country [32,33] (this hypothesis has been also made by Gössling *et al.* [15]). The second hypothesis concerns the uniformity of consumption by tourists from a given country of origin. In other words, the fact that there can be substantial differences between tourists coming from the same country is ignored in order to simplify the calculation. Moreover, the sensitivity of the coefficients for the assessment of uncertainty has not been ignored because the analysis conducted is done *ex post*. Thirdly, we take into account 85% of the tourist arrivals and as Gössling *et al.*[15] it was assumed that for the 15% remaining tourists the consumption corresponds to the average consumption of the others.

Some others coefficients were used to calculate the mean footprint of a tourist for a trip in Tunisia. We used those detailed in the methodology of Gössling *et al.* [15] for the calculation of fossil energy land. It refers to the coefficient described before:β (2.0), α_1 (0.04), α_2 (0.014) and μ (0.03).

On the basis of these hypotheses and using the data of the ONTTT, we obtained the following results (Table 2).

Table 2. Calculation of the ecological footprint per tourist per stay in Tunisia.

EF component in ha		Corrected *EF* component in gha	
$\beta.\alpha_1.d_a$	0.2500	$ef_1.\beta.\alpha_1.d_a$	0.4501
$\beta.\alpha_2.d_v$	0.0084	$ef_1.\beta.\alpha_2.d_v$	0.0151
$\mu.e_{ni}$	0.0114	$ef_1.\mu.e_{ni}$	0.0206
$\delta.s_t$	0.0007	$ef_2.\delta.s_t$	0.0022
$\eta.s_h$	0.0014	$ef_2.\eta.s_h$	0.0046
s_g	0.0002	$ef_2.s_g$	0.0007
c_a	0.005	$ef_3.c_a$	0.0161
c_p	0.0297	$ef_4.c_p$	0.0119
c_e	0.0105	$ef_5.c_e$	0.0010
c_f	0.0036	$ef_6.c_f$	0.0064
Ecological footprint per stay and per tourist			0.5286

This table presents the details of the calculation of the EF by referring to the mathematical expression expressed before. Thus, the mean EF of a tourist staying in Tunisia is of the order of 0.53 gha [34,35] (by considering that a mean stay in Tunisia is 5.1 days according to the ONTT [30]). Transportation has the greatest impact, accounting for something of the order of 86% of the footprint. It should also be noted that the components associated with energy and infrastructures make the greatest contributions to the EF, accounting for nearly 90% of the total footprint.

A comparison of the situations in Tunisia and the Seychelles led us to use the results reported by Gössling *et al.* [15] for the Seychelles alongside the data reported above. The Table 3 below summarises these comparisons (For more details see Appendix 1, 2 and 3).

Table 3. EFA of tourism in Tunisia and in the Seychelles (adapted from [15]) in gha.

Surface area	Seychelles	Tunisia
Fossil energy land	1.7373	0.4857
Built up land	0.0105	0.0075
Arable land	0.0632	0.0161
Pasture	0.0277	0.0119
Sea space	0.0032	0.0010
Forest	0.0145	0.0064
Aggregated footprint per tourist	1.8564	0.5286
Aggregated footprint of all tourists	218,482	1843,039

This table presents the components of the EF of Tunisia and of the Seychelles. First of all, the surface areas are higher for all the components in the case of the Seychelles than in the Tunisian case. Fossil energy land for the Tunisian case represents less than 30% of fossil energy land required for the Seychelles due to the air travel. Conversely, the difference between the built-up land components (land directly used for tourist infrastructure) for both destinations is not so high, because infrastructure required for air transport is similar, only the accommodation has an incidence with a higher surface for the Seychelles. In the two case studies, built-up land is small because roads, airports, accommodation establishments, *etc.* are used by a great number of tourists per year, leading to a rather small per capita built-up area demand. The consumption for pasture and forest for Tunisia corresponds to 40%–50% of the consumption of the Seychelles for the reason that the length of stay here (10.4 days) is doubly more important than for the Tunisian case (5.1 days). Arable land and sea space for the northern destination considered refer to 25%–30% of the consumption for the southern country studied. This percentage is explained by the length of stay and the origin of the tourists with an important share of tourists coming from Eastern Europe which consume less than those from Western Europe.

Thus, the aggregated footprint per tourist is better for a tourist in Tunisia. Nevertheless, this conclusion is not valid when considering the aggregated footprint of all tourists for the two destinations, because the number of tourists in Tunisia is much higher in this country than in the Seychelles. So, comparison of the findings for Tunisia (with a footprint of the order of 0.53 gha per tourist for a length of stay of 5.1 days) and those for the Seychelles (with a footprint of the order of 1.85 gha per tourist for a length of stay of 10.4 days), confirms the extent to which air travel forms a substantial part of the EF of tourists. The shorter the distance tourists travel, the less pressure is brought to bear on the environment at the global level. Furthermore, some of the forms of consumption linked to food, which are expressed by the last four components, also reveal a definite advantage for Tunisia, which draws most of its tourists from Eastern Europe; these tourists consume less than those from Western Europe.

On the basis of this EFA, we can conclude that a tourist journey in Tunisia has a less harmful impact on the environment at the global level than a tourist journey in the Seychelles. Nevertheless, the aggregated EF per year for all categories of tourists combined is clearly lower for the Seychelles than for Tunisia due to the scale effect that results from the huge volume of tourists holidaying in Tunisia. The volume of tourists going to Tunisia is at least 30 times greater than that visiting the Seychelles (for the baseline years chosen for the two countries). In this way, the aggregated EF for the Seychelles corresponds to 12% of the total aggregated footprint for Tunisia, whereas the arrivals in the Seychelles represent only 3% of the total tourism arrivals in Tunisia.

4. DISCUSSION

4.1. Discussion of the Methodology Underlying the EF Applied to Tourism

Of course EFA presents some boundaries [23,36]. First, EF is a stock indicator generally used to evaluate the sustainability of production or consumption of groups or areas. It leaves aside exchanges between areas. However, considering distribution between geographical spaces is necessary for the elaboration of a sustainability indicator [37]. Second, EF is a static measure and comparisons from time to time are not sufficient to understand changes. In particular, they cannot permit dissociation between substitution effects of goods or production processes that have been realised and decreasing or increasing effects in production or consumption. In articles [38,39], their point of view, underlined the necessity to develop a dynamic approach which can highlight stakes of different scenarios. Third, EF ignores technological changes, in particular, the development of clean technologies. Fourth, EF does not take into account freshwater withdrawals [40,41,42]. EF does not consider the freshwater consumed by tourists or the freshwater used by the tourism sector, whereas tourism does not have a neutral effect on freshwater resources at the local scale in arid countries like Tunisia or in islands like the Seychelles. The focus on fresh water and not water in general is important because fresh water is scarce. The water footprint (WF) was introduced in the field of water studies in 2002 [43] in order to measure the human appropriation of the globe's freshwater resources [44,45]. The term was chosen by analogy with the EF concept, but the WF concept has other roots than the EF concept. As a result, the methods to quantify the different indicators show both striking similarities and differences [46]. Results obtained by WWF [47] with the concept of WF shows that in Tunisia the stress on water resources could reach around 60% for freshwater withdrawn (from water bodies that are used and not returned), which corresponds to a severe stress, whereas stress on freshwater withdrawn is under 40% for the most part of the country. The situation is also problematic in the case of the Seychelles because of the high population density induced by tourism.

Nevertheless, EF presents advantages. As Moffatt [48] underlined, it possesses the capacity to give a clear message without ambiguity which can be used by public and policy makers and it is a key point of every good indicator.

Furthermore, as we apply it in the case of tourism, EF is not really concerned by previous critics. On one hand, we do not try to evaluate the evolution of tourism EF, but to compare, at any moment of time, the EF of two forms of tourism. On the other hand, we do not consider technological changes. They are short term. Finally, the distribution between two areas is taken into account in our evaluation by the travel of tourists between the departure area and the arrival area of tourists. In other words, we do not seek to use EF more than it can be used. Of previous boundaries, the only one that remains is the concern for freshwater withdrawals, and it is a true limit of our evaluation. On this point rests a crucial problem for evaluating tourism impact. However in the mean time, it is mainly a local stake and not a global one, which we are trying precisely to evaluate. We have nevertheless discussed this question in our conclusion.

For a decade, numerous sustainable indicators have been proposed and discussed (see for instance [49,50], for a survey of main indicators by countries). Tourism did not escape from this freneticityy of sustainable indicators. As Ceron and Dubois [51] pointed out, most of these indicators are environmental indicators at a local level. Even if recent indicators are indicators that can be used for comparison between countries and are holistic indicators (see for instance [52,53,54]) they remain indicators about the impact of tourism in countries, while not considering travelling between countries. Now, Hunter and Shaw [27] underlines that an undeniable advantage of EF is to give a global indicator of sustainability. In this sense, it is not only an indicator of tourism in a place-country, but an indicator of tourism travelling from one country to another.

However, for a few years, another indicator has competed with EF called carbon footprint. Carbon footprint evaluates carbon emissions from production or consumption activities. Carbon footprint should be restricted to analyses that include only carbon emissions, in practice analyses proceed by including non carbon emissions, then using an equivalent indicator [55]. Carbon footprint is very similar to the Global warming potential (GWP) indicator used in life cycle assessment. Its success is indeed based on its readability concerning the effect of goods consumed on global warming [56].

Whereas carbon footprint measures CO_2 emissions, EF measures the ratio between CO_2 emissions and the biosphere capacity to keep carbon locked up, notably through forests. Some underline (for instance [57]) that, since CO_2 emissions play a major role in EF, it could be judicious to refer only to carbon footprint. It highlights policy stake, *i.e.*, to reduce CO_2emissions.

Meanwhile, such a point of view seems excessive for two reasons. First, if global warming is a crucial stake, it is not due to CO_2 emissions, but to the incapacity of the planet to absorb them. It is then more logical, even if less clear politically, to use EF. Second, the use of carbon footprint could lead to solutions with disputable effects on environment. For example, the development of biofuel is from a carbon footprint a good solution; CO_2 emissions are reduced. But such a conclusion leaves aside the fact that the production of biofuel needs large areas which increase significantly the pressure on forests. EF, comparing emissions and capacity of absorption lead to a more mitigated conclusion. If areas needed to product biofuels contribute significantly in reducing forest which absorbs carbon, then biofuels are no more a **good** solution.

From a general perspective, the major asset of EF is to give an equivalent surface. This equivalent surface is furthermore a relevant indicator in the case of tourism, to an extent touristic activities are done on a given space and cannot be reduced to carbon emissions, even if, as we have noted in previous sections, it is of course a main dimension. EFA thus assesses environmental sustainability of tourism at a global level.

4.2. Discussion of the Results Obtained for the Seychelles and Tunisia

Applying the concept of the EFA to tourism makes it possible to determine the pressure that tourist consumption exerts on the area of productive land. It is measured in surface area (gha per tourist). It is an indicator that can be used to determine the sustainability of the different forms of consumption of tourism by considering the area required to cater for tourists' consumption relative to that naturally available. An average tourist's journey to the Seychelles requires more than 1.8 gha of world average space to maintain the necessary resource flows and to offset the contribution of greenhouse gas emissions to global warming. On the contrary, an average tourist's journey to Tunisia requires less than 1.8 gha of world average space. In the context of tourism, the conclusion is relatively severe for distant destinations, because the footprint for a journey equals or even exceeds the biological capacities required for the activity involved, and this is particularly true of the Seychelles. A necessary condition for development to be sustainable is clearly that the EF must be smaller than the biocapacity available [27].

The question of distance also has a relatively significant impact with regard to the previous observation. The fraction attributable to fossil energy consumption makes up the greatest part of the EF [15], and transportation and accommodation are the main consumptions that involve fossil energy to a substantial extent. Consequently, long-distance travel considerably increases the EF per tourist. According to [5], a linear relation between distance and footprint is proved with a high correlation between these two indicators. In this way, the greater the distance travelled, the heavier the EF. In this context, the heavy footprint of ecotourists travelling to the southern hemisphere appears to be in contradiction to the objective of sustainable development, despite the fact that ecotourism may be sustainable on the local level (in the sense that its puts a minimum threat to local ecosystems through the conversion of lands, collection of species, trampling, *etc.*). Thus, the problem of air travel is particularly striking with regard to the development of tourism, to the extent that the intensification of international tourism, and therefore the multiplication of air miles linked to tourism, can only increase the pressure of tourism on the global environment as a result of the relentlessly increasing emission of greenhouse gasses [58].

The comparison between the two major variables, air travel and length of stay, proves also that length of stay has a marginal impact on the results. For example, if we consider a length of stay of the order of 10 days for both destinations (so twice that of Tunisia), it will of course impact the consumption in the new calculation of the footprint of a tourist in Tunisia with an increase of arable land, pasture, forest and sea space, but not the other elements like built up land and fossil energy land which do not depend on the length of stay or marginally for the energy used for accommodation (approximately less than 5% of the total footprint).

Our comparison of the data available for Tunisia and for the Seychelles confirms the major impact of air travel in the EF of tourists, which is consistent with the results shown above and extracted from the work of [5] on various tourism scenarios. According to [5], a "mass" tourism destination such as Tunisia is less sustainable than an ecotourism destination as in the Seychelles, and our results prove that also if we just focus on the tourism activity at the national level with the higher number of arrivals for Tunisia. Nevertheless, this conclusion has a sense only at the national level because the individual footprint of a tourist is higher for the Seychelles than for Tunisia (around three times more). In this way, 433,581 international tourists in Tunisia have the same impact on the global scale as 117,690 international tourists travelling to the Seychelles. Thus, it limits the sustainability of ecotourism in destinations like the Seychelles because of the air travel. It proves that tourist travel should involve relatively small distances and the development of tourism should be constrained in order to reduce the global pressure on the environment. In order to become more sustainable, destinations should seek to attract clients from close source markets and regional tourism involving short transport distances may often only be sustainable from an ecological point of view.

Finally, one of the clearest conclusions of the EFA is that "mass" tourism near to the countries of origin of most of the tourists will have a less harmful impact at the global level, than a more sustainable form of tourism, such as ecotourism, that takes place at destinations far from the countries of origin of most of the tourists. This conclusion leads to problems, because most of the major areas of high biodiversity, which are the basis of the growth of ecotourism (biodiversity hotspots), are located in the southern hemisphere [59].

5. CONCLUSIONS

The sustainability of ecotourism can be called into question if it takes place in southern countries involving long-distance travel. Thus, ecotourism appears to be only a partially sustainable form of consumption, and can be partially responsible for causing ecological damage at the global level, unless it takes place in countries that are relatively close to the countries of origin of the tourists.

Meanwhile, our conclusion must be put into context for at least three reasons.

First, as we have previously underlined, freshwater withdrawals are not taken into account in EF. According to Mekonnen and Hoekstra [60], the WF of the global average consumer in the period 1996–2005 was 1,385 m^3/year The EF discussion should be enriched with a water-use perspective in addition to a use-of-space perspective. The average consumer in Tunisia has a WF of 2,217 m^3/year while the average citizen in the Seychelles has a WF of 2,192 m^3/year. It is quite similar and above the overall average. Now tourism, by the pressure it exerts constitutes a major factor of stress on freshwater. To evaluate and compare the effect of the two forms of tourism, WF should be taken into consideration. Such a footprint will be particularly relevant in our case. Indeed, Tunisia is an arid country. Pressure exerted by mass tourism could largely contribute to the scarcity of freshwater resources. If impact is mainly local, side effects could be produced, especially through increasing migratory phenomenon. Seychelles, as an archipelago, must tackle concentration of tourism on coastal areas. Small islands are very vulnerable to the pressure exerted by tourism with access to freshwater and land. A strong tourist pressure could affect significantly access to freshwater for the native population. In the case of Tunisia, as in the case of Seychelles, WF is certainly a relevant indicator for tourism impact and should be used in addition to EF.

Second, EF does not take into account irreversibility effects. From such a point of view mass tourism in Tunisia can be different, to ecotourism in Seychelles. Degradations produced by mass tourism could be, to a certain extent, irreversible, whereas ecotourism effects are probably less irreversible, because they are less concentrated.

Third, EF does not take into account technological changes. Technological changes in travel that reduce their impact could significantly modify our results. As we underlined, our study is static and we do not use EF more than it should be used. The dynamics of sustainability is poorly represented by this indicator. It produces only a static picture, at a given moment of time and our results must be considered in this way: a comparison between two forms of tourism at a given moment of time. It does not signify that ecotourism, if it be developed through cleaner travelling means, would not be more interesting than mass tourism at a global level.

A challenge for future research is to bring EFA, WFA and the other types of sustainability analysis together in one framework. A first step would be to harmonise the footprint calculation methodologies and develop ways to use EF, WF and other sustainability estimates in a complementary way to assess the sustainability of the use of the natural environment by human beings.

ACKNOWLEDGEMENTS

We thank anonymous reviewers for helpful comments on an early draft of this paper. The usual disclaimer applies.

REFERENCES AND NOTES

1. Rees, W.E. Ecological footprints and appropriated carrying capacity: What urban economics leaves out. *Environ. Urban.* **1992**, *4*, 121–130.
2. Rees, W.E.; Wackernagel, M. Ecological footprints and appropriated carrying capacity: Measuring the natural capital requirements for the human economy. In *Investing in Natural Capital: The Ecological Economics Approach to Sustainability*; Jansson, A.M., Hammer, M., Folke, C., Costanz, R., Eds.; Island Press: Washington, DC, USA, 1994; pp. 362–390.
3. Chambers, N.; Simmons, C.; Wackernagel, M. *Sharing Nature' Interests: Ecological Footprints as an Indicator of Sustainability*; Earthscan: London, UK, 2000.

4. Hoekstra, A.Y. Human appropriation of natural capital: A comparison of ecological footprint and water footprint analysis. *Ecol. Econ.* **2009**, *68*, 1963–1974.
5. Hunter, C.; Shaw, J. Applying the ecological footprint to ecotourism scenarios. *Environ. Conserv.* **2006**, *32*, 294–304.
6. For detailed descriptions of the procedures involved in EFA see [3,7,8,9].
7. Monfreda, C.; Wackernagel, M.; Deumling, D. Establishing national natural capital accounts based on detailed ecological footprint and biological capacity assessments. *Land Use Policy* **2004**, *21*, 231–246.
8. Wackernagel, M.; Monfreda, C.; Moran, D.; Wermer, P.; Goldfinger, S.; Deumling, D.; Murray, M. *National Footprint and Biocapacity Accounts 2005: The Underlying Calculation Method*; Global Footprint Network: Oakland, CA, USA, 2005.
9. Wackernagel, M.; Onisto, L.; Bello, P.; Linares, A.C.; Falfan, I.S.L.; Garcia, J.M.; Guerrero, A.I.S. National capital natural accounting with the ecological footprint concept. *Ecol. Econ.* **1999**, *29*, 375–390.
10. Ewing, B.; Goldfinger, S.; Wackernagel, M.; Stechbart, M.; Rizk, S.; Reed, A.; Kitzes, J. *The Ecological Footprint Atlas 2008*; Global Footprint Network: Oakland, CA, USA, 2008.
11. Wackernagel, M.; Yount, J.D. Footprints for sustainability: The next steps. In *Environ. Dev. Sustain.*; 2000; pp. 21–42. [Google Scholar]
12. Hall, C.M. *Tourism Planning: Policies, Processes and Relationships*; Pearson Education: Upper Saddle River, NJ, USA, 2008.
13. Cooper, C.; Hall, C.M. *Contemporary Tourism: An International Approach*; Butterworth-Heinemann: Oxford, UK, 2008.
14. Large parts of some coastal areas of Tunisia can be considered as "mass tourism". Tourism is one of the most dynamic sectors of the economy and is a major source of foreign exchange earnings. The tourist arrival figure in 2010 was 6.9 million.
15. Gössling, S.; Borgström-Hansson, C.; Hörstmeier, O.; Saggel, S. Ecological footprint analysis as a tool to assess tourism sustainability. *Ecol. Econ.* **2002**, *43*, 199–211.
16. The tourist arrival figure in 2010 was 175,000. Even if serious doubts have been expressed about the effectiveness of environmental conservation [18].
17. Lindén, O.; Lundin, C.G. The Journey from Arusha to Seychelles. Successes and failures in integrated coastal zone management in

Eastern Africa and island states. In Proceedings of the Second Policy Conference on Integrated Coastal Zone Management in Eastern Africa and Island States, Seychelles, 23–25 October 1997.

18. Froger, G. Le tourisme durable dans les Suds : Solution ou mirage? *Mondes en Dév.* **2012**, *157*, 1.
19. Weaver, D.B.; Lawton, L.J. Twenty years on: The state of contemporary ecotourism research. *Tour. Manag.* **2007**, *28*, 1168–1179.
20. Wackernagel, M.; Rees, W.E. *Our Ecological Footprint: Reducing Human Impact on the Earth*; New Society Publishers: Gabriola Island, BC, Canada, 1996.
21. We retain the terminology used by Cooper and Hall [13], even if several updates in the National Footprint Accounts have been released by the Global Footprint Network [10,23]. Indeed, new factors do not include fossil energy land, arable land, pasture and sea space anymore but CO_2 uptake-land (or carbon footprint, *i.e.*, forest for carbon sequestration), cropland, grazing land and fishing ground. The factors built-up land and forest (for timber and fuel wood) remain.
22. Ewing, B.; Moore, D.; Goldfinger, S.; Oursler, A.; Reed, A.; Wackernagel, M. *The Ecological Footprint Atlas 2010*; Global Footprint Network: Oakland, CA, USA, 2010.
23. World Wildlife Fund, *Living Planet Report*; WWF: Gland, Switzerland, 2000.
24. Wackernagel, M.; Lewan, L.; Borgström Hansson, C. Evaluating the use of natural capital with the ecological footprint: Applications in Sweden and Subregions. *Ambio* **1999**, *28*, 604–612.
25. Bossel, H. *Indicators for Sustainable Development: Theory, Method, Applications*; A Report for the Balaton Group, International institute for Sustainable Development: Winnipeg, MB, Canada, 1999.
26. Hunter, C.; Shaw, J. The ecological footprint as a key indicator of sustainable tourism. *Tour. Manag.* **2007**, *28*, 46–57.
27. Gössling, S.; Peters, P.; Scott, D. Consequences of climate policy for international tourist arrivals in developing countries. *Third World Q.* **2008**, *29*, 873–901.
28. Gössling, S.; Schumacher, K.P. Implementing carbon neutral destination policies: Issues from the Seychelles. *J. Sustain. Tour.* **2010**, *18*, 377–391.

29. Office National du Tourisme Tunisien (ONTT), *Le Tourisme Tunisien en Chiffre*; ONTT Tunisia, Ministry of Tourism: Tunis, Tunisia, 2004.
30. Gössling, S. Tourism, environmental degradation and economic transition: Interacting processes in a Tanzanian coastal community. *Tour. Geogr.* **2001**, *3*, 230–254.
31. This hypothesis is needed to use the same process of calculation as with [15]. Some authors consider that tourists consume more during their stay [33]. Nevertheless, if tourists consume more when they are on holidays in developing countries, this is balanced by the specificities of local products, which have a smaller EF than similar products in developed countries.
32. Cogoy, M. The consumer as a social and environmental actor. *Ecol. Econ.* **1999**, *28*, 385–398.
33. According to [35], in 2009, the EF obtained for the Tunisian case is of the order of 0.5225 gha and the aggregated footprint is of the order of 2,046,092 (for data see Appendix 1, 2 and 3). The difference is not significant between 2004 and 2009 for the individual footprint because of the similar repartition of category of accommodation establishments and the similar repartition of nationality of tourists. The aggregated footprint is higher for 2009 because of the increase of tourist arrivals.
34. Office National du Tourisme Tunisien (ONTT), *Le Tourisme Tunisien en Chiffre*; ONTT Tunisia, Ministry of Tourism: Tunis, Tunisia, 2009.
35. Kitzes, J.; Peller, A.; Goldfinger, S.; Wackernagel, M. Current methods for calculating national ecological footprint accounts. *Sci. Environ. Sustain. Soc.* **2007**, *4*, 1–9.
36. Van den Bergh, J.; Verbruggen, H. Spatial sustainability, trade and indicators: An evaluation of the ecological footprint. *Ecol. Econ.* **1999**, *29*, 61–72.
37. Moffatt, I. *Sustainable Development: Principles, Analysis and Policies*; Parthenon Press: Carnforth, UK, 1996. [Google Scholar]
38. Lange, G.M. How to make progress toward integrating biophysical and economic assessments. *Ecol. Econ.* **1999**, *29*, 29–32.
39. Goldfinger, S.; Wackernagel, M.; Niazi, S.; Peller, A.; Kaercher, M.; Kitzes, J.; Ewing, B.; Silvestri, F.; Hayes, K.; Wakabayashi, T. *Africa—Ecological Footprint and Human Well-Being*; WWF, Global Footprint Network, SDC: Oakland, CA, USA, 2008.

40. David, M.; Dormoy, C.; Haye, E.; Tregouët, B. *Une expertise de l'empreinte écologique; Etudes et documents No. 16* ; Commissariat général du développement durable: Paris, France, 2010.
41. "The EF includes only those aspects of resource consumption and waste production for which the Earth has regenerative capacity, and where data exist that allow this demand be expressed in terms of productive area. For example, freshwater withdrawals are not included in the footprint, although the energy used to put or treat the water is" [40] (p. 28). The EF does not, and is not intended to measure freshwater flows.
42. Hoekstra, A.Y. Virtual water trade. In Proceedings of the International Expert Meeting on Virtual Water Trade, Delft, The Netherlands, 12–13 December 2002.
43. The WF of an individual or community is defined as the total volume of freshwater that is used to produce the goods and services consumed by the individual or community. Though they are often compared and contrasted, EF and WF are, as indicators, fundamentally incapable of being substituted. Whereas the EF denotes the bioproductive area (global hectares) needed to sustain a population (or measures the biological capacity a population uses), the WF represents the freshwater volume (cubic meters per year) required. They each provide a different piece of information in the sustainability puzzle. Instead of being seen as competing metrics, they should be seen as two complementary indicators of natural capital use in relation to human consumption. For more information on the methodology and the applications of WF, see [4] and [45].
44. Hoekstra, A.Y.; Chapagain, A.K.; Aldaya, M.M.; Mekonnen, M.M. *The Water Footprint Assesment Manual—Setting the Global Standard*; Earthscan: London, UK, 2011.
45. Two differences are, for example, that EF are usually calculated based on global average productivities, while WF are calculated based on local productivities, and that EF are often not made spatially explicit, while WF are [4]. Furthermore, the components of an EF are weighted (based on equivalence factors) before adding up to the total EF, while the components of a WF are added without weighting. For more information on the similarities and differences between EF and WF analysis, see [4].

46. World Wildlife Fund. *Living Planet Report*; WWF: Glan, Switzerland, 2008.
47. Moffatt, I. Ecological footprint and sustainable development. *Ecol. Econ.* **2000**, *32*, 359–362.
48. Bossel, H. *Indicators for Sustainable Development: Theory, Method, Applications; A Report for the Balaton Group*; International institute for Sustainable Development: Winnipeg, MB, Canada, 1999.
49. Hass, J.L.; Brunvoll, F.; Hoie, H. *Overview of Sustainable Development Indicators used by National and International Agencies; OECD Statistics Working papers*; OECD Publishing: Paris, France; February; 2002.
50. Ceron, J.-P.; Dubois, G. Tourims and sustainable development indicators: The gap between theoretical demands and practical achievements. *Curr. Issue Tour.* **2003**, *6*, 54–75.
51. Reddy, M.V. Sustainable tourism rapid indicators for less developed islands: An economic perspective. *Int. J. Tour. Res.* **2008**, *10*, 557–576.
52. Bojanic, D. Using a tourism importance-performance typology to investigate environmental sustainability on a global level. *J. Sustain. Tour.* **2011**, *19*, 989–1003.
53. Manning, E.W.; Clifford, G.; Dougherty, D.; Ernst, M. *What Managers Need to Know-A Practical Guide to the Development and Use of Indicators of Sustainable Tourism*; World Tourism organization: Madrid, Spain, 1996.
54. Wiedmann, T.; Minx, J. *A Definition of Carbon Footprint; ISAUK Research Report 2007-01*; Durhau, Centre for integrated Sustainability Analysis, ISAUK Research & Consulting: Durham, UK, 2007.
55. Weidema, B.P.; Thrane, M.; Christensen, P.; Schmidt, J.; Løkke, S. Carbon footprint. A catalyst for life cycle assessment. *J. Ind. Ecol.* **2008**, *12*, 3–6.
56. Ledant, J.-P. *L'empreinte Ecologique: Un Indicateur de…Quoi? Working paper IDD/J-P.L*; Institut Pour le Développement Durable: Ottignies, Belgium.
57. Scott, D.; Becken, S. Adapting to climate change and climate policy: Progress, problems and potentials. *J. Sustain. Tour.* **2010**, *18*, 283–295.
58. Christ, C.; Hillel, O.; Matus, S.; Sweeting, J. *Tourism and Biodiversity, Mapping Tourism's Global Footprint*; UNEP: Nairobi, Kenya, 2003.

59. Mekonnen, M.M.; Hoekstra, A.Y. *National water Footprint Accounts: The Green, Blue and Grey Water Footprint of Production and Consumption; Value of Water Research Report Series No. 50*; UNESCO-IHE: Delft, The Netherlands, 2011.

Chapter 7

Exploring Gamification Techniques and Applications for Sustainable Tourism

Adina Letiţia Negruşa, Valentin Toader, Aurelian Sofică, Mihaela Filofteia Tutunea and Rozalia Veronica Rus

Department of Hospitality Services, Faculty of Business, Babeş-Bolyai University, Cluj-Napoca, 7 Horea Street, 400174 Cluj-Napoca, Romania

ABSTRACT

Tourism is perceived as an appropriate solution for pursuing sustainable economic growth due to its main characteristics. In the context of sustainable tourism, gamification can act as an interface between tourists (clients), organisations (companies, NGOs, public institutions) and community, an interface built in a responsible and ethical way. The main objective of this study is to identify gamification techniques and applications used by organisations in the hospitality and tourism industry to improve their sustainable activities. The first part of the paper examines the relationship between gamification and sustainability, highlighting the links between these two concepts. The second part identifies success stories of gamification applied in hospitality and tourism and reviews gamification benefits by analysing the relationship between tourism organisations and three main tourism stakeholders: tourists, tourism employees and local community. The analysis is made in connection with the main pillars of sustainability: economic, social and environmental. This study is positioning the role of gamification in the tourism and hospitality industry and further, into the larger context of sustainable development.

Keywords: sustainable tourism; gamification; hospitality; applications

1. INTRODUCTION

The relationship between tourism and the environment represents one of the main topics in the agenda of tourism stakeholders, all of them being aware of its importance. Academics and researchers focus on it, emphasizing two approaches [1,2,3,4,5,6]:

- Environment-sensitive tourism: reduce environmental effects and increase sustainability;
- Environment-dependent tourism: the environment represents one of the main inputs for tourism, which must be preserved in order to guarantee long-run success.

Due to its characteristics, tourism is perceived as an appropriate solution for pursuing sustainable economic growth. According to the United Nations Environment Programme and the World Tourism Organisation (UNWTO), sustainable development represents "a process to meet the needs of the present without compromising the ability of future generations to meet their own needs" [7]. This approach can be explained using the three pillars of sustainability: economic, social and environmental. Economic sustainability is related to the efficiency of companies' activities in the long run; social sustainability refers to providing equal opportunities, respecting different cultures, reducing discrimination and poverty, while environmental sustainability is related to a proper management of scarce resources, especially of those that are not renewable. A sustainable tourism process should take "full account of its current and future economic, social and environmental impacts, addressing the needs of visitors, the industry, the environment and host communities" [7].

Despite the fact that almost all tourists declare they support the idea of sustainable tourism, Budeanu [8] emphasizes that only a small fraction of them are adopting a sustainable behaviour during their holidays. Usually, companies and public institutions are more environmentally aware and focused on implementing methods and techniques to make tourism more sustainable. During holidays, tourists' attention is focused on experiencing a high level of comfort, or at least higher than in their everyday life. In this situation, it is difficult to convince them to adopt a more sustainable behaviour, because it may be perceived as a factor with a negative influence on their comfort level. Therefore, tourists need a strong incentive to:

- Choose public transportation or other types of transportation with lower environmental impact, because these are more time-consuming, in the sense that would take tourists more time to reach their destination and, therefore, their stay there would be shorter;

- Opt for a certified sustainable accommodation unit as these units are usually more expensive;
- Consume less water, energy or reduce the amount of waste because they would have to change their habits and might not be ready or willing to adapt;
- Purchase domestic products when there are so many international trusted brands.

Fortunately, tourists do not purposely adopt a behaviour that could prove harmful for the environment, therefore proper incentives and techniques could positively influence a sustainable behaviour, a transition to a more sustainable lifestyle during their holidays. One solution could be the use of the Internet and Internet-based technologies. UNWTO [9] considers tourism and the Internet ideal partners, because the latter represents an efficient way to provide services and information, to target a large number of tourists at lower costs and, most important, to develop adequate communication channels that promote a sustainable lifestyle during holidays.

Lately, the implementation of gamification mechanism has determined a more sustainable approach for tourism development. This is a technique that involves giving rewards and providing entertainment as incentives for different stakeholders, to elicit a specific behaviour from them [10]. This technique has the ability to enhance tourists' experience in an innovative way, combining tourism services with game elements. Due to its characteristics, it increases the added value of the tourism product in a sustainable way [11] and diversifies tourists' experiences at the destination [4].

Gamification is not a new term. It "originated in the digital media industry" [12]. According to Gartner, the interest in gamification started around 2010, "first appearing on Google Trends in September 2010" [13]. The most cited definition of gamification is the one proposed by Deterding *et al.*: "'Gamification' is the use of game design elements in non-game contexts" [12]. Uskov and Sekar define gamification as the "implementation and use of game design elements and techniques, game thinking, game mechanics and analytics, computer game technology in business models, framework activities, processes, procedures, services, *etc.* to improve employee and/or customer/user skills, experience, engagement, effectiveness, and productivity outside pure entertainment area" [14]. Gamification is "not all about giving out badges and points to incentivize behaviour, but needs to apply game thinking to encourage and motivate for further activities" [15].

Gamification can be applied in any industry, yet the hospitality sector is not taking full advantage of gamification as other sectors do. Many well-known companies have already adopted gamification to increase customer engagement, gain customer loyalty, improve employee performance or gain competitive advantages. Companies will continue to adopt gamification and by 2016, it will be "an essential element for brands and retailers to drive customer marketing and loyalty" [13]. The research conducted by M2 Advisory Group estimates that the market expenditures on gamification solutions will reach $2.8 billion by 2016 [16]. Also, a study developed by the Pew Research Center's Internet & American Life Project, in which participated 1021 technology stakeholders and critics, supports the same idea: 53% of the respondents agreed that by 2020 "there will have been significant advances in the adoption and use of gamification" [17].

Some of the technologies that influenced the development of gamification are mobile technologies (mobile applications, location-based services), cloud computing, Web 2.0 and Augmented Reality. Knowing that by 2030 half of the world population will be online and mobile, it becomes vital for the tourism and hospitality industry to integrate Information and Communication Technologies (ICT) in their services [18]. ICT provide strategic tools for sustainable tourism development, creating opportunities to enhance the positive effects of tourism and to reduce its negative impacts at destination [19]. Bennett and Williams [20] are going further stating that there is a synergy between gamification and sustainability, based on the fact that both focus on emotional responses. According to UNWTO [21] game mechanisms have the capacity to create positive experiences in tourism—fun, excitement, arousal, pleasure, sense of achievement—and to provide tourists with both entertainment and information. Furthermore, there are good chances that tourists will adopt a more sustainable behaviour if provided with: a way to measure their progress and reasons to move on—collect points to obtain discount vouchers; goals to achieve—new levels for new facilities; the opportunity to compete with other tourists—leaderboards and special rewards for the best scores.

The gamification market is developing exponentially as companies are becoming more aware of its benefits [13,16,17]. This represents an important incentive for gamification solution providers to increase the quality of the products and develop more efficient solutions. The best proof is the fact that, even if at present gamification is less developed in tourism and hospitality than in other sectors, there are many producers and platforms (online and mobile) on the market (see Table 1).

The purpose of this study is to analyse the areas in which gamification has the potential to improve tourism sustainability. In particular, the paper concentrates on gamification success stories in tourism and hospitality and emphasizes the benefits of using "game design elements". To achieve this goal, the authors use three methods: analysis based on a broad bibliography, content analysis and observation. In the first part, the paper analyses the relationship between gamification and sustainability, identifying the links between these concepts and the mechanism that triggers a more sustainable behaviour in tourism.

Table 1. Gamification platforms used in hospitality and travel.

Product	Vendor/Producer	Use	Game Mechanics Features	Social Features
Badgeville	Badgeville Inc. (Redwood City, CA, USA) (badgeville.com)	Customer engagement Marketing campaign	Competition Feedback Levels & missions Profiles Ranks & badges	Activity feed Ratings & reviews Social media login
Bigdoor	Bigdoor Media Inc. (Seattle, WA, USA) (bigdoor.com)	Customer engagement	Economy/Marketplace Feedback Profiles Progress Bar Ranks & badges	Activity feed Community forums Likes & comments Ratings & reviews Share plugin
Gigya	Gigya Inc. (Mountain View, CA, USA) (gigya.com)	Customer engagement Marketing campaign	Economy/Marketplace Levels & missions Profiles Ranks & badges	Activity feed Likes & comments Live chat Ratings & reviews Share plugin Social login
Keas	Keas Inc. (San Francisco, CA, USA) (keas.com)	Customer engagement	Competition	Likes & comments Ratings & reviews Social login
Bamboo	Manumatix Inc. (Redwood City, CA, USA) (manumatix.com)	Customer engagement Marketing campaign	Competition	Activity feed Social login
Pluck	Demand Media Inc. (Santa Monica, CA, USA) (demandmedia.com)	Customer engagement Marketing campaign	Competition Economy/Marketplace Feedback Ranks & badges	Community forums Likes & comments Ratings & reviews

Source: [22].

In order to draw connections between the main stakeholders and to visually represent the value of gamification for sustainable tourism, an interpretative framework was developed (see Figure 1). The initiators of gamification in tourism are the providers of gamification services from ICT and business consultancy industries. The buyers of gamification services are the organisations with an interest in the tourism industry, like companies (hotel chains, restaurants, tour-operators), local institutions (mayor's office, tourism ministries, public employment offices) and NGOs (concerned with social, environmental and political issues). The providers and the buyers are the developers of the phenomenon. The purpose behind this endeavour is to engage and persuade the players (tourists, employees, community) to change or reinforce their behaviour through enhancing specific organisational activities (marketing, human resource management, public relations), a process carried on with the help of game mechanics (instant rewards, feedback, levels and missions, ranks and badges,*etc.*). The challenge is to do all this in a sustainable way, taking into account economic, social and environmental components.

In the second part of the paper, the focus is on identifying success stories in tourism gamification and analysing the impact of gamification. The paper ends with conclusions and the limitations of the present study.

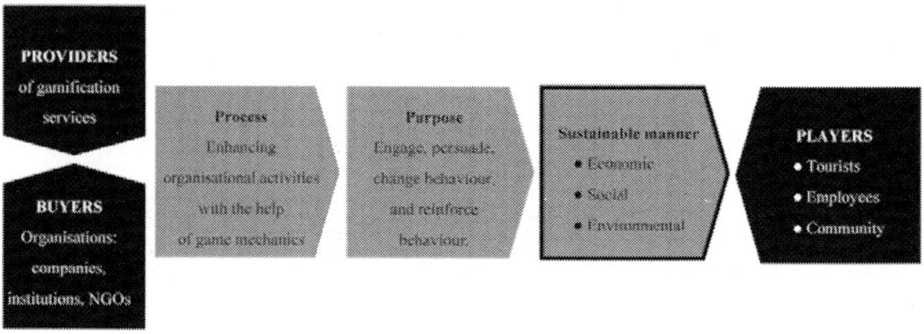

Figure 1. The mind map of a sustainable gamification process in tourism.

2. LITERATURE REVIEW

According to UNWTO, sustainable tourism should use environmental resources efficiently; promote biodiversity and the conservation of natural resources; respect the culture and values of host communities; encourage inter-cultural understanding and tolerance; provide stable employment, income-earning opportunities and social services to residents; reduce poverty and ensure long-term efficiency [7]. Similar approaches are available in the literature: sustainable tourism development is the growth that maintains a state of health necessary for survival at a higher level of quality, where quality means "the state of health in a tourist destination, sustaining the benefits of local community, satisfaction of tourist experience, and conservation of natural resources" [23].

However, tourism, like any other economic activity, generates both positive and negative effects. Being impossible to eliminate the negative ones, it is important to decide which aspects of tourism development can be considered sustainable. The destination's carrying capacity represents the first approach to sustainable tourism—natural resources are scarce, so there will always be a maximum number of tourists a destination could host without causing irreversible damage to the environment. As tourism evolves, the destination may increase its carrying capacity through different methods, like improvement of tourism products, introduction of new products or infrastructure development. Because the industries' interests and resource consumption may be in conflict, the local community should have a say regarding the acceptable level of changes a destination can go through because of tourism development [24]. Therefore, stakeholders represent the main component in the implementation of sustainable tourism. Their perception will influence their involvement in the tourism sustainable development strategy [25]. The decision to focus on more sustainable activities could be influenced by two types of factors:

- Exogenous, like environmental regulation, societal values and norms, and market factors—results of customers, suppliers, competitors and shareholders' decisions/interest regarding sustainability;
- Endogenous or internal factors, like company's strategy and its ability to integrate sustainability, the company's culture and resources [26].

ICT is considered a powerful tool used to improve sustainability in the case of both public and private companies [19,27,28,29]. Among all types of ICT applications, the location-based services are recognized as being one of the main sources of tourism sustainability. This type of applications, including the gamification tools using them, will generate important benefits for the destination community [19,28]:

- Increase the efficiency of resource consumption as a result of the information provided to tourists;
- Improve tourist flow management (*i.e.*, reduce the length of waiting queues, traffic congestions);
- Support an adequate interaction between members of local communities and tourists, and
- Diminish the physical degradation of the environment and support biological diversity.

Furthermore, Ali and Frew [27] consider that the relation between ICT and sustainable tourism is symbiotic, as ICT tools have a significant contribution to the amelioration of tourism negative impacts. In this respect, ICT represents a way to improve the organisation's management and control, facilitates the identification of environmental sensitive areas and provides information for decision makers and offers to community the opportunity to protect and promote its endogenous resources [27]. Schieder, Adukaite and Cantoni [29] see ICT tools as the best option for public institutions to promote cultural enrichment and, at same time, to protect cultural attractions from degradation or tourists' negative influence. They explain the case of UNESCO World Heritage sites, emphasizing the dilemma faced by their administrators, who had to choose between protection and preservation of the sites for the future generation and education and cultural enrichment of the present generation of tourists.

Gamification represents the intentional use of game elements for a gameful experience of non-game tasks and contexts [12,30,31]. Relating the concepts of sustainability and gamification, Seaborn and Fels consider that sustainability applications "seek to support and encourage sustainable behaviours, such as reducing the amount of resources used, investing in recycling initiatives and renewable forms of energy, and reusing material wherever possible" [30]. Therefore, as other authors [31,32,33] mention, gamification can focus on business processes or outcomes with the purpose to change or engage specific behaviour of employees, customers or communities of interest. Tourism and hospitality companies can combine game elements and use them to induce certain types of behaviour. They have the possibility to determine tourists to reduce the level of energy and water waste, to use public transportation or to rate services *etc.* Furthermore, companies may determine their employees to engage in learning and to improve their productivity.

According to Robson *et al.* [31], the key drivers of gamification stay into motivational drivers of human behaviour: reinforcements (extrinsic and intrinsic) and emotions. These elements encourage the players to learn and repeat the desired behaviour due to the fact that most of the gamification solutions are competition-based [33], while the game itself represents an intrinsically incentive for the player, sometimes more powerful than traditional motivation [30]. The developers of gamification applications, together with the managers of the companies using gamification applications play the role of the designers in the game design process. They are the ones who should customize the companies' goals into game mechanics, develop the structure of the gamified experience and create the reinforcements that will motivate players' actions or trigger behaviour changes [31].

Loong sees gamification as a complement of destination authenticity [34]. Together, these two elements create an artificial ecosystem where tourists' experience will be enhanced and shared with their social network. Due to the intangible nature of the services provided, sharing these memorable experiences is very important for the sustainability of tourism [34]. Tourists would like to see different places, taste the atmosphere, know different people or learn about the places they visit. Virtual interaction between stakeholders represent the best way to find and learn about destinations, imagine the services provided, rate experiences and develop tourists' brand awareness [35]. Sharing experiences through virtual interaction will help potential tourists to reduce the intangible nature of their holiday. Moreover, using a gamified application, the tourist has the opportunity to feel the emotion of a game progressing, share his or hers results with friends and even know new people and become a part of that game community. By combing these incentives, tourism and hospitality companies have the opportunity to create a mix of reinforcements to enhance tourists' experience [11,33].

The factors promoting users' involvement in gamification applications can be divided into utilitarian (usefulness, ease of use) and hedonic (enjoyment, playfulness) [36,37]. Enjoyment, usefulness and ease of use stimulate the willingness to use gamification, but the effectiveness of these factors diminishes as the time of using the application rises [36,38]. As a result, to obtain an improvement in the sustainability of tourists' behaviour, both utilitarian and hedonic incentives should be included in the gamification mechanism. Rapp [39] recommends to adjust the game mechanics to the player's motivational stage. At the beginning, the game mechanics should provide extrinsic rewards for tourists, like accommodation discounts, free drinks *etc.*; as the game progresses and the desired behaviour becomes more and more frequent, the rewards should be intrinsic and focus on special titles, social recognition, *etc.*

For a critical analysis, another important aspect is to review the challenges and disadvantages brought about by implementing gamification. According to Meloni [16], Anderson and Rainie [17], too much competition and a too strong game-style approach could determine an overall loss in employees' productivity and other negative outcomes.

Froehlich [40] mentions that in many situations, companies use gamification to build a green or eco profile meant to improve their brand image awareness or to promote new technologies rather than a sustainable behaviour. In addition, companies have to be aware of the "challenges initiated by the over-gamification of life" [15]. Not all the activities in a company can be gamified. Furthermore, it is not recommended to transform everything into competition or to make players expect to collect points for every human move [17]. In the case of gamified loyalty programmes, companies can reach the situation where it is not possible to differentiate loyal guests from "so-called programme bargain hunters" [41]. Another important aspect is that personal profile data are collected thought gamified applications, without users being aware of it, and can be used by companies to gain intelligence or for behavioral manipulation [17,32].

3. METHODOLOGY

Despite the fact that there are many studies trying to measure the level of tourism sustainability, there is no universally accepted method for it. Starting from the UNWTO system of Indicators of Sustainable Development for Tourism Destinations (2004) to different scientific papers or operational systems, different approaches for evaluating sustainability can be identified, at different levels (see [10,23,42], or the Accor Planet 21 system).

This is an exploratory study based on a detailed industry survey of self-identified case studies of gamification applications in the tourism and hospitality industry. According to Yin [43], a case study analysis should be considered an appropriate approach under the following conditions:

- When the research questions focus on "how" and "why";
- When one of the main purposes is to study how contextual conditions influence the phenomenon, because you believe they are relevant to it;
- The boundaries between the phenomenon and context are not clear.

Besides, in exploratory research, case studies offer the possibility to investigate causal mechanisms and the specific contexts in which they are activated [44].

The case study approach is a good choice for this research because it is an important aid for in-depth exploration and for studying the gamification phenomenon in the context of organisation and sustainability. Mapping a range of diverse cases of gamification applications gives the possibility to incorporate all aspects and key players in an overall picture. It outlines the role of gamification, adding sustainability features to organisations activating in the tourism and hospitality sector.

In this respect, the study was designed based on a multiple case analysis, which allows conclusions to be better grounded and offers the advantages of a cross-case comparison. In the case selection process, the best practices of gamification presented in the literature and documentary sources were taken into consideration. This criterion was used because the knowledge incorporated in best practices is usually well-defined and already tested. Best practices are defined as processes and/or methods that were successfully carried out and adjusted, leading to increased productivity and efficiency [45]. In the selection process, other criteria were established: organisations' area of activity and benefits of gamification for stakeholders. Because any example of good practices is described based on a general set of elements, these were used for collecting and analysing data about industry study cases. This approach made possible the uniform collection of data for a multiple case research and ensured the reliability of the cross-analysis carried out. Direct observation of the gamification platforms was useful in providing additional information about the topic.

In order to prove the progress in sustainability connected with gamification implementations, 37 case studies of gamification applications from different sectors of tourism and hospitality were analysed. From the selected case studies, 17 included all the three pillars of sustainability (economic, social and environmental) but to different degrees.

Due to the fact that gamification is a creative endeavour, it was difficult to create a common and completely codified framework of analysis. As in any other qualitative study, the data collection and analysis occur concurrently and in the present study the following elements were taken into account:

- Stakeholders and partners—Who are the beneficiaries or the target group of the good practice? Who are the users of the good practice?
- Context—What is the context (initial situation) and challenge being addressed?
- Validation—Confirmation by the beneficiaries that the practice addresses the needs properly. Has the good practice been validated with the stakeholders/final users?
- Impact—What has been the impact (positive or negative) of this good practice on the beneficiaries?
- Lessons—What are the key messages and lessons learned to take away from the good practice experience?
- Sustainability—What are the elements that need to be put into place for the good practice to be institutionally, socially, economically and environmentally sustainable?

In this study, tourism organisation was used as a generic term which includes, beside private companies, public institutions (city halls, tourism information centres, other public institutions in charge with tourism), NGOs and Destination Management Organisations (DMOs) because examples of gamification were identified both in the public and private sector. Further on, these relations were studied from the perspectives of sustainability changes driven by the implementation of gamification.

To review the applications, a structured analysis was conducted focusing on the relationship between tourism organisations and three main tourism stakeholders: tourists, tourism employees and local community. To emphasize the relation between stakeholders, the analysis was concentrated on a set of questions which had the role to identify what benefits gamification applications have for sustainability:

- Which is the scope or final objective of the implementation of gamified applications?
- Which is the player target group?
- What types of behavioural changes could gamification trigger?

- How much is the tourism experience influenced?
- Does it have the capacity to support local community and environment?

To conclude, the primary defining feature of a case study is represented by the multiplicity of perspectives that are rooted in a specific context [46]. The multiplicity of perspectives in our case deals with clients, employees and local community, and these perspectives are embedded in the specific context born at the intersection of gamification, tourism and sustainability.

4. RESULTS AND DISCUSSION

4.1. Gamification Effects on Tourism Organisation—Tourist Relationship

Implementing gamification for marketing purposes, the majority of organisations from tourism and hospitality perceive the use of game elements as an opportunity to reshape loyalty programmes, innovate advertising and promotion campaigns and customize and develop new products.

Due to a new generation of tourists eager to compete in the virtual environment for status recognition and build-in rewards, the attractiveness of the traditional loyalty programmes based on frequency of staying in hotel chain or serving dinners at restaurants decreased. As a result, international hotels and restaurant chains introduced game mechanics into their loyalty programmes (see Table 2). The following examples are relevant for this benefit: InterContinental Hotels Group, Starwood Hotels and Resorts, BJ's Restaurant and Brewhouse, Caesars Entertainment and Starbucks.

InterContinental Hotels Group added a daily online trivia game in 2011 to the group's loyalty programme. The trivia game awards participants with free miles based on quick and correct responses related to travel destinations. The marketers observed that this "Win It in a Minute" programme determined the attractiveness among younger consumers, a group which was not normally credited with brand loyalty.

Starwood Hotels and Resorts linked the Starwood Preferred Guest (SPG) loyalty programme with Foursquare, a location-based app. As a result, members were able to receive special promotions, and users with confirmed reservations could earn Starpoints by checking into a Starwood hotel on Foursquare or unlock other potential rewards (including a five-night resort getaway) and badges. Each month, Starwood will declare the member with the most check-ins across all properties as the SPG Mayor. Mayoral duties will include sharing favourite insider travel tips and destination hotspots with fellow frequent travellers around the globe.

BJ's Restaurant and Brewhouse developed Premier Rewards programme allowing clients to earn points for every $1 they spend on eligible transactions and for signing up for automatic posting on Facebook, Foursquare or Twitter [47].

Caesars Entertainment developed a guest loyalty programme that boasts more than 40 million participants. They were interested not only in improving the brand awareness but also in identifying the spending patterns to be used in establishing a potential, future value for each customer. The company has taken a reverse approach regarding the process of obtaining standard rewards based on past achievements. Instead, Caesars rewards guests instantly based on projection for forthcoming spending activities (e.g., lodging offers, special food and beverage opportunities based on potential play in casinos).

A similar example comes from Starbucks. The company has recently begun implementing quick response (QR) codes as part of its mobile gamification strategy for new product promotions. By scanning a product's QR code, consumers are directed to the company's mobile website where they can find information on its products, newsletters, coupons, customer service and other similar benefits. In order to intensify the interaction with consumers, the mobile website is linked to social media outlets. A Facebook link is used to create a social platform for consumers to interact and comment on likes and dislikes. From the company's perspective, accessibility via QR code ensures an analytic database (e.g., time of day, gender, location, *etc.*) that can be incorporated in future marketing campaigns and product presentations.

On the other hand, the mobile technology offers innovative tools for improving the promotion strategy (see Table 2). One of this is the Gift Cards App that allows consumers to store and redeem gift cards by accessing store data via the phone display. Connecting to Facebook, Twitter and Google+ allows users to send, re-gift and receive gift cards instantly. Beside its explicit benefits, this approach provides an opportunity for an additional sales channel for hotels and restaurants, to advertise directly and sell gift cards through social media. For example, if a customer holds a hotel gift card and accesses a Gift Cards App, the hotel's marketer is able to gain visibility of the cardholder and offer tailored promotions aimed at powerful revitalization bundled purchasing.

The game mechanics could help companies achieve more than sales and promotion related objectives. This concept used in the development of the basic services could have positive impacts on the product diversification and added value strategies. In this way, using gamification for new purposes will result in creating a competitive advantage for the hospitality industry. A good example is Pizza Hut, which is introducing interactive touchscreen tables allowing customers to customize their pizza before ordering. Clients can post orders and choose a payment method. The interactive table can be used to play games while customers are waiting for their pizza [48]. Even if it is experimental, the implementation introduces some benefits such as: redefining the main product, providing the opportunity to customize the service for each client and influencing the restaurant's production flow (see Table 2). The use of gamification in product customization has a positive impact on the environment because clients can order a pizza of a preferred size, with the preferred ingredients, reducing in this way the quantity of food waste. The time for taking orders decreases; the employees don't have to wait for clients to decide what to order and they can focus on other activities.

Another example of gamification application for reshaping the service and creating some unique characteristics is 4food restaurant, a unit which proposed a "de-junking food" concept for burgers. Because the owner considers social interaction as a part of the business concept, the quick service restaurant provided a web-based application allowing clients to create their own type of burgers, name their own creations and market them via social media. They used a suite of web-based technologies which offer the possibility to build customer's profile and match menu items to a particular profile feature and generate incentives. Through this application the restaurant gained those clients who were undecided regarding the healthier food consumption pattern, by shaping their behaviour using communication and social interactions elements. In this case, the gamification example stresses on the intrinsic motivation of customers interested in a healthy diet.

EpicMix used gamification for diversification strategy. This is a popular mobile app between skiers, now being available at 10 ski resorts. The app captures the ski and snowboard experience of skiers based on Radio Frequency Identification (RFID) tags included in ski passes. Users can connect with other people using social media, compete and earn virtual pins. This location-based application offers the following functions: photo sharing, customized guides, ski and snowboard School experience capture, useful information on ski slopes, weather conditions, *etc.*

Meanwhile, the tourism destinations location-based gamified apps use the benefits of Augmented Reality (AR) in order to take visitors on an individual and interactive walking tour to discover the place. Already available in cities like London, Stockholm and Berlin, these applications combine the tourist interest in discovering the destination (destination's history, events, people and social life) with its competitive spirit and joy of playing.

The analysis of the previous best practices emphasizes four important aspects. First, gamification reinforces the economic sustainability in gaining brand loyalty, especially in the case of young customers. For a company, tourists' loyalty is extremely important for achieving its strategic objectives. Even if loyalty programs have been used by tourism companies since 1981, when American Airlines initiated AAdvantage [41], the implementation of gamification can be considered marketing innovation in the tourism and hospitality sector. Because loyalty is determined by quality, perceived value and satisfaction [49], the companies have to collect data about tourists' behaviour and their perception regarding the services consumed during the holiday. Many gamification applications provide a sustainable way to do that, allowing companies to customize their products in accordance with clients' needs and their economic objectives. This approach responds better to the preferences of a new generation of clients, more oriented towards ICTs.

Table 2. Benefits of gamifying the relationship tourism organisation—tourist.

Type of Gamification Applications by Activities	Economic Effects	Social Effects	Environmental Effects
1. Loyalty programmes (InterContinental Hotels Group, Starwood Hotels and Resorts, BJ's Restaurant and Brewhouse, Caesars Entertainment and Starbucks)	• Increase young customers` loyalty • Sharing travel tips • Improving the knowledge about customers behaviour • Identifying the future values/costumer	• Social recognition—special status (*i.e.*, "mayor") • Increased visibility on social networks	Easier, quicker and less resource consuming interactions between customers and supporting systems
2. Advertising and promotions (gift cards apps)	• Alternative sales channels • Improving brand awareness • Improving promotions strategy	Increased visibility on social networks	Resource consumption reduction
3. Developing or creating new services (Pizza Hut, EpicMix, 4food)	• Product diversification • Increased added value • Improving production flow	Learning about nutrition facts, tourism destinations, *etc.*	• There is no need to print menus or tour guides • Resource consumption reduction

Second, gamification brings new hedonic incentives to engage customers' brand recognition. As Rapp [39] suggests, the game mechanism is customized to the level of brand recognition, providing clients with more appealing rewards as their loyalty increases. Extrinsic rewards and utilitarian facilities are used at the beginning, like discounts for the services consumed and an easier and quicker interaction between customers and supporting systems but also interactions among different users of the same system [50]. As the loyalty increases, virtual interaction and loyalty statutes are used to reward clients and achieve companies' objectives.

By analysing the type of incentives used by gamification mechanism of the studied best practices, there can be identified both intrinsic and extrinsic motivations. Loyalty programmes and advertising and promotion campaigns focus more on extrinsic motivation, providing external benefits to tourists in order to engage them in the desired behaviour, while the development of new products and services in tourism and hospitality focus mostly on intrinsic motivation, including in the tourism products elements of social and environmental responsibility—eating healthier products, buying local products, reducing waste. This represents a successful way to exploit the tourists' awareness for sustainability and to activate the incentives which will bring to a sustainable behaviour.

Third, all the applications of gamification with mechanism focusing on marketing goals imply social interaction between guests. According to Gigya's internal study, the implementation of gamification increases the number of website comments by 13% and social sharing by 22% [51]. The potential of increasing tourism organisations' visibility on social networks is linked to several directions: increasing brand notoriety, getting local prosperity and becoming an important local player with a significant influence role for the community's members.

Finally, the effects of applying gamification at the level of the relations between tourism organisation and tourists are mostly economic. Loyalty programmes are one of the four main reasons for selecting a hotel brand [41]. According to Forgacs [52] brand performance can be linked with marketing and operations. Thus, in operations, a well-managed brand can generate maximum revenue, ensure better training to its employees, and create a consistency in delivering their service. In marketing, the hotel can build its brand more effectively, conduct broader and more coordinate promotion, and develop a more advanced product. In this case, a brand can reach a penetrated market more effectively and offer a higher or premium price [52]. The increase in clients' loyalty determines an increase in overnight stays or service consumption, an improvement in cash flows and a rise of companies' notoriety [53,54]. Moreover, due to the fact that gamification is applied to services, improvements in production flows are possible as a result of clients' involvement in the game mechanism. The increase in the length of stay at the same location also has an important environmental effect, reducing the amount of consumed resources like water, energy, and working time and leads to a stronger socio-cultural interaction between tourists and residents. Gamification is an excellent way to strengthen brands' image as customer-friendly, socially-minded and community-involved. Some companies engage clients into different causes associated with the main purpose of the business unit by donating a part of the price to these types of actions.

The final objectives of gamifying the relationship tourism organisations—tourists are related mostly to the economic dimension of sustainability—more loyal clients, higher consumption frequency, increased brand awareness or new products development. The tourists and the hospitality units' clients represent the target group of players, companies being interested in determining specific behavioural changes that will lead to an increased profitability. Gamification boosts tourists' satisfaction, as they facilitate access to additional opportunities and services at the destination—customized products, further information, social recognition, better ways to use their travel time, increased virtual interaction with other tourists from that destination, *etc.*

Positive externalities at the level of social dimension of sustainability may arise from the tourists' behavioural changes or from the facilities created for them: learning effects for both tourists and residents; involvement of tourists in charitable acts or in tourism activities focused on reducing poverty and bringing positive changes to the places where residents live; improvement of destination ICT infrastructure, the residents thus having the possibility to benefit from it for personal use; the facilities for disabled tourists will be available for disabled residents, *etc.* Furthermore, there is a positive environmental externality, because, in general, digital media reduces the usage of physical resources regarding manufacturing and distribution compared to print media but, on the other hand, at the consumers' level, it implies higher energy due to the use of electronic devices [55].

4.2. Gamification Effects on the Tourism Organisation—Employee Relationship

Gamification is used in HR mainly to facilitate a playful and motivated environment (intrinsic and extrinsic) during an array of HR activities like training, recruiting, monitoring and stimulating productivity, with the help of elements commonly associated with game mechanics. Some of these elements are player profiles, competition between individuals or teams, feedback, levels and missions, ranks and badges. In HR practices, serious gaming is less used than casual gaming. Serious gaming means that the player will immerse in a game situation where the real environment is recreated "with all the complexities of the technical and social interactions typically found there" [56]. Casual gaming "tends to emphasize rote behaviours, puzzle solving, simplistic graphics, and straightforward play mechanics" [56]. HR activities invest less in environment reconstruction and more in the educational and pragmatic component.

Gamification in training can be defined as the addition of elements commonly associated with games to an educational or training programme in order to make the learning process more engaging [56]. One relevant example for gamification in tourism training is Costa Cruise. The company, with the help of Hydra-New Media, used gamification in order to inform travel agents, retailers and sales staff about travel, ships, destinations and special offers, *etc.* They planned to use e-learning for its unifying power and especially to overpass the fact that classical e-learning was considered boring by employees, because of the inherent lack of interaction, feedback, competition and rewards. They developed three games ("Exploring the world with Costa"; "Dream destination"; "Aboard our new cruise ship") each aimed at different levels of learning. Some e-learning topics were designed to blend e-learning with classroom training and other events, to integrate the virtual learning space with the real one. The following successes were reported: the average time spent by users on the platform was 20 min; most of them visited the platform several times; most of the tasks were completed successfully; content-related questions were answered correctly [57].

Gamification in recruiting can be defined as the addition of elements commonly associated with games to the hiring process (job analysis, recruiting and selection) in order to market the job openings to the potential employees in a funnier and more engaging way. One of the first companies to use gamification in the tourism industry was Marriott International Inc (Bethesda, MD, USA). Their online game My Marriott Hotel, was launched in June 2011 [58]. The game was developed as a Facebook app in English, Spanish, French, Arabic and Mandarin to address the rapid expansion outside the US and to reach out for new recruiting targets, such as the Millennials. Users from 120 countries were simultaneously caught in the game executing tasks and 30% of the users clicked on the "try it for real" button that redirected them to the Marriott's career website [59].

Gamification in stimulating and monitoring productivity can be defined as the addition of elements commonly associated with games to the stimulation and monitoring process, in order to increase competition and performance. The third example deals with a delicate area of gamification. The case study reflects the unintended consequences of gamification or what can go wrong when companies don't "zoom in" on ethical details. Such is the case of using scoreboards or leaderboards in Disneyland where public monitors displayed the work speed of housekeeping employees, for instance who was the quickest at loading pillow cases, sheets and other items into a laundry machine. Their names were displayed with efficiency numbers: green if the performance was high and red if it was lower than expected. The purpose of the gamification process was to measure and increase employees' productivity, but fierce competition between the employees started. Some employees skipped bathrooms and breaks out of fear; low-paid workers felt they were controlled; co-workers felt badly for a pregnant employee and the system got the name "electronic whip" [60].

The analysis of these three examples of using gamification in HR practices brings to the foreground some important issues in terms of gamification and sustainability.

First, from the three pillars of sustainability, the most important for tourism are the economic and social. As mapped by UNWTO, they merge together under the concept of "*employment quality*", which aims "to strengthen the number and quality of local jobs created and supported by tourism, including the level of pay, conditions of service and availability to all without discrimination by gender, race, disability or in other ways" [7].

Table 3. Benefits of gamifying the relationship tourism organisation—tourism employees (HR).

Type of Gamification Applications by Activities	Economic Effects	Social Effects	Environmental Effects
1. Training (Costa)	• Improving learning, better knowledge regarding facts, rules, procedures at lower costs • Improving clients' experience as a result of observable skills improved in daily activities.	• Increasing employee's confidence in their knowledge and skills. • Increase the employability on the labour market for trained employees. • Provide an element of fun.	Reduce resource consumption for trainings in terms of money, traveling (carbon footprint) and manpower.
2. Recruiting (My Marriott Hotel)	Innovative way to market job openings bypassing classical methods. Increase brand awareness and target performance. Targeting specific recruiting sources (the Millennials).	Gamification practices in recruitment and selection can promote equal opportunities by focusing on performance and not on race, experience, education, *etc.*	Targeting simultaneous potential employees at international level, lower resource consumption.
3. Stimulating & monitoring productivity (Disneyland)	Increased labour productivity. Ability to monitor performance and compare "players". Inject adrenalin into the company.	Delicate area considering the potential adverse effects: fierce competition between employees—they tend to work individually not as a team members; discrimination (e.g., for pregnant employees).	Reduction of resource consumption.

Second, when the training is delivered in disadvantaged areas and positively contributes to the development of the human capital, the sustainable benefits of gamification diffuse into society to a greater level (see Table 3). In the majority of cases, training and gamification can be business-centric, aimed at the needs of the business, but considering the next level, they can become sustainable if the developers, organisations and gamification service providers integrate the process into the bigger picture of the local community and personal development. Personal development pursued in the organisational setting, but disconnected from the local community's needs, lacks the sustainability component and social legitimacy. In the majority of cases, the employers are reluctant to and disengaged with the training process because they see the company as interested only in the "bottom line" and not in the genuine development of the main stakeholders. Gamification creates a learning environment which activates the employees' intrinsic motivations to learn and to progress—they may satisfy their curiosity on different topics, compete with other colleagues and gain self-esteem and colleagues' recognition for their achievements.

Third, gamification in recruiting goes beyond the classic requirements of ensuring and enforcing labour regulation and surpasses the classical approach to recruiting, delivering results at three levels: employer branding, targeting performance and providing equal opportunities (see Table 3). Employer branding can be skilfully used in gamification as another layer of game mechanics, because the company culture (visible artefacts, norms, values and basic assumptions) are embedded in the game. The Marriott case study presented above was one of the first examples of merging employer/industry branding and gaming with the purpose of attracting young people to the hospitality and tourism industry. In terms of targeting performance in the recruitment and selection process, gamification as a business case competition allows the best performers to step out and be noticed and avoids, at the same time, biases like school name, years of experience, age, race,*etc.* The main aim of these practices is to identify the best performers with initiative and problem solving skills and, how the candidate will behave in terms of organisational fit. It is interesting that these practices challenge the classical approach to recruitment, job advertising and candidate selection. The order is reversed: first the people with interesting and relevant skills are tested or selected without knowing their demographic or professional background and only then are they recruited or informed by the organisation's interest in employing them. Therefore, the classical recruiting methods are bypassed. Furthermore, the focus is on player's intrinsic motivation, its' incentives to play being fully related to the fact that the player likes tourism and hospitality as a field of activity or he/she enjoys similar tasks with the one performed in the game.

Finally, stimulating and monitoring productivity is the most delicate area when it comes to the use of gamification. If the other areas of HR and gamification are not questioned almost at all, a lot of ethical questions arise when it comes to stimulating work and productivity (see Table 3). From a realistic perspective, a performance-focused rhetoric covers the fact that gamification is nothing else than a specific persuasive technology and procedure aimed at reinforcing behaviour with the help of tunnelling, self-monitoring, surveillance and conditioning [61].

Providing mostly extrinsic incentives to employees, gamification could be seen as an unfair mechanism used to increase productivity, with no real costs, via playfulness, but with the cost of increasing pressure on the employees, creating unnatural competition, disregarding work diversity, *etc.* There is an "exploitation-ware" which actually means to use gamification in order to motivate staff to do more than what their job requires [62]. The genealogy of this capitalist process is not a flattering one, gamification having roots in the "socialist competition". Factories were competing for a medal in order to motivate workers in a society alternative to capitalism [63]. It is difficult not to take into account all these "postmodern" perspectives when it comes to gamification of productivity and sustainability, because ethics is an inherent component of the company's sustainable development. Sometimes, like in the Disneyland case presented above, the mechanics of the game degenerates. This case and other raised the question of the "thin red line" that gamification can cross.

The objectives of using gamification in HR activities are primarily concerned with partnering, engagement and alignment of a targeted audience with the organisation (private sector, public sector, NGO sector). Partnering means identifying potential external organisations with a higher purpose, ideally NGOs, and involving them in aligning their mission with the society. Engaging means to increase the involvement rate of internal and external stakeholders on various frequencies associated with the organisation. Aligning means to adjust all HR processes to resonate with the values of a sustainable business [64].

The targeted audience or the players are the employees of the company and the potential candidates on the labour market, or members of the community interested in various sustainability dimensions. In the cases aforementioned, the players are the potential employees or candidates on the labour market, the organisation's employees and, in the future, maybe the community. On the pragmatic side, gamification wants to reshape specific behaviours depending on the targeted group. An organisation might want to:

- Stimulate people to apply for a job—candidates or potential employees from the labour market or other companies;
- Increase the assimilation of knowledge and skills of the internal stakeholders—employees;
- Increase productivity levels—employees;
- Increase the degree of control over employees' activities.

These types of behavioural changes are already targeted in tourism gamification practices, but another one, proposed with the specific purpose of enhancing the social and ecological component of sustainability, might enhance employees' willingness to volunteer in social or ecological programmes by using a gamification process or platform.

The touristic experience can be definitely influenced, indirectly, by introducing gamification in HR activities. The influence is hard to be quantified, but critical events with the tourism experience, be they positive or negative can provide some guidelines. Gamification allows performance-oriented recruitment and selection, ensuring a qualified pool of candidates. Besides, gamification of the training process boosts the learning engagement and the employee development. The three HR dimensions ensure the necessary social capital that provides high quality services. Regarding the controlling and performance enhancement component of gamification, the tourists' experience might be influenced in a negative way if the control and performance stimulation programme fails to be balanced and counteract extreme competition, work-life balance, pregnant employee protection and any other equity dimension the job involves. Stressed, overworked employees, lack of empathy, all can backfire and reflect in the tourism experience or in the employer's branding image.

In order to gain legitimacy, the company needs to be a good corporate citizen and address the "triple bottom line": profit, people and planet. Gamification in the HR activities of the tourism industry definitely has the capacity to support certain aspects of the local community in terms of career development, employment opportunities, selection and employer branding. It can develop if companies create and support volunteering programmes for employees with a gamification layer targeting environmental issues.

Gamification in HR is not necessarily about what it is happening presently but about what it can be in an organisational environment, educated, ethical and playful enough to construct the next level of connections and transparency between stakeholders. It can create an elegant, casual and funny infrastructure, connecting in a proactive manner employees working in tourism with the local community, in a coherent effort to solve social and environmental challenges.

4.3. Gamification Effects on the Tourism Organisation—Community Relationship

Some of the gamification applications were targeted to influence visitors or tourists' behaviour and to promote sustainable consumption. The use of game mechanics can encourage the interest in sustainability and tourism sustainability. Studies comparing resource consumption (water, energy, food, gas, *etc.*) during holidays to everyday life emphasize an accentuated level of consumption and waste during holidays [65,66,67]. Therefore, raising awareness about sustainable consumption during holidays and creating tools to improve or change tourists' consumption behaviour could represent an important contribution of gamification to the development of sustainable tourism.

With respect to this, a very good example is a platform concept for the city of Eindhoven (Netherlands) for reducing carbon emissions at a destination [68]. This requires involvement of the local community, companies and tourists. The system implies social interaction through game-play and incentive points. The idea is very simple: visitors who use public transportation will gain points and will use them to obtain discounts in restaurants, bars, museums or a reduced tourist's tax, *etc.* The platform provides visitors and citizens superior service solutions for sustainable tourism and, at the same time, acts as a tool to connect people and enable them to act virtuously as a social community. This concept can be implemented likewise for sensible issues like water consumption, waste reduction or energy saving.

Related to energy saving, an example is the Nest smart thermostat [69] which tries to control and manage home heating and cooling systems. This application combines three data resources: built-in motion and temperature sensors, local weather data and observations of the home occupant. In about a week Nest creates a personalized heating and cooling schedule for each household. According to Matsuoka [70], Nest is aesthetically pleasing, uses an awarded mechanism with a virtual Leaf for setting temperature to energy-efficient levels and started to impact on behaviour, 92% of the Nesters use check for the Nest Leaf at least once a week.

The energy-saving behaviour is important not only for single family homes, but mainly for shared residences like residence halls, student' dormitories or accommodation units [71] because the inhabitants themselves have little awareness about energy usage, while construction managers or owners have a high level of interest in controlling it. Because the occupants do not have incentives to conserve energy or reduce the resource waste, gamification represents a way to raise their awareness about environmental issues. The game mechanics were used successfully for engaging students from Oberlin College [72] and University of Southern California [73] in a competition for lowering resource consumption in residence halls. As result of their activities, an eco-feedback was provided on the web or on shared public displays. According to Froehlich [40], these types of applications are rather artificial because they have to be intensively promoted and conduct to good results only for a short period of time. Knowing the importance of social interaction in gamification, a platform based on the previous idea, with some additional incentives, could transform saving energy or other types of resources in a behavioural pattern for tourists and the local community (see Table 4).

Table 4. Benefits of gamifying the tourism organisation—community relationship.

Type of Gamification Applications by Activities	Economic Effects	Social Effects	Environmental Effects
1. Residents reactions toward tourism and tourists (TripAdvisor)	• Empower local businesses • Facilitating networking among local enterprises • Destination branding	Smooth cultural interactions between tourists and local community.	Equitable use of resources.
2. Stimulating sustainable consumption (Nest, students dorms)	Increased efficiency in the use of resources	• Changes in the local community and tourists' behaviour • Social interaction.	Interest for environmentally sustainable products or services.
3. Charity (Hotel Prinz Luitpold)	Clients' loyalty	Increase life quality	Funds for protecting the environment and endangered species
4. Information, education, tourists guiding (Tourism Ireland, Tourism Australia, Travel Plot Porto, Sighter, Spy Denton, Catch the Kid, History Hero, APA, SFMOMA Families, Race Against Time)	• Identifying new segments of clients • Promoting local businesses	• Learning effects • Connecting destinations with their visitors	Management of tourist flows

Hotel Prinz Luitpold (Bavaria, Germany) uses gamification to engage their guests in a charitable act. Starting from the name of the hotel, they developed a game mechanism through which they reward their loyal clients. If a guest stays for at least 21 nights, the king will ennoble him. In order to qualify for this honour, the guest has to make a good deed in the king's name. Guests had the feeling of being part of a higher mission, being important, which increased the connection with the hotel [74]. Moreover, the local community and not only, had many benefits because of the good deeds implemented by hotel guests. Some clients took care of sick people, others donated to mountain rescue or raised support for kindergarten projects in Africa.

TripAdvisor, the world's largest travel portal, is one of the best-known examples of gamification in the tourism industry. The company offers a gamification platform on its website, which allows tourists to express their reviews on different destinations, hotels, restaurants, *etc.*, gaining various rewards. This stimulates the interaction among tourists, creating a TripAdvisor community, promoting travellers' choice destinations, attracting tourists on website, increasing the time spent on the website, and developing tourists' loyalty (see Table 4). TripAdvisor was used to show how "funware is developed for motivating travellers to engage with its website and for improving the travellers' online experiential values and trip planning processes" [75]. As a result, TripAdvisor increased the experimental values (customer ROI, social, enjoyment-immersion) and generated some important behavioural outcomes, like: "user engagement with the TripAdvisor website, impacts on processes, outcomes and social interactions, destination cognition and behaviour, emotional-relational experiences, overall experiences" [75].

To have the complete picture of the gamification use in tourism and its benefits in terms of sustainability, it is necessary to consider two other special types of organisations, namely destination management organisations (DMOs) and cultural institutions, like museums, art galleries, exhibitions, castles, memorial houses, *etc.* The role of DMOs in sustainable tourism is "to develop the whole product and include the interests of all stakeholders" [76]. DMOs have to concentrate both on tourists and residents, providing tourism experiences in a sustainable manner and enhancing the social and economic well-being of the residents [77].

Tourism Ireland and Tourism Australia implemented gamification in 2011 in order to promote Ireland and Australia as holiday destinations [78]. Tourism Ireland adopted a gamification solution which allowed tourists to create their own city and pub and to make virtual visits of the island. The solution has generated a significant increase in Ireland's visibility on social networks and an increasing number of visitors interested in this tourism destination. Tourism Australia, through the campaign "Best Jobs in the World", launched a gamification solution to attract young people around the world to a dream job in a working holiday in Australia [79]. The participants in this campaign were selected based on images or videos sent and shared on social networks and the winners were rewarded with offers of trips to Australian destinations. The results were overwhelming—more than 330,000 persons from 196 countries applied for the six jobs they offered, while the number of persons actively seeking information about working in Australia reached 430,000 people in 2013 [79].

From the transmedia storytelling category in gamification, which integrates various playing techniques with social networks and other collaborative platforms, Travel Plot Porto, represents an interesting gamification solution in tourism approached by DMOs. The tourists, in their mission to save Port Wine, must identify various hidden treasures. Achieving the goals, they are rewarded with various prizes and discounts and have the opportunity to create and share their own story related to this touristic destination [74]. The app provides tourists an opportunity to discover and learn about Porto's history in a funny and engaging way, to experience the local gastronomy and atmosphere and to share their experience with their friends and families.

Due to their roles to attract and educate tourists, DMOs must consider integrating social networks with simple and intuitive social games or apps. A well-known gamified application is a Hungarian application, an urban treasure hunt game for different destinations and cities—Sighter. The app allows the users to capture and send images of urban locations to everyone using mobile devices.

I Spy Denton, a digital scavenger hunt application that promotes the city of Denton (TX, USA), rewards the users involved in the transmission of images and creates an important community on social networks with the final purpose of promoting the destination. The New Mexico Tourism Department has launched a wide application, Catch the Kid, where tourists seek different locations where Billy the Kid could be hidden. Besides attracting tourists to the destination, the application has an important economic impact, the rewards granted to users stimulating local entrepreneurship [80].

The analysis of gamification solutions adopted by cultural tourism institutions must start from their role of attracting and educating visitors. Gamification solutions from this category have to guide visitors, facilitate interaction between them and attract funds for conservation and development of the destinations (see Table 4). The Dallas Museum of Art uses gamification incentives for visitors for recognizing artworks by sharing images on social networks, without the need of sophisticated scanning devices [81].

A specific solution for gamification based on digital curation can be considered History Hero, which was the first interactive mobile guide for visitors from important museums, churches and historical sites in the world. It was adopted by almost 40 institutions in the US and Europe, including the Metropolitan Museum of Art, St. Peter's Basilica, Versailles Castle and National Mall [82].

Another example of using gamification applications is APA that incorporates solutions of serious game with VM (virtual museum) techniques, which allows users to explore a specific historical time—the 13th century in Roman Bologna. The educational aim of the solution is to attract visitors to the destination Bologna, known for its preserved medieval heritage preserved [83]. The San Francisco Museum of Modern Art has its own application called SFMOMA Families, which is aimed at attracting families to the museum and encouraging conversations and comments on the exhibits through gamification techniques. In addition, Tate Modern, UK adopted gamification applications (Race Against Time) using mobile devices as a tool for the presentation of art history by incursions of modern art [84]. Other well-known gamification apps adopted by museums are Meanderthal by Smithsonian, Eduweb's augmented reality app MoonWalking or LaunchBall by the Museum of Science in Boston [85].

The analysis of the previous gamification best practices resulted in the following issues.

First, there is an important difference in the amount of benefits identified in this case and the previous two situations. The benefits of gamifying the relation between community and tourism organisations are concentrated mainly on social and environmental pillars of sustainability. Social interaction, cultural understanding, resource saving behaviours and the increase in the quality of life of both tourists and residents represent the main objectives and desired behavioural changes.

The messages promoted through gamification engage two types of positive behaviours [86]:

- Motivate tourists to involve in a new activity or to visit a new destination they never did previously or;
- Motivate visitors to make permanent changes in their holiday behaviour.

Examples for the first type of behaviour could be the approaches of Tourism Australia, Hotel Prinz Luitpold or Travel Plot Porto, while for the second type of behaviour, representative are Nest or the platform for the city of Eindhoven.

Second, even if the economic benefits of gamification in this case are smaller, they are concentrated on the marketing area. Specific for DMOs is that they have to "engage visitors before, during and after the travel" [74], gamification having to provide an adequate method to achieve that. The benefits of gamification in the case of marketing objectives of DMOs are similar to those of private companies, but it is more difficult for a non-profit organisation or a public institution to engage the tourists because they are sending a less tangible message [86].

Gamification complements the marketing strategies DMOs may use in order to build the destination's brand. Furthermore, the adoption and implementation of gamification solutions by DMOs determine an increase in the number of inbound and outbound tourists to the destination, higher visibility and more virtual interaction on social networks, quicker feedback and increase in tourists' loyalty [78].

Third, both intrinsic and extrinsic motivation can be identified in the previous best practices. The focus is more accentuated on intrinsic incentives engaging the tourists in activities which bring them self-esteem and social recognition, like charities, honest reviews for the services consumed, reduction in the waste of resources or knowledge development. The extrinsic incentives act as a complement in the game mechanism, providing strong enough reasons for tourists to reduce waste or to learn things about the history and the life of destinations' community which otherwise may be boring or not so appealing, especially for the younger tourists.

The impact of visitors' feedback on local communities, through the reviews' mechanism, is significant. By writing reviews, travellers are paying back to the local community an "honest thanks" for the experience they had. The information about their experience represents the best promoting tool of the services provided, for the entire destination and its endogenous resources. From the community's point of view, the reviews received increase their awareness about tourism effects and represent an incentive for improving and promoting tourism activities.

Finally, the use of gamification by museums and other cultural sites represent an opportunity to attract new visitors and in particular younger visitors. Gamification enhances the learning experience, providing visitors with the opportunity to discover and learn gradually the aspects about the tourist destination. On the other side, in the case of cultural attractions sensitive to visitors, gamification represents a way to facilitate access to these sites [29]. Through gamification, the learning effects and the learning experience do not end when the visitor leaves the site. Tourists may use these apps before they are starting the journey to see what points of interest are worth visiting and learn about, during the journey to plan the visits and to find information about the destination and after the journey to remember the experience they had and to share with their friends and relatives the knowledge they gained.

Looking to the previous best practices, from the point of the methodological approach, the main objectives of gamification was to promote, develop and educate a sustainable tourism framework for the entire community. Due to the fact that sustainable development cannot be achieved without the implication of a larger number of the society's members, the tourists, the residents and the employees became the target group that local entities, DMOs or cultural institutions tried to engage in different actions through gamification. The virtual interaction and the game mechanism represent a perfect environment to engage all the tourism stakeholders in responsible behaviours. The main behavioural changes targeted through gamification were:

- Raise awareness about the necessity for sustainable consumption among tourists and local residents;
- Responsible consumption during the holidays;
- Save energy and reduce waste;
- Involve in charitable acts.

The tourism experience is directly influenced by these types of gamification applications because there are new funny tools that enriched the level of education and awareness of tourists. Moreover, in this case, the tourism experience is enriched by an inner satisfaction the tourists may have as a result of their involvement in the communities' problems. Making donations, helping poor people to improve their lives, knowing and understanding the beliefs and behaviours of residents, and contributing to the protection of some endangered species of plants and animals create a sense of self-esteem which makes the holiday unforgettable. In conclusion, gamification represents an adequate support for the interaction between local communities' members and tourists, making them more aware of the environmental impacts of both parties.

5. CONCLUSIONS

The paper contributes to the literature by reviewing the benefits of gamification in tourism and hospitality based on the pillars of sustainability: economic, social and environmental. The result is a complete grid with the main implications of gamification techniques for distinct types of stakeholders. This study is positioning the role of gamification for tourism and hospitality into a larger context of sustainable development.

Gamification contributes to the overall tourist experience in different ways. Gamification can be used to act as an interface between tourists, organisations and local communities, to leverage responsible and ethical behaviour. Regardless of some approaches that consider gamification only another way of attracting tourists or clients, there is a visible sustainable feature. Even if this technique is without question very efficient to ensure economic benefits in the best practices analysed, it showed also significant benefits in terms of social and environmental sustainability. Of course, the most important advantage of gamification is to increase brand loyalty. However, the results should not be evaluated only in terms of quantity of products or services consumed but also in terms of tourists' willingness to adopt the sustainable behaviours recommended by the gamification providers. As a result, some values created for reducing the negative environmental and social impact, especially in the case of younger generations, will be implemented.

On the other hand, gamifying tourism services has a positive impact on communication and social interaction, leading to a collective consciousness extremely important for the development of local communities. The resulting changes—better informed and more skilled personnel, increased satisfaction, a more accurate entrepreneurial strategic thinking, locals' more positive attitudes toward tourism and sustainability—lead to a satisfied client, ensuring the sustainability of the tourism organisation. The direct connection with tourists helps local communities to learn more about, promote and integrate their culture and cultural values.

Together with better qualified employees, gamification contributes to improving the overall image of the city or location, a development of the cooperation between companies that can have a positive impact on the well-being of the entire community. The best results in terms of sustainability improvement were noticed in the case of gamification tools meant to develop or create new products, train employees and educate tourists and residents. Through gamification, the new tourism products integrate the economic objectives with social and environmental positive externalities, the training environment becomes more engaging and the tourists discover the history, the life and the destinations' identity, all these things with a minimum negative impact on the local community and the environment.

To aspire to a sustainable approach, gamification needs not only to tighten the relation between tourists and employees but also to include the local community in the equation. Gamification can help to shift from a business-centric perspective of gamification to a truly sustainable perspective. The best proof in this regard is the fact that almost all the best practices analysed incorporated the players' intrinsic motivations in the game mechanism. Self-esteem, social recognition and the opportunity to create new friends have the power to raise the tourism stakeholders' awareness about their holiday impact and to engage them in sustainable behaviours. Even if Seaborn and Fels [30] cite some authors who don't recognize the role of intrinsic motivations in gamification, the present conclusion supports the idea of the necessity of intrinsic incentive inclusion in the game mechanism, like other studies did [39].

The next step of undeniable difficulty would be applying a unifying approach to gamification that ensures the engagement of all the stakeholders (tourists, employees and community) in the same gamified space, where everybody is aware of everybody's interests, power, and status. Above all, sustainability means transparency between the developers and players, as gamification can go in two opposite directions:

a) Tunnelling, manipulating, distracting attention, squeezing more than appropriate, or

b) Synergy, responsible consumption, work or life balance, genuine fun, authenticity.

Considering the global trends in the use of gamification solutions, some aspects of the overall profile of key players in the gamification market can be highlighted:

- Mostly, both providers and buyers of these solutions are often part of large companies that already have a recognized brand on the market, financial resources and development strategies that make possible these innovations;

- The majority of organisations that have adopted these solutions are part of a specific offline community, geographically bounded, or of an online one, purpose bounded and have a special supportive relationship with the stakeholders.

Thus, for small companies or new entrants to the market, the adoption of these solutions can become an important challenge for top management, and a possible positive decision in this regard must start from a well-founded cost-benefits analysis. There is a bottom-up approach stressing the important role the leadership vision has when it comes to the successful implementation of gamification.

The limitations of this study may arise from the inability to fully cover the gamification applications dedicated and used in the tourism and hospitality industry. However, the study results can provide a very important basis for the development of other comparative studies both in the same and other industries that have adopted gamification solutions.

Another limitation of the paper is that it is focused on the online component of gamification (ICT based), without analysing the offline component, the use of gamification mechanics and philosophy in the organisational activities or the infusion of gamification values in the organisational culture. For example, gamification of the organisational learning process can be present offline in training sessions designed specifically to create competition, a certain degree of fun, provide feedback, monitor performance and offer instant rewards without the help of ICT, in a socio-professional gamification process. The practical implication of this alignment of online and offline gamification are important in terms of authenticity and sustainability and provides a new research direction.

AUTHOR CONTRIBUTIONS

All authors contributed equally to this work. All authors read and approved the final manuscript.

REFERENCES

1. Aall, C. Sustainable Tourism in Practice: Promoting or Perverting the Quest for a Sustainable Development?*Sustainability* **2014**, *6*, 2562–2583.

2. Petrescu, D.C. Sustainability, Tourism and Consumer Behavior. *Qual. Access Success* **2012**, *13*, 243–247.
3. Erdem, B.; Tetik, N. An environmentally-sensitive approach in the hotel industry: Ecolodges. *Int. J. Responsib. Tour.***2013**, *2*, 22–40.
4. UNWTO. Sustainable Tourism for Development Guidebook, 2013. Available online: http://dtxtq4w60xqpw.cloudfront.net/sites/all/files/docpdf/devcoengfinal.pdf (accessed on 19 July 2015).
5. Tang, Z. An integrated approach to evaluating the coupling coordination between tourism and the environment. *Tour. Manag.* **2015**, *46*, 11–19.
6. Stefănica, M.; Butnaru, G.I. Research on Tourists' Perception of the Relationship between Tourism and Environment.*Proc. Econ. Finance* **2015**, *20*, 595–600.
7. UNEP; UNWTO. Making Tourism More Sustainable—A Guide for Policy Makers, 2005. Available online: http://www.unep.fr/shared/publications/pdf/DTIx0592xPA-TourismPolicyEN.pdf (accessed on 10 April 2015).
8. Budeanu, A. Sustainable tourist behaviour—A discussion of opportunities for change. *Int. J. Consum. Stud.* **2007**, *31*, 499–508.
9. UNWTO. E-Business for Tourism. Practical Guidelines for Tourism Destinations and Businesses, 2001. Available online: https://pub.unwto.org/WebRoot/Store/Shops/Infoshop/Products/1210/1210-1.pdf (accessed on 15 July 2015).
10. Torres-Delgado, A.; Palomeque, F.L. Measuring sustainable tourism at the municipal level. *Ann. Tour. Res.* **2014**, *49*, 122–137.
11. Kim, S. Interdisciplinary Approaches and Methods for Sustainable Transformation and Innovation. *Sustainability***2015**, *7*, 3977–3983.
12. Deterding, S.; Dixon, D.; Khaled, R.; Nacke, L. From game design elements to gamefulness: Defining gamification. In Proceedings of the 15th International Academic MindTrek Conference: Envisioning Future Media Environments, Tampere, Finland, 28–30 September 2011; pp. 9–15.
13. Burke, B. Gamification 2020: What Is the Future of Gamification? Available online: https://www.gartner.com/doc/2226015/gamification--future-gamification (accessed on 1 March 2015).
14. Uskov, A.; Sekar, B. Serious games, gamification and game engines to support framework activities in engineering: Case studies, analysis, classifications and outcomes. In Proceedings of the 2014

IEEE International ConferenceElectro/Information Technology (EIT), Milwaukee, WI, USA, 5–7 June 2014; pp. 618–623.

15. Xu, F.; Weber, J.; Buhalis, D. *Gamification in Tourism*; Springer International Publishing: New York, NY, USA, 2013; pp. 525–537.
16. Meloni, W.; Gruener, G. Gamification in 2012 Market Update. Consumer and Enterprise Market Trends, 2012. Available online: http://gamingbusinessreview.com/wp-content/uploads/2012/05/Gamification-in-2012-M2R3.pdf(accessed on 10 March 2015).
17. Anderson, J.; Rainie, L. The Future of Gamification. Available online: http://www.pewinternet.org/2012/05/18/the-future-of-gamification/ (accessed on 2 March 2015).
18. Euromonitor International. Half the World's Population Will Be Online by 2030. Available online: http://blog.euromonitor.com/2015/04/half-the-worlds-population-will-be-online-by-2030.html (accessed on 2 June 2015).
19. Touray, K.; Jung, T. Exploratory study on contributions of ICTs to sustainable tourism development in Manchester. In*Information and Communication Technologies in Tourism*; Springer: Wien, Austria, 2010; pp. 493–505.
20. Bennett, G.; Freya, W. Mainstream Green: Moving Sustainability from Niche to Normal, Ogilvy Earth. Available online: https:/..ogilvy.com/truffles_email/ogilvyearth/Mainstream_Green.pdf (accessed on 15 July 2015).
21. UNWTO. Handbook on E-Marketing for Tourism Destinations, 2008. Available online: https://pub.unwto.org/WebRoot/Store/Shops/Infoshop/48EA/148E/19AC/BB41/5091/C0A8/0164/579A/080926_handbook_e-marketing_excerpt.pdf(accessed on 15 July 2015).
22. SoftwareInsider. Compare Gamification Platforms. Available online: http://gamification.softwareinsider.com/(accessed on 10 June 2015).
23. Ko, T.G. Development of a tourism sustainability assessment procedure: A conceptual approach. *Tour. Manag.* **2005**,*26*, 431–445.
24. Saarinen, J. Critical Sustainability: Setting the Limits to Growth and Responsibility in Tourism. *Sustainability* **2014**, *6*, 1–17.
25. Waligo, V.M.; Clarke, J.; Hawkins, R. Implementing sustainable tourism: A multi-stakeholder involvement management framework. *Tour. Manag.* **2013**, *36*, 342–353.

26. Schrettle, S.; Hinz, A.; Scherrer-Rathje, M.; Friedli, T. Turning sustainability into action: Explaining firms' sustainability efforts and their impact on firm performance. *Int. J. Product. Econ.* **2014**, *147*, 73–84.
27. Ali, A.; Frew, A.J. *Information and Communication Technologies for Sustainable Tourism*; Routledge: London, UK, 2013.
28. Mareba, M.S. The Role of Information and Communication Technology in Supporting Sustainable Tourism: In-Trip Tourists Perspectives. Available online: http://etheses.qmu.ac.uk/1615/1/ 1615. pdf (accessed on 15 July 2015).
29. Schieder, T.; Adukaite, A.; Cantoni, L. Mobile Apps Devoted to UNESCO World Heritage Sites: A Map. In Proceedings of the International Conference in DublinInformation and Communication Technologies in Tourism 2014, Dublin, Ireland, 21–24 January 2014; pp. 17–29.
30. Seaborn, K.; Fels, D.I. Gamification in theory and action: A survey. *Int. J. Hum. Comput. Stud.* **2015**, *74*, 14–31.
31. Robson, K.; Plangger, K.; Kietzmann, J.H.; McCarthy, I.; Pitt, L. Is it all a game? Understanding the principles of gamification. *Bus. Horiz.* **2015**, *58*, 411–420.
32. Owen, P. *How Gamification Can Help Your Business Engage in Sustainability*; Do Sustainability: Oxford, UK, 2013.
33. Burke, B. *Gamify: How Gamification Motivates People to Do Extraordinary Things*; Bibliomotion: Brookline, MA, USA, 2014.
34. Loong, B.L.S. Tourism and simulacrum: The computational economy of algorithmic destinations. *Procedia* **2014**, *144*, 237–246.
35. Barreda, A.A.; Bilgihan, A.; Nusair, K.; Okumus, F. Generating brand awareness in Online Social Networks. *Comput. Hum. Behav.* **2015**, *50*, 600–609.
36. Hamari, J.; Koivisto, J. Why do people use gamification services? *Int. J. Inf. Manag.* **2015**, *35*, 419–431.
37. Bilgihan, A.; Bujisic, M. The effect of website features in online relationship marketing: A case of online hotel booking.*Electron. Commer. Res. Appl.* **2014**, *14*, 222–232.
38. Koivisto, J.; Hamari, J. Demographic differences in perceived benefits from gamification. *Comput. Hum. Behav.* **2014**,*35*, 179–188.
39. Rapp, A. Designing interactive systems through a game lens: An ethnographic approach. *Comput. Hum. Behav.* **2015**.

40. Froehlich, J. Gamifying Green: Gamification and Environmental Sustainability. In *The Gameful World*; MIT Press: Cambridge, CA, USA, 2014; pp. 563–596.
41. Laškarin, M. Development of loyalty programmes in the hotel industry. *Tour. Hosp. Manag.* **2013**, *1*, 109–123.
42. Franzoni, S. Measuring the sustainability performance of the tourism sector. *Tour. Manag. Perspect.* **2015**, *16*, 22–27.
43. Yin, R.K. *Case Study Research: Design and Methods*, 3rd ed.; Sage: Thousand Oaks, CA, USA, 2003.
44. George, A.L.; Bennett, A. *Case Studies and Theory Development in the Social Sciences*; MIT Press: Cambridge, CA, USA, 2005.
45. Negrusa, A.L.; Petrescu, D.C.; Gică, O.A.; Bota, M.; Rus, R.V. *Perspective Asupra Clusterelor, Inovării și Bunelor Practici din Domeniul Științelor Vieție. Perspectives on Clusters, Innovation and Best Practices in Life Science*; Risoprint: Cluj-Napoca, Romania, 2013.
46. Ritchie, J., Jane, L., Eds.; *Qualitative Research Practice: A Guide for Social Sciences Students and Researchers*; Sage: Thousand Oaks, CA, USA, 2003.
47. BJ's Premier Rewards. Available online: http://www.bjspre mierrewards. com/ (accessed on 15 June 2015).
48. Sayeed, I.; Faheem, P. Game on: Restaurants Winning with Gamification. Available online: http://www.digitalsignagetoday .com/articles/game-on-restaurants-winning-with-gamification/ (accessed on 2 March 2015).
49. Campon, A.M.; Alves, H.; Hernandez, J.M. Loyalty measurement in tourism: A theoretical reflection. In *Quantitative Methods in Tourism Economics*; Matias, Á., Nijkamp, P., Sarmento, M., Eds.; Springer: Berlin, Germany, 2013; pp. 13–40.
50. Schiano, D.J.; White, S. The first noble truth of cyberspace: People are people (even when they MOO). In Proceedings of the ACM CH'97 Conference on Human Factors in Computing Systems, Los Angeles, CA, USA, 18–23 April 1998; pp. 352–359.
51. Kuo, I. Gigya's Social Gamification Boosts Content Discovery by 68%. Available online: http://www.gamification.co/2013/03/28/gigya-boosts-content-discovery/ (accessed on 28 March 2015).
52. Forgacs, G. Brand Asset Equilibrium in Hotel Management. *Int. J. Contemp. Hosp. Manag.* **2003**, *15*, 340–342.

53. Nunes, J.C.; Drèze, X. Your loyalty program is betraying you. *Harv. Bus. Rev.* **2006**, *84*, 124–131.
54. Lee, J.J.; Capella, M.L.; Taylor, C.R.; Gabler, C.B. The financial impact of loyalty programs in the hotel industry: A social exchange theory perspective. *J. Bus. Res.* **2014**, *67*, 2139–2146.
55. Carli, D. Print *vs.* Digital Media: False Dilemmas and Forced Choices. Available online: http://www.sustainablecommunication.org/downloads/whitepapers/printvsdigital_false-dilemmas-and-forced-choices.pdf (accessed on 15 July 2015).
56. Landers, R.N.; Callan, R.C. Casual Social Games as Serious Games: The Psychology of Gamification in Undergraduate Education and Employee Training. In *Serious Games and Edutainment Applications*; Ma, M., Oikonomou, A., Jain, L.C., Eds.; Springer: London, UK, 2011; pp. 399–423.
57. Gamification Case Study—Costa Travel Company, 2014. Available online: http://www.hydra-newmedia.com/references/case-study-gamification/?lang=en (accessed on 25 May 2015).
58. Marriott. My Marriott Hotel Opens Its Doors on Facebook. Available online: http://news.marriott.com/2011/06/my-marriott-hotel-opens-its-doors-on-facebook.html (accessed on 2 March 2015).
59. Maister, J. Gamification: Three Ways to Use Gaming for Recruiting, Training and Health & Wellness. Available online: http://www.forbes.com/sites/jeannemeister/2012/05/21/gamification-three-ways-to-use-gaming-for-recruiting-training-and-health-amp-wellness/ (accessed on 26 May 2015).
60. Lopez, S. Disneyland Workers Answer to "Electronic Whip". Available online: http://articles.latimes.com/2011/oct/19/local/la-me-1019-lopez-disney-20111018 (accessed on 19 May 2015).
61. Llagostera, E. On Gamification and Persuasion. In Proceedings of the SBGames, Rio de Janeiro, Brazil, 2–4 November 2012; pp. 12–21.
62. Shahri, A.; Hosseini, M.; Phalp, K.; Taylor, J.; Ali, R. Towards a Code of Ethics for Gamification at Enterprise. In *The Practice of Enterprise Modeling*; Frank, U., Loucopoulos, P., Pastor, Ó., Petrounias, I., Eds.; Springer: Berlin, Germany, 2014; pp. 235–245.
63. Nelson, M. Soviet and American Precursors to the Gamification of Work. Available online: http://www.kmjn.org/publications/Gamification_MT12.pdf (accessed on 15 May 2015).
64. HRM's Role in Corporate Social and Environmental Sustainability, 2012. Available online: http://www.shrm.org/about

/foundation/products/documents/4-12%20csr%20report%20final%20for%20web.pdf (accessed on 15 July 2015).

65. Schipper, L.; Bartlett, S.; Hawk, D.; Vine, E. Linking Life-Styles and Energy Use: A Matter of Time? *Ann. Rev. Energy* **1989**, *14*, 273–320.
66. Bin, S.; Hadi, D. Consumer Lifestyle Approach to US Energy Use and the Related CO_2 Emissions. *Energy Policy* **2005**, *33*, 197–208.
67. Tribe, J. *The Economics of Recreation, Leisure and Tourism*; Routledge: London, UK, 2012.
68. Caines, M. Review: Visioning Workshop on Sustainable Tourism. Available online: http://www.cluster.eu/2011/12/08/review-visioning- workshop-on-sustainable-tourism/#sthash.HvUH8cpk.dpuf (accessed on 15 June 2015).
69. Nest. Nest: Saving Energy. Available online: http://www.nest.com/saving-energy/ (accessed on 10 June 2015).
70. Matsuoka, Y. Nest Learning Thermostat's Auto Schedule Feature Reduces Residential Energy Usage. In Proceedings of the Behavior, Energy, and Climate Change Conference, Sacramento, CA, USA, 12–14 November 2012.
71. Pierce, J.; Odom, W.; Blevis, E. Energy Aware Dwelling: A Critical Survey of Interaction Design for Eco-Visualizations. In Proceedings of the 20th Australasian Conference on Computer-Human Interaction: Designing for Habitus and Habitat, New York, NY, USA, 8–12 December 2008.
72. Petersen, J.E.; Shunturov, V.; Janda, K.; Platt, G.; Weinberger, K. Dormitory Residents Reduce Electricity Consumption When Exposed to Real-time Visual Feedback and Incentives. *Int. J. Sustain. High. Educ.* **2007**, *8*, 16–33.
73. Sintov, N.D.; Desario, G.; Prescott, C. Effectiveness of a Competition-Based Intervention in Promoting Pro-Environmental Behavior in a University Residential Setting. In *ACEEE Summer Study on Energy Efficiency in Buildings*; ACEEE: Washington, DC, USA, 2010; pp. 322–336.
74. Weber, J. Gaming and Gamification in Tourism. Available online: http://thinkdigital.travel/wp-content/uploads/2014/05/Gamification-in-Tourism-Best-Practice.pdf (accessed on 10 June 2015).
75. Sigala, M. The application and impact of gamification funware on trip planning and experiences: The case of TripAdvisor's funware. *Electron. Markets* **2015**.

1.2. Problem statement

At the beginning of the present millennium, the United Nations General Assembly committed itself to "Spare no effort to free fellow men, women and children from the abject and dehumanizing conditions of extreme poverty, to which more than a billion of them are currently subjected" [3]. More than one decade after this commitment, the International Fund for Agricultural Development (IFAD) reveals that there are still about 1.4 billion people living on less than US$ 1.25 a day and close to one billion suffering from hunger, with 70 percent of these living in rural areas [4]. IFAD further states that "Rural poverty results from lack of assets, limited economic opportunities and poor education and capabilities, as well as disadvantages rooted in social and political inequalities" [4]. The problem here is the perpetuation of this untenable situation despite the fact that research since 1973 [5-11] indicates that tourism can be used as a means of orchestrating development.

1.3. Significance of the pilot project

The significance of this study lies both in improving the quality of life of community members and raising awareness on issues of global concern such as the protection of endangered species, climate change, health and so on. More so, by tailoring training and education to relevant community development projects, the challenges posed by jobless graduates will be curbed, if not eliminated. Other specific gains will include job creation, social cohesion, peace and foreign exchange flows to the local economy.

2. BACKGROUND TO THE STUDY

2.1. The Cameroon tourism situation

Cameroon is commonly described in tourism circles as Africa in miniature as most of the attractions found on the continent can be seen in the country. However, the contribution of tourism to economic development in Cameroon can be described as a missed opportunity as the country occupies an unenviable position 148 out 181 in world rankings in terms of the relative contribution of tourism to national economy [12]. This situation is quite unfortunate considering the high unemployment rate in the country (estimated at between 30 to 50.2 percent) on the one hand, and the great potential of tourism to create jobs and generate incomes on the other.

In his address to the Cameroon National Tourism Board on December 28, 2010, the Cameroon Minister of Tourism, Baba Hamadou revealed the government's recognition of the strategic role that tourism has to play in the country's vision of becoming an emerging country by 2035. The Minister also indicated that

Cameroon received 572,728 international visitors in 2010, contributing FCFA 28.64 billion through entry visas to Cameroon, FCFA 2.66 billion through airport stamps and FCFA 59.142 billion from taxes and duties. Tourism direct employment in Cameroon currently stands at over 60,000.

By 2035, tourism is projected to contribute 13 percent to the Gross Domestic Product (GDP), generating FCFA 250 billion to the state treasury. Over the same period, the accommodation sector is expected to grow to 5689 hotels with a capacity of 156,023 rooms and 170,260 beds.

According to the World Travel and Tourism Council [12], key projections on tourism in Cameroon in the year 2011 are as follows:

- GDP: Direct contribution

  The direct contribution of travel and tourism to GDP is expected be FCFA 263.6 billion (2.3% of total GDP) in 2011, rising by 5.4% pa to FCFA 447.5 billion (2.4%) in 2021.

- GDP: Total contribution

  The total contribution of travel and tourism to GDP, including wider economic impacts is forecast to rise by 5.6% pa from FCFA 579.3 billion (5.0% of GDP) in 2011 to FCFA 1,000.4 billion (5.3%) by 2021.

- Employment: Direct contribution

  Travel and tourism is expected to support directly 88,000 jobs (1.9% of total employment) in 2011, rising by 2.6% pa to 115,000 jobs (2.0%) by 2021.

- Employment: Total contribution

  The total contribution of travel and tourism to employment, including jobs indirectly supported by the industry, is forecast to rise by 2.8% pa from 200,000 jobs (4.4% of total employment) in 2011 to 264,000 jobs (4.7%) by 2021.

- Visitor exports:

  Travel and Tourism visitor exports are expected to generate FCFA 153.1 billion (4.3% of total exports) in 2011, growing by 8.9% pa (in nominal terms) to FCFA 288.1 billion (4.2%) in 2021.

- Investment:

  Travel and tourism investment is expected at FCFA 45.2 billion or 1.9% of total investment in 2011. It should rise by 7.6% pa to reach FCFA 93.9 billion (or 2.4%) of total investment in 2011.

2.2. Site and population of the study

Buea, capital of the South West region of Cameroon (Appendix 1) is the site of this experimental study, hence the base of the two operating structures: Achas Higher Institute of Tourism, Hospitality and Business Management and Achas centre of sustainable Community Development. The choice of Buea as the study site is strategic considering that the region is endowed with two flagship attractions, namely the volcanic Cameroon Mountain (the highest peak in West Africa -Appendix 2) and the coastline of the Atlantic Ocean flowing through the coastal town of Limbe (Appendix 3). With a vibrant population of 90, 088 inhabitants [13], this University town also has huge political and historic significance, hence its historic colonial significance as the capital of German Kamerun (1901-1919-Appendix 4) and capital of Southern Cameroons from 1949 to 1961. The annual mount Cameroon race in February of every year, the amazing mountain trails, the indigenous Bakweri culture (Appendix 5) among others, beautifully complement and complete the variety of attractions in this area.

3. LITERATURE REVIEW

3.1. Community tourism

Hall [14] describes tourism as an expression of lifestyle "identified either through voluntary travel or a voluntary temporary short-term change of residence". The motives of such travel are increasingly wideranging, including among others education, business, health and leisure. Following on this, Cooper et al. [15] indicate that the core elements of tourism arise out of the movement of people to and their stay in various destinations outside their usual environments and places of work. This in turn fuels the demand for various facilities and services designed to provide for and ease the transition and temporary residence of visitors in the host community. The supply of direct and support services such as transport, accommodation, catering, souvenirs and other shopping to tourists should help stimulate the local economy, create jobs, bring additional income and hence improve the quality of life of community members.

In this light, Spenceley [16] defines community tourism as "tourism which is owned and/or managed by communities with the aim of generating wider community benefit". Choi and Sirakaya [2] further argue that the main aim of community tourism should be to improve the residents' quality of life by maximizing local economic benefits, protecting the natural and built heritage and providing a high quality of experiences for the visitors. However, taking a slightly different view, Petric [17] emphasises that community tourism should aim at giving visitors personal contact with the physical and human environment of the countryside and allow them to participate in the activities, traditions and lifestyles of the local people.

3.2. Sustainable tourism

The concept of sustainability gained prominence after the publication of the World Commission on Environment and Development (WCED) report "Our Common Future" which defined Sustainable Development as development that "Meets the needs of the present without compromising the ability of future generations to meet their own needs" [18]. Keyser [19] points out that the attainment of sustainable development necessitates a balance between economic efficiency, environmental integrity and social justice. According to Butler [20], sustainable tourism refers to "Tourism which is developed and maintained in an area (community or environment) in such a manner and at such a scale that it remains viable over an infinite period and does not degrade or alter the environment (human or physical) in which it exists to such a degree that it prohibits the successful development and well being of other activities and processes". The United Nations World Tourism Organization's [21] provides a comprehensive definition of sustainable tourism development when it states that sustainable tourism development "meets the needs of present tourists and host regions while protecting and enhancing opportunity for the future. It is envisaged as leading to management of all resources in such a way that economic, social and aesthetic needs can be fulfilled while maintaining cultural integrity, essential ecological processes, and biological diversity, and life support systems".

3.3. Tourism and sustainable community development

The study of tourism business development has gained impetus in recent times. According to Lee and Chang [10], "A general consensus has emerged that it (tourism) not only increases foreign exchange and incomes, but also creates employment opportunities, stimulates the growth of the tourism industry and by virtue of this, triggers overall economic growth". Sebele [22] indicates that the concept of community development was introduced as an approach to rural development in the 1950s and 1960s. This approach calls for the more active involvement of locals in development issues. In the same vein, tourism literature since the 1980s has advocated the inclusion and involvement of local communities in tourism as local residents have been identified as being instrumental in sustaining the product [23]. Furthermore, Tosun [24] asserts that apart from being instrumental in tourism sustainability, community participation also ensures that a number of benefits accrue from tourism, namely: employment opportunities for local people, positive local attitudes, conservation of local resources and physical development within the community.

It is evident from the above that there is a considerable overlap between community tourism, sustainable tourism and sustainable development hence the need to explore a strategy that would ensure that local communities benefit from all three.

Considering the rapid growth of tourism and the subsequent impacts (economic, environmental and socio-cultural), its strong association with issues of sustainability is hardly surprising. Tourism is "intimately connected with issues of sustainable development. Indeed, since the late 1980s there has been an explosion in the number of texts and articles, plus courses and consultants, concerned with sustainable forms of development and mitigating or managing tourism's undesired effects" [14]. The World Tourism Organization (WTO) defines sustainable tourism as tourism development that "meets the needs of present tourists and host regions while protecting and enhancing opportunity for the future. It is envisaged as leading to management of all resources in such a way that economic, social and aesthetic needs can be fulfilled while maintaining cultural integrity, essential ecological processes, and biological diversity, and life support systems" [21]. Three types of impacts are identified as originating from tourism development:

- Impacts on the environment (both natural and man-made);
- Socio-cultural impacts (Effects on host communities through direct and indirect relations with tourists, and interaction with the tourism industry); and
- Economic impacts.

In the light of the foregoing, the World Tourism Organization envisages that a sustainable tourism strategy should:

1) "Make optimal use of the environmental resources that constitute a key element in tourism development, maintaining essential ecological processes and helping to conserve natural heritage and biodiversity;
2) Respect the socio-cultural authenticity of host communities, conserve their built and living cultural heritage and traditional values, and contribute to inter-cultural understanding and tolerance; and
3) Ensure viable, long-term economic operations, providing socio-economic benefits to all stakeholders that are fairly distributed, including stable

employment and income-earning opportunities and social services to host communities, and contributing to poverty alleviation.

Sustainable tourism development requires the informed participation of all relevant stakeholders as well as strong political leadership to ensure wide participation and consensus building. Achieving sustainable tourism is a continuous process and it requires constant monitoring of impacts, introducing the necessary preventive and/or corrective measures whenever necessary" [24].

The rationale behind this experiment therefore lies in assessing the extent to which tourism development theory can be incorporated into community development strategies redress the huge imbalance between the economic, socio-cultural and environmental opportunities presented by the tourism industry and the abject poverty suffered by many people in local communities.

4.4. Maximising Tourism Gains for Poor Communities: The Operation of Achas Centre for Sustainable Community Development

Achas centre for sustainable community development (ACSCD) operates alongside Achas Higher institute of tourism, hospitality and business management (Achas Higher Institute of Management – AHIM in short) in Buea, South West region of Cameroon. The rationale behind having the two centers operate side by side is to ensure that while AHIM deals with issues of training and education, the sustainable development centre (ACSCD) commutes the skills acquired into practical gains for the community through developmental projects (Figure 1). In order to maximize the benefits from these projects, community focus areas are identified. This will ensure that skills and efforts are properly channeled to impact in specific areas of community life. Training is pivotal to the success of this model because it provides the basic tourism awareness, information and education and the quality human resource needed to operate tourism and its support industries. Women empowerment is considered a focus area because this is a vital resource that is often neglected in many communities. Women generally constitute a greater part of the population and so putting them at the fore of community development can have significant positive impact. Conservation projects will ensure that resources and the quality of the experience do not deplete or degenerate beyond repair. Food security is essential for the livelihood of community members and provides for visitors as well. Research and innovation ensures there are improvements both in the way of doing things and the quality of the product offering as well. Of course, all these will only flourish if conditions in the local environment (enabling environment) are favorable for investment and the operation of businesses.

Achas Higher Institute of sustainable tourism, hospitality and business management is strategically positioned to coordinate, lead and empower local communities to take advantage of the development opportunities presented by the travel and tourism industry, through:

- The training of employees, entrepreneurs, government sectors and community residents through certification programs (National diploma, professional degrees and Master's degrees (in the long run) and Short Learning Programs (SLPs).
- Empower women (through workshops and short Learning programs), especially in the informal economic sector to better manage their small business initiatives;
- Alert women to entrepreneurial opportunities (gap identification) in the tourism industry;
- Conducting research on tourism-related impacts: economic, environmental and socio-cultural;
- Training employees and raising awareness on the importance of conservation, especially in Cameroon and the Central African region;
- Initiating networks, collaboration and partnerships with conservation Non-Governmental Organizations (NGOs) with the aim of tailoring our curriculum to practical needs and participating on conservation programs;
- Increase capacity-building in tourism and conservation via various short learning programs (SLPs);
- Working with public and private sector groups to formulate policies that will lead to improvements in sustainable tourism planning and development;
- Developing best practice guides and training manuals;
- Training employees for tourism support sectors such as banking, hospitality, human resources management, and so on;
- Initiating collaboration, networks and interactions with institutions and organizations that focus on conservation and sustainable tourism;
- Facilitating collaboration between industry partners through workshops and conferences;
- Attract international attention to the tourism potential of Cameroon, and support to improve its planning and development;
- Harness all the above towards community development in Cameroon and improve the quality of life of the local people.

4.1. From theory to practice

Though still in the early stages, the application of this theory can be seen in operation of Achas Higher institute of sustainable tourism, hospitality and business management and Achas centre for sustainable community development as follows (Figure 2):

4.2. Transforming the skills' base into sustainable community development

Empowered community members shall contribute towards the development of the community by serving at Achas centre for sustainable community development whose specific objectives are the following (Figure 3):

4.2.1. Women for sustainable community development

Goal: Empower women to play a leading role in sustainable community development; through:

1. Identifying the needs of community women;
2. Providing basic education for healthy and balanced social development;
3. Organize women into common-initiative groups so that the synergy can further enhance the possibility for success;
4. Provide seed-funding to local women for small business development;
5. Create job and income opportunities for women;
6. Provide training in small business management skills through Short Learning Programmes (SLPS), workshops and conferences;
7. Identify potential markets for finished products;
8. Monitor the progress of women in their respective economic activities.

4.2.2. Conservation for sustainable livelihoods

Goal: Work towards the preservation and sustainable use of flora and fauna species; through:

1. Providing education on conservation through accredited courses;
2. Provide training in conservation work through SLPs, workshops, conferences, etc.;

3. Work with existing conservation NGOs to identify endangered species;
4. Raise awareness within communities on the importance of conservation;
5. Produce basic EIA reports (Environmental Impact Assessment Reports);
6. Work towards re-habilitating endangered species;
7. Use conservation as an opportunity for job creation for community members;
8. Contribute towards enhancing the HR (Human Resource) needs of the conservation movements.

4.2.3. Food security

Goal: Work towards ensuring that an increasing number of community members have access to their daily nutritional needs; through:

1. Providing training in the better management of existing farms;
2. Providing training in improved farming methods;
3. Ensure a better distribution of available food stock;
4. Conduct information sessions of healthy eating;
5. Provide financial assisting to existing and new farmers;
6. Encourage cooperation among farmers.

4.2.4. Research and Innovation

Goal: Contribute to and encourage the use of scientific and innovative methods to enhance community development; through:

1. Training more people to improve the HR needs of the research sector;
2. Get community members to be active and participate in data collection;
3. Educate community members in the benefits of data collection and research in general;
4. Work with the public sector, especially local government departments on research projects;
5. Encourage the use of new and innovative methods to resolve individual and community challenges;

6. Encourage and facilitate research work among community members by availing literature and data sources such as the internet, libraries, and so on.

5. CONCLUSIONS

- The possibility of job creation: Although this experimental study is still in its infancy, fifteen direct jobs have already been created. Ten of these jobs have been in various positions within the training centre and five in the sustainable community development projects. This is evidence of the possibility of this community development model to create employment for community members.
- Opportunities for empowered communities to participate in their own development: This model further illustrates the possibility for community members to engage in their own development. There is ample opportunity for multiple talent development and exploitation, either in direct tourism services or in the support industries.
- The enhancement of corporate social responsibility: Also evident in this model is the opportunity to engage other corporate partners to contribute towards community development, either by creating opportunities for internship programmes for young graduates within their business environment, create permanent employment or donate resources towards community projects.
- Community benefit from available natural, heritage and other resources: Hallmark attractions such as the Cameroon Mountain and the beach front by the Atlantic Ocean in Limbe can be further exploited by the community for its own development.
- Greater social cohesion and synergy among community members: Through working together on various community development projects community members energize each other and improve the social bonding therein.
- Income generation within the community: This sustainable community development model also demonstrates opportunities for income generation to fast-track community development. For the year 2011, the budget for the training institute (Achas Higher Institute of sustainable tourism, hospitality and business management stands at XAF 49,595,000 ($ 99,190). The fact that most of this money will be flowing from outside into the local economy is a great boost to the economy. In addition, income will come from external sources such as foreign exchange earned from tourism and foreign donors, and internally through the income from community projects. Other spin-

offs and multiplier effects are expected to trickle these gains throughout the community.

- Finally, it is evident from this study that tourism development, just like training requires effective partnership with other industries and economic sectors in order to produce real gains for local communities. Education/training without relevant projects on which to apply the acquired knowledge will likely generate frustration and resentment. In the same vein, development that does not involve the local people breeds a sense of alienation and is therefore not sustainable.

5.1. Limitations

Despite the foregoing benefits of this experimental project, the following limitations to its successful implementation need to be noted:

- Lack of expertise among local people to manage both the training centre (higher institute) and the sustainable development centre. Delivering lectures requires seasoned academicians, just as much as experienced managers are needed to coordinate the projects at the sustainable community development centre. However, expatriate labour is being sought to train trainers within the communities.
- Lack of seed funding to kick-start both the training centre and the sustainable development centre. While the financial sustainability of this development model is assured in the medium and long term through tuition fees and the profitability of its projects, acquiring seed funding remains a major challenge.
- Effective stakeholder participation. In principle, all stakeholders are in favor of community development. However a major limitation to the success of this community development model will be the effective collaboration among the stakeholders. Government processes are usually burdened by bureaucracy and delays. This could leave the other parties (community members, NGOs, investors, etc) frustrated and disinterested.
- Lack of training facilities and resources. The success of this community development model will depend on the availability of training facilities and resources such as a well equipped library and a computer laboratory. This will enable community members access information and empower themselves on how to deal with various challenges.
- Finally, one limitation to this study lies in the fact that it is still in its infancy so the full impact cannot be determined at this stage. It is however important to note that these impacts will be measured with regards to the number and extent of women empowerment, conservation activities, food production and distribution and research output.

5.2. Recommendations

Considering the foregoing challenges, the following recommendation would further ensure the sustainability of this project:

- The active participation of all stakeholders such as government, Non-governmental organizations (NGOs), community members, investors and so on.
- The provision of seed funding to kick-start the projects. This could come from private individuals such as the case of Achas centre for sustainable development, but this funding could also come from the corporate community as part of their social responsibility packages, international development agencies and government grants. Considering recent unfortunate happenings such as the famine in East Africa, climate change pressures, popular uprisings around the world, and illegal migration, one can only hope that the international community has learned to apply the common adage that "prevention is better than cure". Assisting communities before a crisis should cost less and save more lives.
- The voluntary transfer of skills, especially from the developed to the developing communities. This could take the form of working holidays
- Donations of equipment in the form of used or new computers, books and conference material (also used for workshops). Transport facilities in the form of four-wheel-drive vehicles and buses to move people and goods around will also be very useful.
- It is also recommended that further research be conducted to establish the medium and long term impacts (economic, environmental and socio-cultural) of this project on the development of the local communities and the quality of life of the residents.

REFERENCES

1. World Travel and Tourism Council (WTTC) (2011) Travel and tourism economic impact: World. London, UK: World Travel and Tourism Council.
2. Choi HC, Sirakaya E (2006) Sustainability Indicators for managing community tourism. Tourism Management 27: 1274-1289.
3. United Nations (2000) Global summit on peace through tourism. Amman, United Nation.

4. International Fund for Agricultural Development (2010) Rural poverty report 2011: The challenge of ending rural poverty. Oxford: Oxford University Press for IFAD.
5. Erbes R (1973) International tourism and the economy of developing countries. Paris, France: OECD.
6. De Kadt E (1979) Tourism: Passport to Development? New York, USA: Oxford University Press.
7. Jenkins CL (1994) Tourism in developing countries: The privatization issue. Chichester: John Wiley.
8. Sinclair MT (1998) Tourism and economic development. The Journal of Development Studies 34: 1-51.
9. Oh CO (2005) The contribution of tourism development to economic growth in the Korean economy. Tourism Management 26: 39- 44.
10. Lee CC, Chang CP (2008) Tourism development and economic growth: A closer look at panels. Tourism Management 29: 180-192.
11. Tao CH, Wall G (2009) Tourism as a sustainable livelihood strategy. Tourism Management 30: 90-98.
12. World Travel and Tourism Council (WTTC) (2011) Travel and tourism economic impact: Cameroon. London: World Travel and Tourism Council.
13. United Nations Development Programme (2006) Beyond scarcity: Power, poverty and the global water crisis. Human development report 2006. New York United Nations Development Programme, Cameroon.
14. Hall CM (2008) Tourism planning: policies, processes and relationships. (2ndedn), Edinburgh Gate: Pearson.
15. Cooper C, Fletcher J, Fyall A, Gilbert D, Wanhill S (2008) Tourism: principles and practice. (3rdedn), Edinburgh Gate: Pearson Education Limited.
16. Spenceley A (2008) Responsible tourism: critical issues for conservation and development. London, UK: Earthscan.
17. Viljoen J, Tlabela K (2007) Rural tourism development in South Africa: Trends and challenges. Human Sciences Research Council Press.
18. World Commission on Environment and Development (WCED) (1987) Our common future, 1987. Oxford: Oxford University.
19. Keyser H (2009) Developing tourism in South Africa: towards competitive destinations. (2ndedn), Cape Town, South Africa: Oxford University Press.

20. Butler RW (1991) Tourism, Environment, and Sustainable Development. Environmental Conservation 18: 201-219.
21. World Tourism Organization (WTO) (2004) Sustainable development of tourism: conceptual definition. Madrid: World Tourism Organisation.
22. Sebele LS (2010) Community-based tourism ventures, benefits and challenges: Khama rhino sanctuary trust, central district, Botswana. Tourism Management 31:136-146.
23. Hardy A, Beeton RJS, Pearson L (2002) Sustainable tourism: an overview of the concept and its position in relation to conceptualisations in tourism. Journal of Sustainable Tourism 10: 475-496.
24. Tosun C (2006) Expected nature of community participation in tourism development. Tourism Management 27: 493-504.

Chapter 9

Is Romanian Rural Tourism Sustainable? Revealing Particularities

Daniela Ruxandra Andrei [1], Rodica-Manuela Gogonea [2], Marian Zaharia [3] and Jean-Vasile Andrei [1,4,*]

[1]IT, Statistics and Mathematic Department, Romanian American University, 18 Expozitiei St., Sector 1, Bucharest 012101, Romania

[2]Statistics and Econometrics, Bucharest Academy of Economic Studies, 15-17 Dorobanti St., Sector 1, Bucharest 010552, Romania

[3]Ciberenetics, Economic Informatics, Finance and Accountancy Department, Petroleum-Gas University of Ploiesti, No.39 Bucuresti St., Ploiesti 100680, Romania,

[4]Business Administration Department, Petroleum-Gas University of Ploiesti, No.39 Bucuresti St., Ploiesti 100680, Romania

ABSTRACT

Research on sustainable tourism involves developing an appropriate framework to highlight the interdependences of economic, social and environmental systems. The interdependence is based on the entropy of the system while respecting the principle of holism and diversity of rural tourism sustainability. In this context, sustainability in general and rural tourism in particular can be considered a complex system of development, which in some ways can be studied by statistical and econometric methods that allow the analysis of the interdependences between the variables of rural tourism at county level and at the level of rural communities. Conducting such studies involves identifying the rural communities where rural tourism has reached significant levels. Based on this consideration, this paper aims to identify the development regions and counties of Romania where the trends of development of rural tourism are significantly above the average recorded at country level, as a first step towards particular studies of sustainability in rural communities.

prices was 31,300 or 34,500 Euro, in Romania it was only 7180 Euro [16], the second lowest in the UE28. Under these conditions, in some counties, especially in rural communities, in many cases, according to the conclusions of the studies of Walpole [17], the low level of development and the poor material circumstances have a negative influence on environmental conservation activities. On the other hand, Mehta [18], and Walpole [17,19], conclude in their studies that between the attitude towards tourism and the attitude towards environmental conservation there is a direct but weak link [20]. It should be emphasized, however, that the actions undertaken by institutions and the concerns of rural communities led to a positive attitude towards sustainable development, as outlined by [21].

3. METHODOLOGY, RESULTS AND DISCUSSION

Assuming that rural tourism is an element that determines growth and that, in turn, economic growth contributes to the development of tourism, between the two elements there is a positive feedback loop; the objectives of the research were, on the one hand, the analysis of tourist traffic (as evidenced by the number of arrivals and overnight stays in agrotouristic boarding houses) and, on the other hand, to highlight how the fluctuations of the GDP (Gross Domestic Product) affect tourist traffic developments, with implications in the sustainable development of rural areas.

To achieve the research objectives we chose to develop an econometric model of the panel type [22], based on data taken from official documents [20,23] and the official data series provided online by the National Institute for Statistics of Romania (2014) on the above-mentioned indicators, for the period from 2000–2012, and for the following economic development regions of Romania: North West, the Central region, North-East, South-East, South-Muntenia, Bucharest-Ilfov, South-West, West.

The choice of methodology for the analysis of this interdependence between rural tourism quantification indicators and economic growth through the panel-type econometric model was based on the latter's capacity for flexibility in modeling the differences registered in the individual components of the influence factors, as well as on the degree of trust in the results obtained, which leads to an increase in the efficiency and consistency of econometric estimates.

The analysis unfolds starting from the following regression model [6]:

$$Y_{it} = a_0 + a_1 X_{1it} + a_2 X_{2it} + \ldots + \alpha_i + \beta_t + e_{it} \tag{1}$$

where:

- x_{it}—the value of the exogenous (explanatory) variable for unit "*i*", at time "*t*"; in this research the exogenous variable is GDP.
- e_{it}—deviation variable, specific to unit "*i*", at time "*t*";
- a_0, a_1—coefficients (model parameters)
- $i = \overline{1,n}$, where n is the number of units (in the application n = 8, the number of regions);
- $t = \overline{1,T}$, with T—number of time periods (T = 13, number of years 2000–2012);

The process of analysis of the individual effects starts from the decomposition of the discrepancy (deviation) e_{it} in three components [22]:

- α_i—the variable that estimates the *specific, unobservable individual effect*, and is invariable in time: it estimates the effect of variables that are unenclosed in the model on the endogenous one, in unit *i* (the specific individual effect);
- β_t—the variable that estimates the *specific temporal effect* invariable in transversal structures: it estimates the effect of variables unenclosed in the model on the endogenous one, at time *t* (fixed time effect).

The research methodology has been contoured in such a way as to lead to a better description of results that would reflect the achievement of the objective set.

For the development of the model and the identification of its parameters, a general procedure developed by Hsiao [24] was adopted, the Toda [25] version. The support software in processing the data it was used EViews. For statistical hypothesis testing, a significance threshold of 5% was chosen [26].

Similarities and Regional Particularities in the Development of Rural Tourism

The modeling of the process that manifests between the variable of rural tourism and the one that reflects economic growth (GDP) has in view the establishment of a causality relation between regional structures between 2000 and 2012; it starts with an analysis of the statistic features of the data series, to the aim of applying the Granger Test [27,28].

The testing of the stationarity of the data series "overnight stays" (OS) and GDP (Null Hypothesis: Unit root), for the chosen significance threshold (Table 1), revealed the non-stationarity of both the OS (Probabilities for

Fisher test = 0.9994 > 0.05) and GDP (Probabilities for Fisher test = 0.1710 > 0.05) series, which required transforming them through differentiation.

The results for the differentiated series d(OS) and d(GDP) are shown in Table 2.

Table 1. Stationarity test results for overnight stays (OS) and economic growth (GDP) series.

Series: OS			Series: GDP		
Method	Statistic	Probability **	Method	Statistic	Probability **
PP—Fisher Chi-square	3.65317	0.9994	PP—Fisher Chi-square	21.1995	0.1710
PP—Choi Z-stat	5.94868	1.0000	PP—Choi Z-stat	−1.76443	0.0388

Notes: Null Hypothesis: Unit root (individual unit root process); Sample: 2000, 2012; Exogenous variables: Individual effects; Newey-West automatic bandwidth selection and Bartlett kernel; Total number of observations: 96; Cross-sections included: 8; ** Probabilities for Fisher tests are computed using an asymptotic Chi-square distribution. All other tests assume asymptotic normality.

Table 2. Test results for the existence of a unit root in the d(GDP) and d(OS) differentiated series.

Series: d(OS)			Series: d(GDP)		
Method	Statistic	Probability **	Method	Statistic	Probability **
PP—Fisher Chi-square	35.3455	0.0036	PP—Fisher Chi-square	112.384	0.0000
PP—Choi Z-stat	−3.02592	0.0012	PP—Choi Z-stat	−8.83166	0.0000

Notes: Null Hypothesis: Unit root (individual unit root process); Sample: 2000, 2012; Exogenous variables: Individual effects; Newey-West automatic bandwidth selection and Bartlett kernel; Total number of observations: 88; Cross-sections included: 8; ** Probabilities for Fisher tests are computed using an asymptotic Chi-square distribution. All other tests assume asymptotic normality.

The results indicate the fact that the error associated with the rejection of the hypothesis according to which the series d(OS) has a unit root is 0.36%, lower than the 5% standard threshold (in fact lower than that of 1%). In this context, we accept the hypothesis that the d(OS) series is stationary. Also, testing the existence of a unit root in the d(GDP) differentiated series leads to an error associated with the rejection of the hypothesis according to which the d(GDP) series has a unit root (it does not have the first significant decimal digits). We accept the hypothesis that the d(GDP) series is stationary.

Since the GDP and OS series, calculated in regional structures between 2000 and 2012, are not stationary, the potential causality relation between the two series cannot be determined directly through the Granger test. Hence, we use the Toda-Yamamoto version of the respective test.

The applicability of the test with the Toda-Yamamoto version requires the observation of the following procedure, which presupposes these stages:

- Determine the order of a VAR (p)-type process between the series in the level;
- Estimate the VAR model of a p + q type, where d is the maximum order of integration for the OS and GDP series. As we have shown above, both series are I(1), so d = 1;
- Calculate the Toda-Yamamoto (Wald type) statistics and compare them with the theoretical value of the distribution $\chi^2$with (p + d – 1) degrees of freedom.

The first results are presented in Table 3.

With the exception of the Akaike information criterion, which indicates a lag 8 order, all the other criteria suggest an order of the VAR process q = 7. Therefore, taking into account that the stationarity of the d(OS) and d(GDP) series was obtained by a 1-order differentiation of the OS and GDP series, it is clear that the chosen process will be VAR (7 + 1).

Table 3. Results regarding the establishment of the order in the VAR process.

Lag	LogL	LR	FPE	AIC	SC	HQ
6	−709.3583	44.47138	6.68×10^{14}	39.74910	40.88110	40.14818
7	−699.0731	**12.23103 ***	**4.93×10^{14} ***	39.40936	**40.71551 ***	**39.86984 ***
8	−694.3501	5.105994	4.99×10^{14}	**39.37028 ***	40.85058	39.89215

Notes: VAR Lag Order Selection Criteria; Endogenous variables: OS GDP; Exogenous variables: C; Sample: 2000, 2012; Included observations: 37; * indicates lag order selected by the criterion; LR: sequential modified LR test statistic (each test at 5% level); FPE: Final prediction error; AIC: Akaike information criterion; SC: Schwarz information criterion; HQ: Hannan-Quinn information criterion.

Finally, we test the existence of the influence of the GDP on OS (Null hypothesis: GDP influence on OS is insignificant) using Toda-Yamamoto statistics, the Wald-type (Table 4), which is compared with the theoretical χ^2 distribution with 7 degrees of freedom.

Table 4. Results of the Granger causality tests s in Toda-Yamamoto.

Excluded	Chi-sq	df (Degree of freedom)	Probability
GDP	23.22066	7	0.0016
All	23.22066	7	0.0016

Notes: VAR Granger Causality/Block Exogeneity Wald Tests; Sample: 2000, 2012; Included observations: 37; Dependent variable: OS.

The probability estimated for the hypothesis according to which the GDP does not influence OS is 0.16%, inferior to the 5% standard threshold (and to that of 1%), which means that the respective hypothesis cannot be accepted at a reasonable threshold. Consequently, we accept the alternative hypothesis, according to which GDP significantly influences the OS variable (GDP is the cause, in a Granger-Toda-Yamamoto sense, for OS). In the established context, we continue with the methodology for the elaboration of an econometric model, starting from this result, through the test of two hypotheses:

- H_1: GDP is an explanatory variable for OS;
- H_2: On short term, the relation between the GDP and OS variables is distorted by circumstantial factors, but there is a stable relationship between them long-term (econometrically, the GDP and OVERNIGHTS variables are co-integrated).

To test the H_1 hypothesis, a VAR model was built between the stationary variables d(OS) and d(GDP). Due to the small number of observations over time (T = 13), the order of the VAR process could not be greatly increased. The applied information criteria led, however, to a 9-order VAR process, which suggests the existence of long-term links between the analyzed variables.

Regarding the H_2 hypothesis, the test results are presented in Table 5.

The negative value of the coefficient of co-integration (−0.027281), which is also significantly different from zero (*t*-Statistic = −2.10969), indicates the presence of a stable long-term relationships between the variables OS and GDP which allows the development of a panel-type analysis. The test is of the type:

$$OS_{it} = a_0 + a_1\, OS_{i,t-1} + a_2\, d(GDP_{it}) + \alpha_t + \gamma_i + e_{it} \qquad (2)$$

where:

- a_0—is the common effect;
- a_1—estimates the inertial effect;
- a_2—measures the influence of the GDP modification on the endogenous variable (OS);
- α_t—is the specific effect in time (t = 2000, ..., 2012);
- γ_i—represents the individual effect (the specificity of every region);
- e_{it}—is the idiosyncratic error.

Table 5. Results of the application of the test to build a VAR model between the stationary of differentiated series for overnight stays d(OS) and also differentiated series for Gross Domestic Product d(GDP) variables.

Cointegrating Equation:	**CointEq1**	**Standard. Error**	***t*-Statistic**			
OS(−1)	1.000000					
GDP(−1)	−3.074897	0.31233	−9.84502			
Error Correction:	d(OS)	Standard. error	*t*-Statistic	d(GDP)	Standard. error	*t*-Statistic
CointEq1	−0.027281	0.01293	−2.10969	0.173205	0.01827	9.48280
D(OS (−1))	0.511978	0.12130	4.22083	−0.265240	0.17133	−1.54814
D(OS (−2))	−0.475473	0.14447	−3.29115	−0.027695	0.20406	−0.13572
D(OS (−3))	0.206023	0.14414	1.42930	−0.138526	0.20360	−0.68040
D(GDP(−1))	−0.043463	0.05006	−0.86816	−0.236349	0.07071	−3.34240
D(GDP(−2))	−0.096805	0.04702	−2.05885	−0.192904	0.06641	−2.90465
D(GDP(−3))	−0.042931	0.03914	−1.09695	−0.170733	0.05528	−3.08860

d(OS)		d(GDP)	
R-squared	0.188150	R-squared	0.724480
Adjusted R-squared	0.109584	Adj. R-squared	0.697817
Sum squared resids	2.46×10^{10}	Sum sq. resids	4.91×10^{10}
F-statistic	2.394794	F-statistic	27.17149
Log likelihood	−777.2895	Log likelihood	−801.1171
Akaike AIC	22.73303	Akaike AIC	23.42368
Schwarz SC	22.95968	Schwarz SC	23.65033
Mean dependent	9825.725	Mean dependent	−14544.88

Notes: Vector Error Correction Estimates; Sample (adjusted): 2004, 2012; Included observations: 69 after adjustments.

If at least one of the values a_0, a_1 and a_2 is significant, then the panel is homogenous, in the sense that, beyond the elements that are specific to every region, or any moment in time, in the connection between the two variables (OS and GDP) elements that are common to all the regions and years under analysis can be found. The results of model testing are presented in Table 6.

The values of the determination ratio, R-squared = 0.955780, and of the F-statistic = 0.955780, which corresponds to an insignificant probability of error, Probability(F-statistic) = 0.00, lead to the conclusion that the model is valid, and the dependencies between the variables OS and GDP in the eight regions developed in Romania during 2004–2012 are well approximated. Consequently, the model can be used to estimate the evolution of overnight stays by GDP evolutions.

Table 6. Test results of the model.

Variable	Coefficient	Standard Error	*t*-Statistic	Probability
C	22411.05	12366.56	1.812230	0.0755
OS(−1)	0.819036	0.191551	4.275814	0.0001
d(GDP(−1))	0.094438	0.032669	2.890741	0.0055
Fixed Effects (Cross)				
_NV--C	9120.251			
_CE--C	37498.26			
_NE--C	15831.54			
_SE--C	−11543.02			
_SM--C	−7787.306			
_BI--C	−20115.99			
_SV--C	−11875.66			
_VE--C	−11128.09			
Fixed Effects (Period)				
2004--C	−10047.28			
2005--C	7200.238			
2006--C	−2974.132			
2007--C	3955.534			
2008--C	9231.927			
2009--C	−15320.37			
2010--C	−16160.52			
2011--C	8814.043			
2012--C	15300.56			
Effects Specification				
Cross-section fixed (dummy variables)				
Period fixed (dummy variables)				
R-squared	0.955780	Mean dependent var		75105.10
Adjusted R-squared	0.941859	S.D. dependent var		73379.15
Sum squared resid	1.69×10^{10}	Akaike info criterion		22.61211
Log likelihood	−796.0359	Schwarz criterion		23.18127
F-statistic	68.65650	Hannan-Quinn criter.		22.83870
Prob(F-statistic)	0.000000	Durbin-Watson stat		1.645208

Notes: Dependent Variable: OS; Method: Pooled Least Squares; Sample (adjusted): 2004, 2012; Included observations: 9 after adjustments; Cross-sections included: 8; Total pool (balanced) observations: 72.

Coefficients a_0(22,411.05), a_1(0.819036) and a_2(0.094438) are significantly different from zero, as shown in Table 6, and the corresponding variables OS(−1) and d(GDP(−1)) are statistically significant for a probability of error of less than 1% even for a bilateral t test. Regarding the coefficient C (constant term a_0), it is statistically significant for a bilateral t test for a probability of 90%.

However, given that the test used to test the validity of model coefficients is a *t*-Student unilateral test, where *t*-statistic = 1.812230 corresponds to a probability of error of 3.77%, it can be said, with a 95% probability, that the panel is homogeneous, which means that in all development regions there are common elements in time and space, which explains the dynamics of the OS variable, over which regional specific effects overlap.

At a general level, considering the coefficients a_1 and a_2 what results is, on the one hand, that the evolution of changes in the volume of the GDP, expressed in the variable factor d(GDP) directly influences, in the same sense, evolutions of overnight stays in the agrotouristic boarding houses, in a proportion of 9.44% ($a_2 = 0.094438$); on the other hand, it forms a positive feedback loop between overnight stays (OS) in two consecutive years; the value of the dependent variable OS in the current year depends on its value in the previous year, the influence being significant ($a_1 = 0.819036$).

Above this general trend, the individual effects of each region overlap. The data from Table 6 and Figure 1 show that, in the North-West, North-East and Central regions, the level of overnight stays in agrotouristic boarding houses exceeds the average, the higher correction corresponding to the Central region (37,498.26), which means that the pressure of this kind of tourism on sustainability in its counties is much higher.

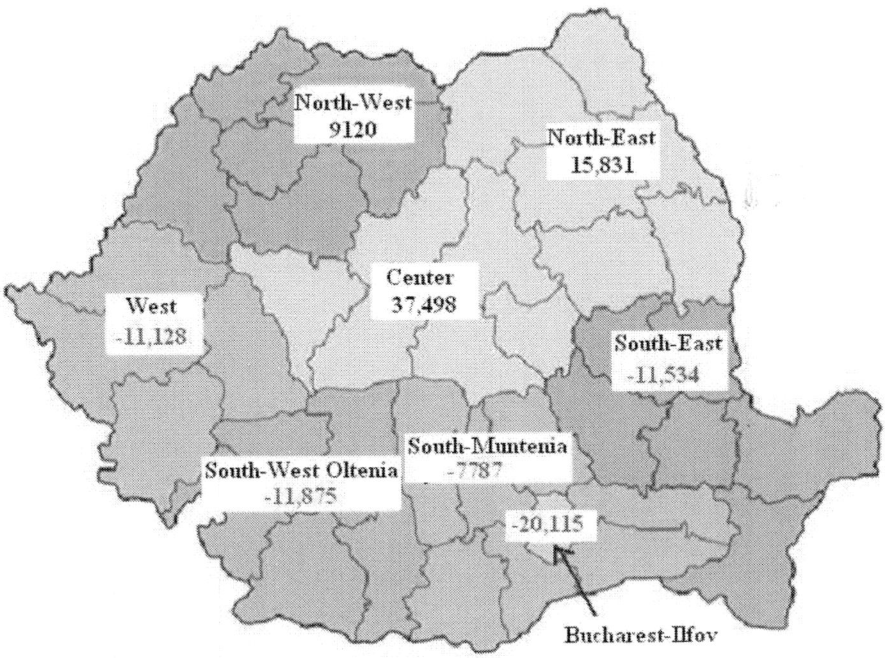

Figure 1. The effects of the specificity of regions on OS.

In the other five development regions, specific effects manifest in reverse. The evolutions of the number of overnight stays in agrotouristic boarding houses are lower compared to the average. A significant decrease was recorded in South-West Oltenia, the South-East and the West, which means that the pressure of rural tourism on sustainability in this region is much lower.

Regarding the Bucharest-Ilfov development region, although the specific correction is −20,115.99, here the pollution and environmental degradation are caused mainly by the urban agglomeration, by the very large number of motor vehicles and by some of the existing commercial and industrial complexes.

Analyzing the temporal effects, we see that corrections are generally positive. However, the corresponding values of 2009 (−15,320.37) and 2010 (−16,160.52) are worth mentioning, as they highlight the effects of the economic crisis on rural tourism in Romania. Since 2011, values which highlight favorable circumstances have been recorded.

4. CONCLUSIONS

The particularities of rural tourism developed in the context of sustainable development are generated by multiple cultural features, biological diversity and the ecological capital as a whole. These must be preserved, along with the need to develop local economies in order to increase the living standards of rural residents.

In Romania, due to the cultural diversity determined by specific traditions inherited over the centuries, and due to the natural heritage of the eight regions, the studies concerning ways of conducting sustainable rural tourism involve approaches, if not punctual, at least at a regional level. On the other hand, we must take into account the fact that this form of tourism has evolved differently over the last decade and therefore the pressures on the sustainable development of counties and local communities are different.

Given the density of rural tourism traffic in the last decade in the eight development regions of Romania, the general trend on the development of rural tourism was highlighted using an econometric model based on panel data series, and quantitative assessments that differentiate development regions from one another were made as well.

Although, at the country level, in Romania rural tourism is a form of tourism which has increasingly drawn more interest, both from Romanian citizens and from foreigners, the analysis of the results shows significant differences between the eight development regions of Romania. In the Central development region, which includes counties from the heart of Transylvania (Alba, Brasov, Covasna, Harghita, Mures and Sibiu) tourist flows are significantly higher than in the other development regions, the Cross Effects having values two times higher than the next ranked. Although the results obtained in the development of the road infrastructure that ensures access to this region are relatively modest, the natural and anthropogenic potential of the counties highlights a good preservation of the medieval age sights (fortified churches, fortresses Fagaras, Alba-Iulia, Bran and others) and the existence of rural civilization areas that have been preserved unchanged, such as the area Viscri, all of these being significant tourist attractions. However, on the other hand, the tourist flows urge the adoption of immediate measures to ensure sustainability.

The North-East Region, comprising the counties of Northern Moldavia, of which Neamt, Suceava and Iasi have a special historical significance, and which, through its monasteries, is an important attraction for religious tourism, also records significant rural tourist flows. Although the natural and anthropogenic potential of this region is no less than that of the Central region, the road infrastructure linking this region with Central and Western Europe is relatively poor, significantly limiting tourist access.

A third region, which has positive values of the Fixed Effects Cross indicator, is the North-West development region, which includes the counties of Bihor, Bistrita-Năsăud, Cluj, Maramures, Satu Mare and Salaj. Among the factors leading to the recording of major tourist flows in this region, there is the proximity of the western border of Romania, which facilitates the access of foreign tourists who use road transport, as well as the particularly rich tourist supply from the Maramures, Cluj and Bistrita-Năsăud counties.

The other five development regions were characterized, in the period under review, by tourist flows that were below the average of the country level. Among these, a special situation is registered in Bucharest-Ilfov, where rural tourism is less developed because of the specificity of this area.

The South-Muntenia region is characterized by significant differences between counties in terms of their rural tourism. Thus, while Prahova, Arges and the north of Dâmboviţa have developed both mountain tourism and rural tourism, counties such as Calarasi, Ialomita, Giurgiu, Teleorman and a large part of Dâmboviţa are plain counties, where tourism is relatively limited, and in which, due to the decline of agriculture that led to a low standard of living, rural tourism offer is quite limited.

Although the tourism offers of the counties of Constanta (the Black Sea) and Tulcea (the Danube Delta) are significant, the attracted tourist flows do not compensate those of the other counties of the South-East development region. An increase in the density of tourist flows in this region, in the fishing villages and the Black Sea villages, can be obtained by the construction of at least a fast road between Pitesti and Sibiu to link the sectors of the A1 and the A2 motorways.

Similar situations are found in the developing regions of West and South-West Oltenia. Within these, although there are areas with special natural and anthropic potential, the relatively weak promotion measures, on the one hand, and poor maintenance of some traditional tourist areas (Baile Herculane), on the other hand, negatively influence the density of rural tourism flows.

Based on these findings, the future research directions will aim at the determination of the specific values of indicators that characterize the aggregate Statistical System at the level of each county. Also, this being given, we aim to identify their similarities and particularities, in order to define an aggregate system at the level of each development region.

ACKNOWLEDGMENTS

This paper has been financially supported within the project entitled "Horizon 2020—Doctoral and Postdoctoral Studies: Promoting the National Interest through Excellence, Competitiveness and Responsibility in the Field of Romanian Fundamental and Applied Scientific Research", contract number POSDRU/159/1.5/S/140106. This project is co-financed by European Social Fund through Sectoral Operational Programme for Human Resources Development 2007–2013. All the authors gratefully thank the reviewers and editors.

AUTHOR CONTRIBUTIONS

Daniela Ruxandra Andrei and Rodica-Manuela Gogonea has designed the research methodology and carried out the calculus involving econometric tools and instruments. Marian Zaharia and Jean-Vasile Andrei have written the literature review, also they contributing in analyzing and debating results. All the authors has read, approved and submitted the paper.

REFERENCES

1. Aall, C. Sustainable tourism in practice: Promoting or perverting the quest for a sustainable development?*Sustainability* **2014**, *6*, 2562–2583.
2. Moscardo, G.; Murphy, L. There is no such thing as sustainable tourism: Re-conceptualizing tourism as a tool for sustainability. *Sustainability* **2014**, *6*, 2538–2561.

3. Andrei, D.R.; Sandu, M.; Gogonea, R.M.; Chiriţescu, V.; Kruzslicika, M. Modeling of rural tourism towards sustainable development from the perspective of specifically organic food. *Sci. Pap. Ser. Manag. Econ. Eng. Agric. Rural Dev.* **2012**, *12*, 21–28.
4. Rodriguez, L.G.; Perez, M.R.; Yang, X.; Geriletu. From Farm to Rural Hostel: New Opportunities and Challenges Associated with Tourism Expansion in Daxi, a Village in Anji County, Zhejiang, China. *Sustainability* **2011**, *3*, 306–321.
5. Gao, S.; Huang, S.; Huang, Y. Rural tourism development in China. *Int. J. Tour. Res.* **2009**, *11*, 439–450.
6. Ailenei, D.; Jula, D. Diminuarea Inegalităţilor Condiţie Esenţială a Coeziunii Economice şi Sociale. Available online: http://coeziune.ase.ro/files/Sinteza.pdf (accessed on 14 July 2014). (In Romanian).
7. Daly, H.E.; Cobb, J.B., Jr. *For the Common Good: Redirecting the Economy towards the Community, the Environment and a Sustainable Future*; Greenprint: London, UK, 1990.
8. Pearce, D.W.; Warford, J.J. *World without End. Economics, Environment, and Sustainable Development*; Oxford University Press: Oxford, UK, 1993.
9. Done, I.; Chivu, L.; Andrei, J.; Matei, M. Using labor force and green investments in valuing the Romanian agriculture potential. *J. Food Agric. Environ.* **2012**, *10*, 737–741.
10. Mebratu, D. Sustainability and sustainable development: Historical and conceptual review. *Environ. Impact Assess. Rev.* **1998**, *18*, 493–520.
11. Bartelmus, P. Dematerialization and Capital Maintenance: Two Sides of the Same Coin. *Ecol. Econ.* **2003**, *46*, 61–88.
12. Common, M.; Stagl, S. *Ecological Economics. An Introduction*; Cambridge University: Cambridge, UK, 2005.
13. Hamilton, K.; Withagen, C. Savings growth and the path of utility. *Can. J. Econ.* **2007**, *40*, 703–713.
14. Popescu, G.; Andrei, J. From industrial holdings to subsistence farms in the Romanian agriculture. Analyzing the subsistence components of the CAP. *Zemedelska Ekonomika* **2011**, *57*, 555–564.
15. Ruta, G.; Hamilton, K. The Capital Approach to Sustainability. In *Handbook of Sustainable Development*; Atkinson, G., Dietz, S., Neumayer, E., Eds.; Edward Elgar Publishing Inc.: Cheltenham, UK, 2007.

16. Eurostat. GDP and main components—Current prices. Available online: http://epp.eurostat.ec.europa.eu/portal/page/portal/statistics/search_database (accessed on 30 September 2014).
17. Walpole, M.J.; Goodwin, H.J. Local attitudes towards conservation and tourism around Komodo national park, Indonesia. *Environ. Conserv.* **2002**, *28*, 160–166.
18. Mehta, J.N.; Kellert, S.R. Local attitudes toward community-based conservation policy and programs in Nepal: A case study in the Makalu-barun conservation area. *Environ. Conserv.* **1998**, *25*, 320–333.
19. Walpole, M.J.; Goodwin, H.J. Local economic impacts of dragon tourism in Indonesia. *Annal. Tour. Res.* **2000**, *27*, 559–576.
20. European Commission. Statistics Explained, Rural Development Statistics by Urban-Rural Typology. Available online: http://epp.eurostat.ec.europa.eu/statistics_explained/index.php/Rural_development_statistics_by_urban-rural_typology (accessed on 14 July 2014).
21. Emptaz-Collomb, J.-G.J. *Linking Tourism, Human Wellbeing and Conservation in the Caprivi Strip (Namibia)*; University of Florida: Gainesville, FL, USA, 2009.
22. Baltagi, B.H. *Econometric Analysis of Panel Data*; John Wiley & Sons Ltd.: West Sussex, UK, 2005. [Google Scholar]
23. European Commission. Towards Quality Rural Tourism. Available online: http://ec.europa.eu/enterprise/sectors/tourism/files/studies/towards_quality_tourism_rural_urban_coastal/iqm_rural_en.pdf (accessed on 16 July 2014).
24. Hsiao, C. *Analysis of Panel Data*, 2nd ed.; Cambridge University Press: Cambridge, UK, 2003.
25. Toda, H.Y.; Yamamoto, T. Statistical inference in Vector Autoregressions with possibly integrated processes. *J. Econ.***1995**, *66*, 225–250.
26. Gogonea, R.M.; Zaharia, M. *Econometrie cu Aplicaţii în Activitatea de Comerţ-Turism-Servicii*; Editura Universitară: Bucureşti, Romania, 2008. (In Romanian)
27. Granger, C.W.J. Investigating Causal Relations by Econometric Models and Cross-spectral Methods. *Econometrica***1969**, *37*, 424–438.

28. Granger, C.W.J. Time Series Analysis, Cointegration, and Applications. *Am. Econ. Rev.* **2004**, *94*, 421–425.

Chapter 10

The Emerging Global Tourism Geography—An Environmental Sustainability Perspective

Paul Peeters and Martin Landré

NHTV Breda University of Applied Sciences, P.O. Box 3917, 4800 DX Breda, The Netherlands; Email:

ABSTRACT

The current development of tourism is environmentally unsustainable. Specifically, tourism's contribution to climate change is increasing while other sectors are reducing their greenhouse gas emissions. This paper has two goals: reveal the main structural cause for tourism's emission growth and show the consequences thereof for (mitigation) policies. It is reasoned that the main cause for tourism's strong emission growth is the time-space expansion of global tourism behavior. Contemporary tourism theory and geography fail to clearly describe this geographical development, making it difficult to understand this expansion and develop effective policies to mitigate environmental impacts. Therefore, this paper explores some elements of a 'new tourism geography' and shows how this may help to better understand the causes of the environmentally unsustainable development of tourism with respect to climate change and devise mitigation policies.

Keywords: tourism; geography; transport; climate change; sustainable development

1. INTRODUCTION

If climate change is a proxy for the environmental issues related to tourism, then it has been clearly demonstrated that tourism is developing unsustainably. Several studies have shown that in the 2050s, tourism related greenhouse gas emissions will be larger than global emissions and will surpass the global emission ceiling that safeguards against 'dangerous' climate change [1,2,3]. Proposed mitigation of tourism's greenhouse gas emissions focuses mainly on technological improvements such as improving the energy efficiency of accommodations and transport modes, using low carbon fuels (mainly bio-fuels for aviation and electric cars for road transport), and off-setting the emissions through funding reductions in other sectors [4,5,6]. Using these measures, the sector envisages reducing total emissions by 50% by 2035 [4], but this ambition is deemed un-realistic [1,7], unless a strong change in tourism transport modes and distances is effected [8,9]. This implies that the trend towards globalization of tourism patterns should be reversed, including reversal of the trends towards travelling farther, more often, for shorter stays and more frequently by air transport.

The aim of this paper is to develop some elements of a 'new tourism geography' and, based on this, show the importance of distance in tourism and its sustainable development. The main questions addressed in this paper are the following: (1) if current tourism is developing sustainably; (2) if geographies/studies are well equipped to assist policy makers in creating sustainable tourism development; and (3) how the geography of tourism has to change to develop tourism sustainably? To find answers, first some geographical mechanisms/relations are explored, providing tools to fill in data gaps in global databases of international and domestic tourism flows. This database is used to find the implications of changes in travel distances on greenhouse gas emissions. Transport modes are related to distances as travel over more than 1,500 km tends to be almost exclusively air transport, while at shorter distances all modes compete, though under 200 km air transport's share approaches zero. Distance is the most important driver for CO_2 emissions as at long distances the large volume of travel (passenger kilometers) combines with higher average emissions per pkm.

2. TOURISM AND TRANSPORT GEOGRAPHY

Definitions of geography, range from rather general—"the study of the world, its people, and the interactions between the two" [10]—to more specific—"the study of the earth and its features and of the distribution of life on the earth, including human life and the effects of human activity" [11]). The main elements in these definitions are the interactions between the physical and living world, including humans. Within geography, Williams defines tourism geography as part of human geography [12] but observes that it has never gained a serious position for academic investigation as tourism is considered to be "a fun-related activity" (p. 19 in [12]). According to Mitchell [13], tourism geography has its focus on the spatial and environmental development of destinations, including access, but not specifically mobility. This description corresponds with the assertion that the main element of human (tourism) geography is places that are defined as "not only merely bounded spaces or locations, but ... also settings (or locales) in which social relations and identities are constituted and in which is developed a sense of place" (p. 185 in [12]). Currently, the destination seems to be the main focus of tourism geography, which is also clear from the contents of some major texts on tourism geography [12,14].

In 1976, Miossec published a geography-based tourism theory that gave a prominent place to transport [15]. He stressed the role of transport, transport technology, travel time, travel distance and mode choice in destination choice. His mapping of tourism from Germany, France, and the United Kingdom showed clearly that for larger distances the number of trips is lower. On page 19, Miossec presents a distance decay graph for Germany [15]. He also presents schematic graphs showing the relations between distance, travel time and travel cost for different transport modes (p. 37, Figure 12 in [15]). Another early description of tourism geography was published by Williams and Zelinsky in 1970 [16]. They reject a rather basic hypothesis for tourism flows, the 'hypothesis of indifference' (p. 556 in [16]), that assumes that if a destination receives X percent of global international tourists, it should receive the same X percent of all departures for each country in the world. This hypothesis fully ignores the role of transportation, distance, cost, and travel time, and it is not surprising that it did not hold against empirical evidence [16]. Williams and Zelinsky also reject the hypothesis of reciprocity, which assumes that a strong tourism flow from A to B will always generate a "reflexive, or counterflow" (p. 564 in [16]). A third hypothesis, the idea that destinations with a lower cost of living will generate the most arrivals, is only partly confirmed by empirical analysis (p. 566 in [16]). Only two hypotheses appear to be valid to them. The first positively relates tourism flows to tourism attraction (e.g., 'desirable climatic characteristics, scenic attractions, cultural and historical features, sports, shopping facilities, night life'). The second hypothesis positively relates tourism flows to international connectivity (e.g., commercial, business, migration, cultural relations and political linkages) (pp. 564–565 in [16]).

The obvious roles of transport and distance have not been tested because of "the absence of data on *points* of origin for tourists from large countries (e.g., the United States and Scandinavia) and a similar lack of data for destination points" (pp. 563–564 in [16]). Unfortunately, this theoretical work has not been further developed into a strong theoretical base [14]. An exemption is the tourism area life cycle model [17] that originated in geography but which is often treated in an aspatial fashion [18]. The lack of theoretical progress on spatial interaction properties of tourism has arguably resulted in tourism geography being mainly focused on the development of destinations, while generally ignoring the (places of) demand and the links between the two. This skewed focus has produced a lack of recognition within the wider scientific community of the importance of tourism geography [12,19,20]. Most textbooks on tourism management and economics see transport as a derived demand with a cost in terms of money and time (e.g., [21]). These books often ignore the fact that transport is an intrinsic part of any travel, also mentally (e.g., [22]). Furthermore, these books also ignore the many interactions between the transport system and the (in-)ability of certain destinations and certain forms of tourism to develop. Even some tourism geography texts pay little attention to the role of transport and transport infrastructure [12,14,19]. An exemption forms a textbook by C. Michael Hall [23], that extensively discusses tourism mobility and models taken from transport and behavioral geography.

Climate mitigation studies conclude that policies may increase costs of tourism and reduce its economic growth in case of carbon taxes [24,25,26,27] or oil price increases [28], though other studies find no significant impacts [29] or believe that second order effects like a mode shift from aviation to the car, may even increase overall emissions when taxing aviation. Unfortunately, most of these studies fail to include important parts of the tourism system, e.g., by just dealing with air transport or international tourism thus failing to acknowledge shifts to other transport modes, or domestic tourism. For instance, Becken *et al.* [28] state on page 135 that the main challenge "that global interactions within the tourism sector (e.g., substitution between destination countries) are not well accounted for" because of "current limitations of global datasets". We provide such a geographical database.

The most fundamental problem of current tourism geography seems to be the unit of geographical scale. Generally, geographical data are based on the nation. Unfortunately, that causes problems as nations' sizes in terms of population, land area and economy vary by up to eight orders of magnitude (e.g., Monaco *versus* China). This problem of differing scales already has been observed in the 1970s [15,16] but has yet to be satisfactorily solved. Derived from this geographical scale problem are two other issues: the divide between international and domestic tourism and the preoccupation with air transport. These two issues lead to many misunderstandings and biases. For instance, the idea that Europe is the most important tourism destination [30,31] is incorrect as China receives annually far more tourists if 1.6 billion domestic tourists [32] were included. The different scales of nations cause that almost all tourist trips by citizens of Monaco are 'international', and most trips by Australians are 'domestic'. But this large seemingly large difference is not representative for the overall difference in travel patters. Most likely, overall travel patterns will be relatively similar in terms of trips, distance distribution, costs and travel times, with the main distinctions being caused by differences in transport infrastructure and the geographical distribution of tourism attractions. More generally, the divide between domestic and international tourism tends to cause an overvaluation of relatively easily measurable international tourism and ignorance of more difficult to measure domestic tourism [14]. Even though UNWTO (United Nations World Tourism Organization) has recently updated the standard of their statistical compendium to include a section on domestic tourism in 2010 (see [33]), this has only resulted in domestic tourism data for just 14 of approximately 210 countries/protectorates recognized in the Tourism Factbook [34].

An important caveat of the focus on international tourism is the overvaluation of the role of aviation in tourism. For example, a recently published advanced tourism-energy model [35] ignores transportation by defining the tourism system in terms of energy system, number of tourists, waste, revenues, and accommodations (supply system) only. Also, a well-known textbook about the tourism economy dedicates only one chapter to aviation [36], and ignores the role of other modes. This is unfortunate as these other modes serve over 80% of all tourism trips in the world [8]. Transport geography is seen as "a sub-discipline of geography concerned about movements of freight, people and information: It seeks to link spatial constraints and attributes with the origin, the destination, the extent, the nature and the purpose of movements" (Rodrigue [37], p. 5). Several concepts from transport geography will be used to overcome the main problems of current tourism theory and geography. In this way, a Global Tourism Transport Database (GTTD, see Section 3) has been developed that covers *all*tourist trips (international and domestic) and provides information on travel distances. Furthermore, the GTTD not only contains all flows between countries, but also aggregates these to 18 world regions as an alternative to the problematic geographical unit of the 'nation state'.

3. THE GTTD AND ELEMENTS OF A NEW TOURISM GEOGRAPHY

Only a few attempts have been made to create a global tourism flow database; one example is a database developed by Bigano *et al.* [38]. As this database uses rather old data––generally 1995 data from the World Resources Institute—NHTV CSTT created one with 2005 as base year: the Global Tourism Transport Database (GTTD), which is based on UNWTO data [39]. In this database, the UNWTO 'Arrivals of non-resident tourists in all types of accommodation establishments, by country of residence' are used when available. These arrivals are given per country of destination. The number of countries of origin per country of destination distinguished by UNWTO data varies over the destination countries between just a few up to more than 100. Remaining countries of origin are given as aggregates per world region (e.g., 'Other Europe', 'Other South-East Asia' or 'Other World'). However, it appears that these 'other' places of residence represented just 3% of the total number of arrivals in the database and we have simply removed them instead of trying to redistribute them over the designated regions. In some cases (e.g., Comoros, Libya), no data for 2005 were available, and 2004 data were used instead. Also, corrections (reductions) were applied to flows between neighboring countries where in some cases large numbers of same-day arrivals were incorrectly counted as 'tourist' arrivals, largely inflating numbers. We consulted national data for this correction where available. The database contains 12,118 country-pairs for a total of 223 countries with international flows totaling to 735 million, which is 2.6% short of the officially published figure of 755 million for 2005 [40]. Furthermore, the data are presented by using world regions as defined in Figure 1, to create more equal geographic entities as nations.

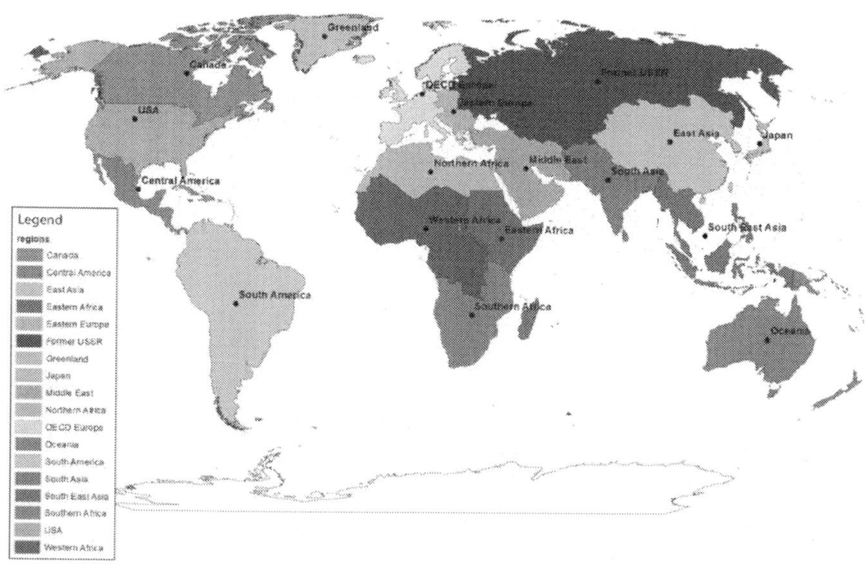

Figure 1. The world regions (source IMAGE-team [41]).

The UNWTO data fail to provide consistent and complete information about domestic tourism, transport mode and distances travelled. The 'transport' element of the GTTD was added by calculating the 'great circle distance' between all countries using the "Great Circle Mapper" [42] and the airport codes for the main airport of each country. For country pairs between neighboring countries with at least one very large country, we replaced the airport in the large country with one nearer to the border between the two countries. This correction kept large numbers of arrivals from being assigned to too large distances. A total of international distances (round-trip) of 3.720 × 10^{12} (pkm) was found, or an average one-way distance of 2,531 km. This distance is 23% higher than the only other published global estimate of 2,051 km [2]. Significant uncertainties remain in both sets of numbers.

Finally the GTTD has been enriched with domestic tourism. The methods used for the domestic number of trips are described in Section 3.1, while the estimates for distance travelled are given in Section 3.4. Section 3.2 describes time and money budgets and Section 3.3 distance decay and power curves. Both of these sections give some new (or renewed) elements for tourism geography and are both used for the domestic travel distance estimates in Section 3.4.

3.1. Trips as Function of GDP

Tourism trips per capita appear to be strongly related to GDP per capita (e.g., by Bigano *et al.* and Peeters *et al.* [8,38]). We assume that this relation is valid for both international and domestic trips and is linear but with a certain maximum number of trips. Hence, domestic trips for all countries were calculated using the following equations:

$$\tau_{T_i} = C_{cy} + \alpha_{cy} \cdot \overline{GDP_{cap_i}} \quad (1)$$

and

$$T_{dom_i} = \tau_{T_i} \cdot P_i - T_{int_i} \quad (2)$$

The number of departures per capita τ_{T_i} are a maximised linear function of GDP per capita (PPP, purchase power parity) as given by Equation (1). Equation (2) gives the number of domestic tourist arrivals T_{dom_i} for country *i* as a function of the total departures from this country minus the (already known) international trips T_{int_i}. The coefficients are given inTable 1. GDP data generally were taken from IMAGE [41] but, in case image did not provide data, also from databases such as the CIA World Factbook [43,44].

Table 1. Baseline values for the parameters determining trip generation (Peeters *et al.* [8]).

Tourism market	C_{cy}	α_{cy}	T_{max}
Total trips	0.2888	0.00007343	5.0

We used this method to estimate domestic tourism. The total global number of trips in the GTTD increased from $3.89 \times 10^9$to 4.625×10^9.

Figure 2 shows a comparison of the results of our method with the results from the method by Bigano *et al.* [38]. Our result is just 2.6% lower than the formerly published number of tourist trips of 4.75×10^{12} [2]. The equation given by Biganoet *al.* would underestimate this total of 4.750×10^{12} by about 17%.

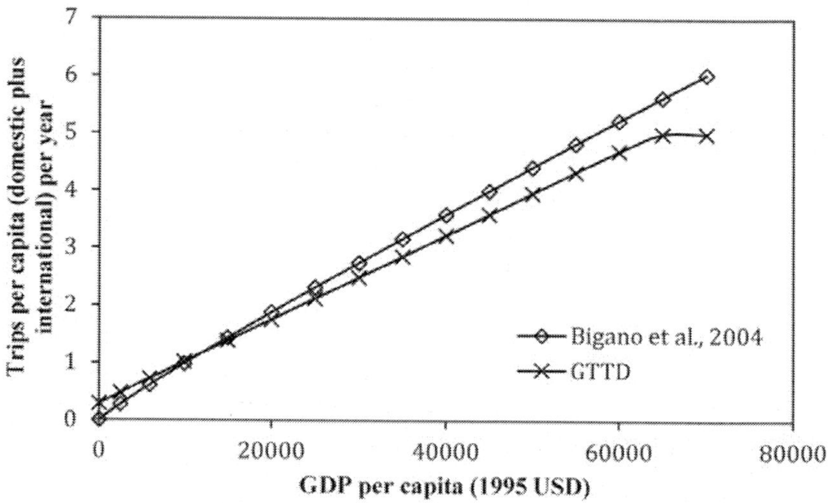

Figure 2. Relations between GDP and trips per capita for the GTTD and Bigano [38].

3.2. Time and Money Budgets and Speed of Travel

The amount people can travel is limited by time and money constraints: the travel money budget (TMB) and the travel time budget (TTB). The TMB hypotheses says that the share of income spent on transport and travel is constant when measured at the population level, while the TTB hypotheses claims that the average travel time is a constant when measured at the population level. Zahavi [45,46,47] and, to a lesser extent, Hägerstrand [48], were among the first to suggest these ideas, but they have been developed by others as well [49,50]. There are time budget constraints that have been successfully modeled in travel demand dynamic models [51]. The most extensive recent work on TTBs and TMBs has been published by Schafer [52,53], who also developed a global mobility model including long distance travel [54], inspiring others [55]. Furthermore, Metz [56] and Banister [57] give strong arguments for the existence of TTB and its use in transport policy, including a plea to leave the flawed transport policies paradigm that says that increasing travel speed saves time.

The constancy of the TTB is challenged as it is only found at the highest aggregation level [58]. Most of these critiques start, oddly enough, by disaggregating the population and showing differences in travel time among different population groups. This approach ignores that the 'constancy' is assumed to exist only at the population level and not for individuals of small population segments.

Kölbl and Helbing [59] relate travel patterns to the energy exerted by the traveler (the individual energy, not the energy consumption of the vehicles used) and show that this budget is a constant. It shows that all different transport modes follow the same distribution of energy used. Lin *et al.* [60] study the experience of travel time and find that cognitive distances are systematically underestimated at short ranges (<150 km) and overestimated at long ranges (>250 km). This result may support the idea that travel times are important as short distances of up to 150 km will involve trips near the average daily TTB of 60–70 minutes [52], making the trip 'feel short', while distances above 250 km will take much longer making their 'feel' to be rather long.

Unfortunately, global travel time data for tourism travel are non-existent. Therefore, a relationship between travel distance and travel is suggested to impute speed to the most common travel modes, as already proposed by Miossec in the 1970s [15]. Using the Mobility Survey Netherlands (MON [61]), Figure 3 has been constructed. MON contains detailed data for 200,000 trips per year generated by 60,000 subjects. The focus of this survey is daily transport and most long international trips are missed. However, it is assumed that the way tourists use transport modes will not fundamentally differ from daily travel. The measured travel speeds were best represented (*i.e.*, highest R^2) by the power law model given by Equation (3), where V is the speed in km/hr, d the distance in km and C_V and b_1 are the model constants estimated with PASW Statistics 18.0.

$$V = C_V \cdot d^{b_1} \tag{3}$$

For the modes private boat, ferry, sea- and river-cruise, and local public transport (bus, tram), we assumed a constant speed because the number of trips in MON was too low or the scatter was too large to accurately calculate the speed. For cycling and walking we assume that these also have a constant speed independent of distance. Figure 3 shows the result.

(a)

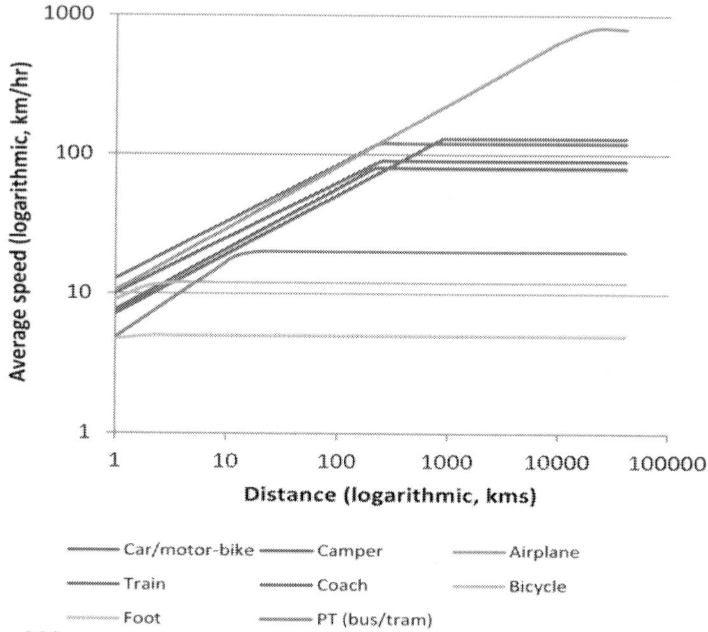

(b)

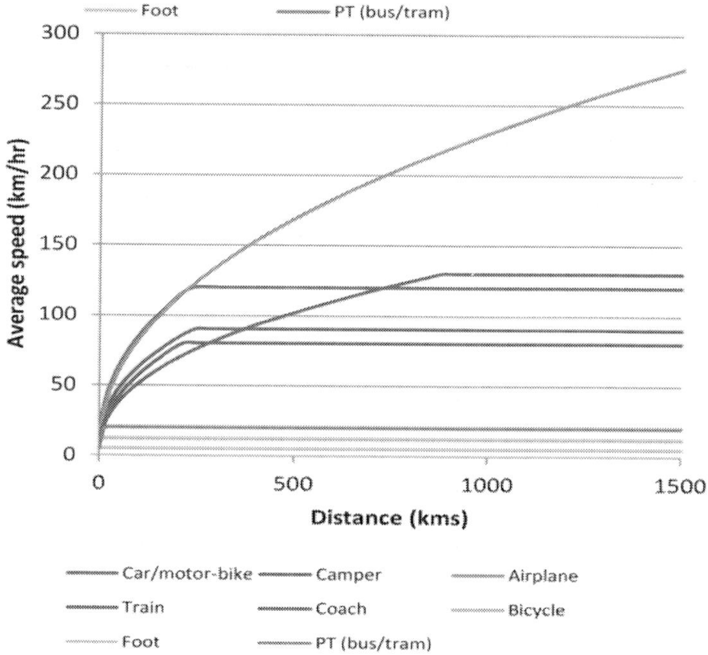

Figure 3. Transport speeds per transport mode as modeled from the MON database [61]. Power curve fit: (a) logarithmic plot for all distances; (b) linear plot for 0–1500 km.

The train data are based on conventional speed trains (less than 180 km/hr maximum speed). MON does not include high speed rail data, but we expect these will show comparable or better average speeds on trips up to 1,500 km.

3.3. Distance Decay and Power Curves

Early attempts to develop tourism geography [15,16,62] related transport system qualities such as cost, distance and speed to travel patterns in a conceptual manner, were inspired by gravity models. These models assume transport flows to be proportional to the 'masses' of origin and destination and inversely proportional to the distance between them [23,63,64,65,66], where distance is defined as a mix of physical distance, cost, travel time and discomfort. This idea is closely related to the 'distance decay' [62,64,67,68,69] theory that assumes the frequency of trips to be inversely proportional to physical distance. Distance decay has even been formulated as the 'first law of geography' (p. 236 in Tobler [68]), which says that "everything is related to everything else, but near things are more related than distant things". Only a few authors have published distance decay for tourism [62,70,71,72,73,74].

For global travel, McKercher [71] finds a surprisingly good fit for 2002 international arrivals, with just a small secondary maximum for the distance class between 5,001 and 6,000 km. For 2005, our GTTD shows the same maximum at 5,500 km and two more at 9,500 km and at 16,500 km (see Figure 4). The maximum at 5,500 km is mainly caused by transatlantic travel (63% between Europe and North America), while the maximum at 16,500 km shows most (79%) traffic between Oceania with either Europe or the USA. The 9,500 km maximum has no specific origin.

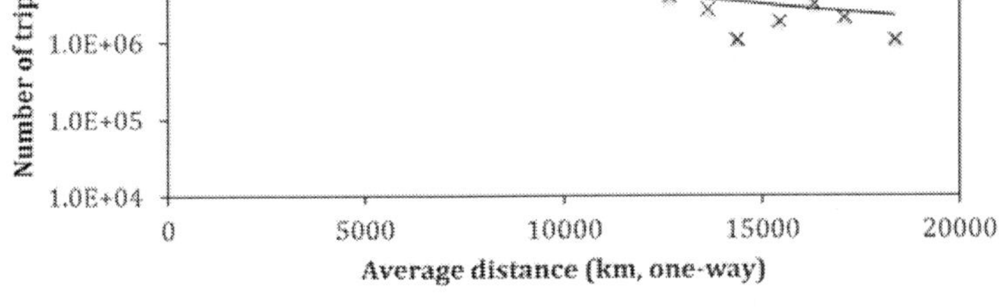

Figure 4. Distance decay and power curve for international arrivals in 2005 (data per nation; source: GTTD).

Many phenomena in nature and society follow a bell shaped probability curve (e.g., the weight distribution of people, which has a clear average and reduced probability the farther away from this average). However, there is also a class of phenomena that follow a much different probability curve that has a very high probability at one end of the range of occurrences and very low at the other end. These power curve probabilities are often suggested for varying physical, natural, economic and social phenomena [75,76,77,78,79,80,81]. Examples of phenomena following a power distribution are earthquakes [75], the number of species in general [79], the use of words [77] and the distribution of wealth [82]. Generally, power curves do not cover the whole dataset but only from a certain minimum cut-off value of the phenomenon. Newman*et al.* [79] mention seven mechanisms in phenomena that may generate power curves in nature of which at least one, the random walk, may be applicable to tourism travel. The random walk literally means that the amount of time before a subject that makes random moves returns to his original position has a power probability (very frequent short times and very infrequent long times). These random walks, also called 'Lévy Flights' [83,84] have been shown to occur in human travel [85,86]. They are also shown [87,88] and disputed [89] for foraging animals. Section 4.1 highlights Lévy Flights for Dutch travel and Section 4.2 for global travel.

3.4. Country Size and Domestic Tourism Travel Distances

The average distance domestic tourists travel will mainly be a function of the size of the country and to some extent the quality of the infrastructure. Based on data from the MuSTT study [90] we drafted a relation between the average domestic distance and the size (the area of land surface in km^2) and fitted a curve to these data (see also Figure 5):

$$\overline{d_i} = \varepsilon_q \cdot 1.6792 \cdot A_i^{0.4387} \quad (4)$$

where d_i is the average domestic one-way distance in km and A_i is the country's land area in km^2. The factor ε_q is an indicator of the quality of the road network, which we consider to be of a better quality in developed countries, enabling higher travel speeds. To be consistent with former work, we set ε_q to 1.0 for developed countries, designated OECD90 in the IMAGE models [41] and 0.8 for all other countries.

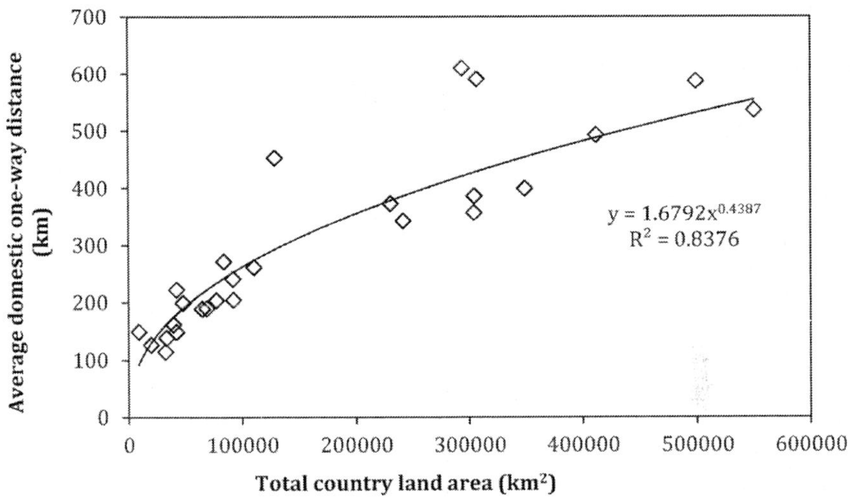

Figure 5. Relation between land area (km^2) and average one-way distance for domestic tourism (source: [90]).

With the method described above, we know the number of domestic trips and the average distance travelled at the scale of nations. This information increases the total distance travelled in the world from 3.72 × 10^{12} billion pkm to 11.36 × 10^{12}billion. The new distance decay results for global tourism are given in Figure 6.

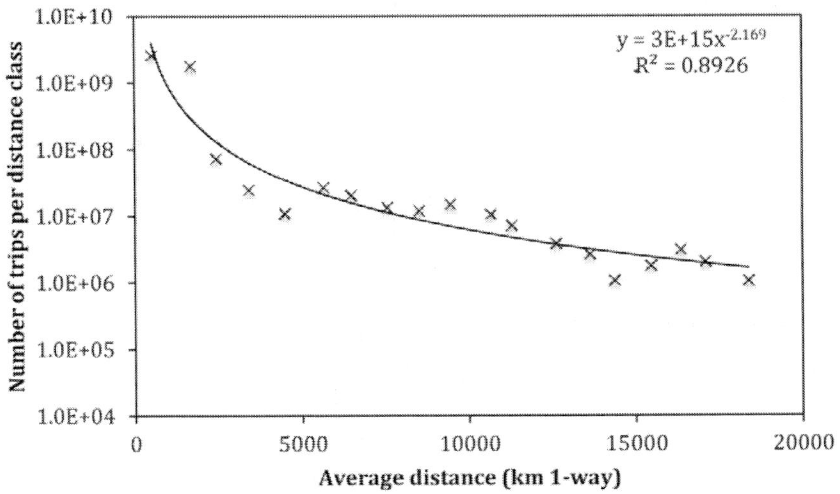

Figure 6. Distance decay and power curve for global tourism (domestic and international; data per nation; source: GTTD).

All domestic trips ended up in one of the first two distance classes. This distorts the figure, as, domestic trips will also be distributed in the same way as international trips, thus with most trips in the lower distance classes and fewer trips for medium to, in the largest countries, even for long haul trips. In the GTTD, the 15 largest domestic flows (0.1% of all data points in GTTD) represent 61% of all trips. Therefore, we have redistributed the 13 largest (all domestic) data points plus two additional large points that are in geographically large countries (Australia and Canada). These 15 data points represent 60.3% of all trips in the GTTD. For this analysis we assumed a power curve distribution for one-way distance classes with averages of 200, 400 and so forth up till 7,000 km. The power curve is as follows:

$$n_T = C \cdot d^{-\mu} \qquad (5)$$

where n_T is the number of tourists, d is the average distance of a distance class, C is a constant and μ is the power coefficient. The minimum value d_{min} is defined by the following relation between C and this minimum value [78]:

$$C = (\mu - 1) \cdot d_{min}^{\mu-1} \qquad (6)$$

Table 2. Overview of the power law constants for the 15 countries with the largest domestic tourism demand.

Country	μ	Distance (km, one-way distance)		
		Maximum	Minimum	Average
Australia	1.123	3,600	800	1,757
Brazil	1.303	4,300	400	1,286
Canada	1.155	5,200	600	1,908
China	1.394	5,000	400	1,342
France	2.698	1,200	400	536
Germany	1.260	1,000	200	399
India	1.235	3,500	200	813
Indonesia	1.682	5,300	200	654
Italy	1.781	1,200	400	610
Japan	1.731	2,500	200	469
Mexico	1.408	3,300	200	679
Pakistan	1.612	1,800	200	450
Russia	1.269	6,000	600	1,971
United Kingdom	1.884	1,300	200	343
United States	1.175	7,000	400	1,902

We used Equation (6) and the 'known' average value and maximum distance (measured using the Google Maps distance tool) to find a value for μ using Excel's Solve add-in and fitting the trip numbers per distance class to the total known domestic trips. Table 2 shows the results. Note that the power law exponents are relatively low with the exception of France. Further research is needed on this issue.

Clearly the 'power curve' fit improved (see Figure 7), although it appears to be following two separate power curves: one in the domestic domain of distances and the other in the medium to long haul international domain. Without better national tourism data it remains unclear if the imputed domestic distance class distribution is realistic.

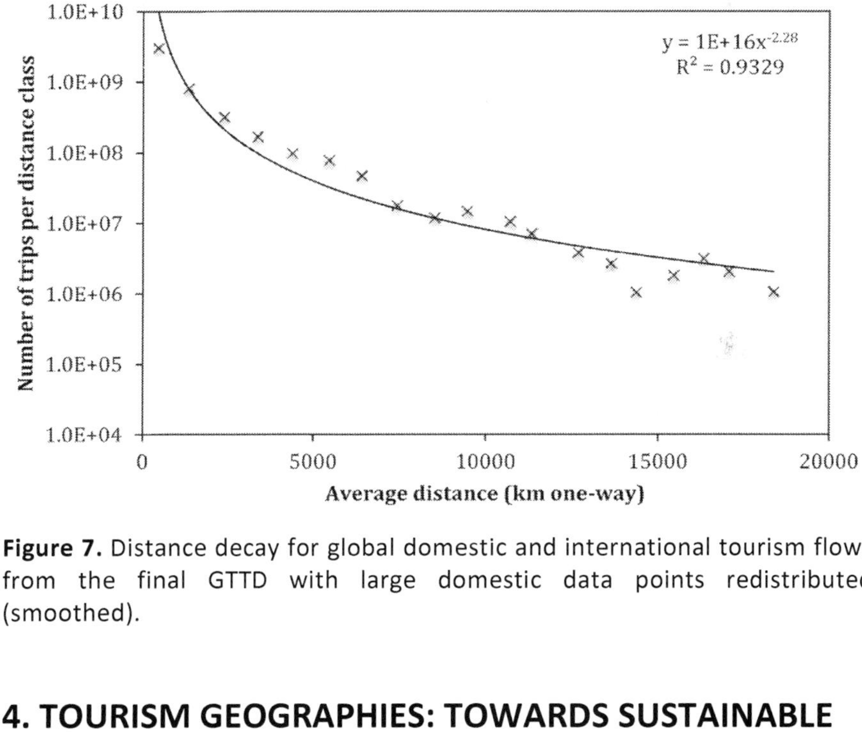

Figure 7. Distance decay for global domestic and international tourism flows from the final GTTD with large domestic data points redistributed (smoothed).

4. TOURISM GEOGRAPHIES: TOWARDS SUSTAINABLE DEVELOPMENT

In Section 4.1 the Dutch case will be introduced, as The Netherlands seems the only country with a sufficiently detailed database for Dutch outbound tourism (excluding business). This database includes information about the time and duration of the trip, transport modes, spending, and (imputed) distances. This information is needed to give some statistical evidence for the ideas presented in Section 3. In Section 4.2 we return to the global level of analysis. Some of the findings from the Dutch case are applied globally, and the flaws in contemporary global tourism geography, which is based on international travel, are discussed.

4.1. Dutch Holiday Travel

Dutch holidaymakers travel all over the world. Figure 8 is based on data from the Dutch Continuous Holiday Survey (CVO [91]) and shows all 104 of the largest international flows. The CVO survey distinguishes 299 destinations with a high resolution at short distances (e.g., at the province level in The Netherlands, Belgium, UK, Germany and France). Travel distances were calculated based on the great circle distances to these destinations [42]. These distances were increased by a certain factor per transport mode to accommodate for detours (see the Travelling Large report [92] for further details about the methods used). Furthermore, travel times were generated that are based on average travel speeds as a function of distance (see Section 3.2).

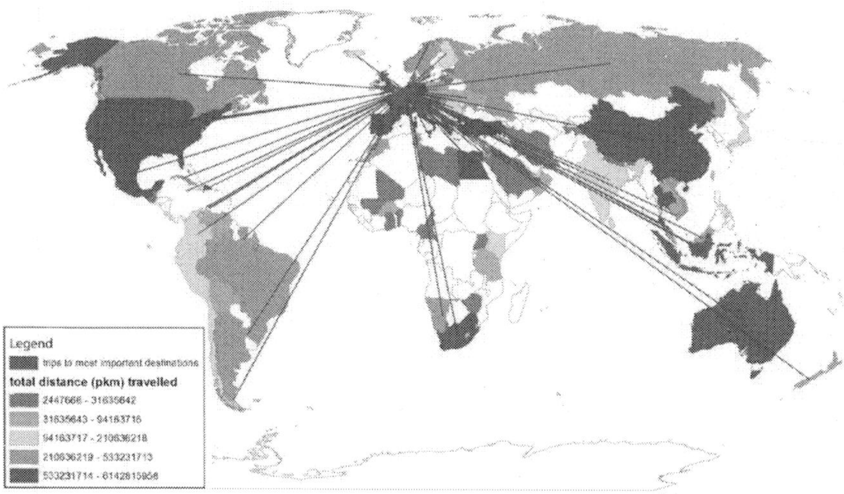

Figure 8. Dutch holiday makers 104 largest international flows for 2009. Background shading gives the total distance (pkm) travelled to the destination country, while the wedge width is a measure of the number of trips.

We tested the distribution frequency of trips per distance class (in classes of 1,000 km one-way distance from 0 up to 20,000 km) and found a power law best represented (largest R^2, though still just 0.71) and a coefficient of 2.06. Most of the scatter appears in the tail of the distribution, where numbers per class are low (see Figure 9(a)). This result gives some support for the power law relation of tourism travel by Dutch holidaymakers, but the result is much weaker than for the global case as shown in Figure 7.

When a TTB is a valid proposition, it might be that travel time is the quantity measured by the human brain, acting as a reference to humans' travel decisions. In that case, the power law for travel time should result in a better fit than for travel distance. Figure 9(b) shows that total travel time has a much better power curve relation than distance (exponent of 2.685 and $R^2 = 0.94$). The idea that travel time is the main parameter in travel decisions has support in the literature concerning the TTB [45,49,50,56,57,93]. Kölbl *et al.* [59] tried to find a physical mechanism for TTB and found it in canonical energy use during travel. The idea that travel time might be related to daily energy use has been convincingly found for animals, as exemplified by the airborne time and energy use of kestrels [94,95,96]. For example, it appears that kestrels have a rather constant activity pattern with approximately 1 hour daily of flight/hunting, except during breeding season when this time increases to approximately 4 hours. However, 4 hours is still far short of the maximum activity time the birds physically could achieve. When the number of chicks in the nest is manipulated, those kestrels confronted with more chicks than they chose to raise showed a much lower survival rate the next season. This reduction might be a strong evolutionary mechanism that creates daily energy constancy of which daily movement (flight) is a derivative, giving support to Kölbl *et al.* [59].

Based on calculations of the carbon footprint of Dutch holidaymakers between 2005 and 2010, we found an increase in carbon emissions from holidaymakers, except for a dip in the economic crisis year 2009. Both the growth and 'dip' are associated with the share of air travel and the average distance travelled; the number of trips was almost constant, giving support to the idea that travel time is more or less a constant. Thus, distance is the dependent and travel speed the independent variable. Both of these variables are strongly related in 2009 for length of stay, cost per day and travel speed (Pearson correlation in bivariate of distance is highest for average speed (0.818), followed by cost per day (0.464) and length of stay (0.402). The carbon footprint calculations include technological energy efficiency improvements calculated for air transport based on Peeters and Middel [97] and measured by the Dutch Statistical Office (CBS) for other transport modes in the Netherlands [98]. As speed and concomitant distance growth is a strong driver for carbon footprint growth, it is necessary to explore what impacts distance restrictions might have on global tourism.

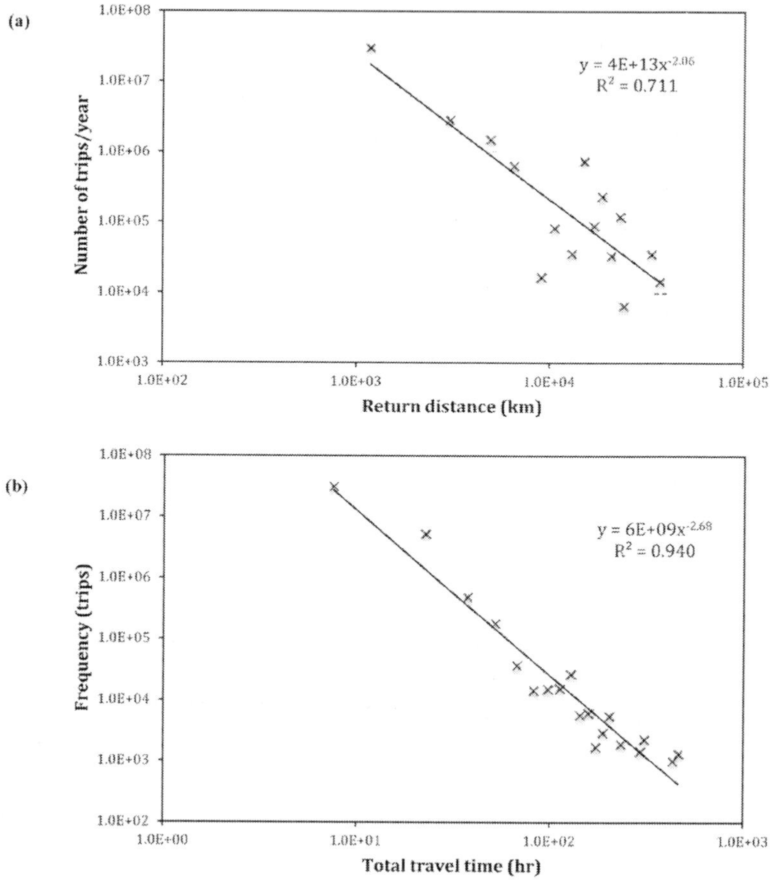

Figure 9. Trip distribution for distance travelled (**a**) and travel time (**b**) by Dutch holidaymakers (source: CVO [91]).

4.2. Global Tourism Flows

Figure 10 shows a geography based map on nations for the 168 most important relations of the 12,000 plus relations in the GTTD. These flows represent 90% of all international trips. Clearly, international tourism is not evenly distributed over the world, and there is a strong bias to the more developed parts of the world. Figure 10 also shows the 'fatter' wedges are those at shorter distances between large countries (such as the USA and its neighbors).

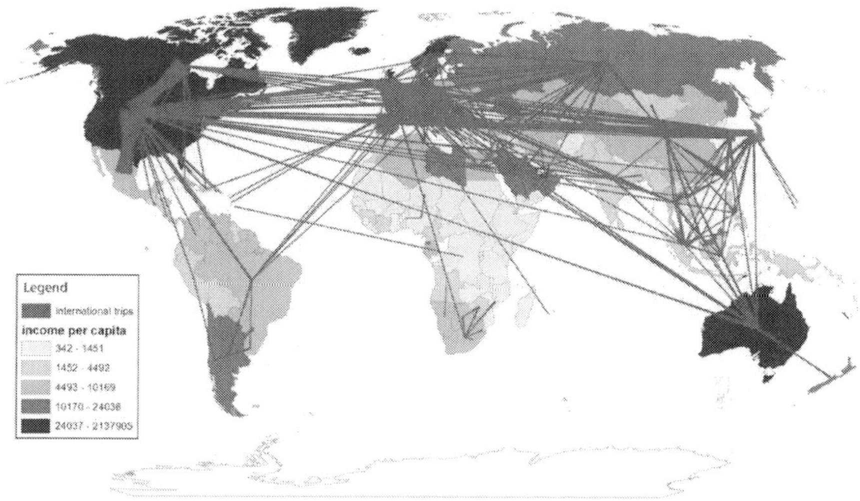

Figure 10. The main international tourist flows in 2005 and income per capita (for 2003 in 2000 USD; based on GTTD).

Many developing countries in Africa, Asia and Latin America do not show any international arrivals because their international tourism flows do not belong to the 168 largest flows we selected. This map also reinforces the perception that Europe is the most important (international) tourism destination.

However, these impressions are incorrect due to some serious caveats underlying the use of only international flows, as outlined in Section 3. First, international flows represent only 17% of all tourism trips as they ignore the 3.89 billion domestic trips (see Section 3.1). Second, international flows are based on flows between very uneven geographic entities, nation states, which are the main cause of the high density of flows in Europe and relatively small flows in North America and Asia. Therefore, we introduced a geographical division of the world that is based on the IPCC scenarios and gives a fairly equal mix of global population, economy and surface area [41].

Figure 11 uses the GTTD to show all of the relations between 17 out of the 18 world regions (Antarctica has no data) for both international and domestic tourism. This map looks rather different from the one in Figure 10. The short but thick intra-regional flows now dominate over the numerous thin inter-regional flows. Also, tourism is now more attached to the richer parts of the world, and the 'special' position of Europe has almost been lost.

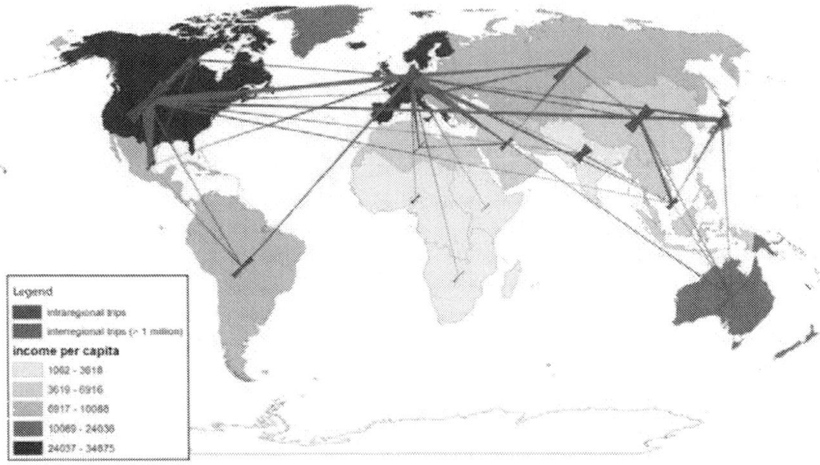

Figure 11. Global map of all flows between and within the IMAGE regions in 2005. The flows are measured in arrivals and include international and domestic tourists (income per capita for 2003 in 2000 USD; source GTTD).

4.3. Mitigating Climate Change by Reducing Tourism Transport

As shown in the literature, a large gap exists between emission reduction goals and mitigation options, such as improved technology, energy efficiency and switching from air and car transport to low-carbon transport modes (e.g., coach and train) [1,2,7,99]. The main, but not only, problem is that the growth in aviation volume outpaces the benefits of technological efficiency improvements [100,101,102]. One way to achieve substantial absolute emission reductions often suggested is to switch aviation to biofuels [103,104,105]. However, there are doubts about the life cycle reductions to be achieved with alternative fuels and problems such as space use, conflicts with food production, and potential loss of biodiversity [106]. It also appears that initially promising biofuels as Jatropha seeds [103], depend greatly on where the seeds are planted [107] and may have serious environmental, social and economic problems associated with them [108,109,110,111,112]. Some researchers recommend using caution when suggesting biofuels such as Jatropha will contribute to large scale greenhouse gas emission reductions [111,113]. In addition, technological solutions have, historically, often not lived up to their expectations [7,114,115].

These problems mean that another solution appears inevitable: decoupling transport volume growth from tourism growth, or, in other words, reducing the average distance people travel per tourist day [116,117]. Figure 12 shows Lorenz curves based on the GTTD for the relation between the share of trips and the share of total distance travelled, with trips ordered from short to long distances. The point labels show the 'cut-off distance' for the case of the fully developed GTTD, including the 15 largest domestic flows redistributed over different distance classes. For example, the '2000 km' point shows that 83% of all trips are shorter, while these trips cover 42.5% of all distances travelled. Thus, if by some measure people were not able to travel farther than 2000 km, this would reduce tourism trips by 17% while reducing total distances travelled by 57.5%.

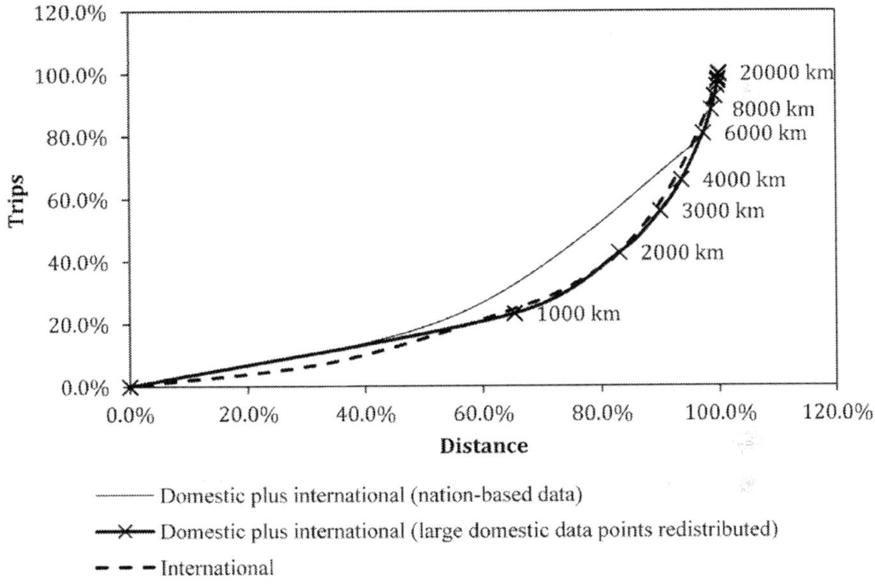

Figure 12. Some Lorenz curves for trips and distance development; trips sorted from low to high distance; labels refer to the one-way distance associated with the share of trips and the total distance of that data point (source: GTTD).

Reducing the number of medium and long haul trips may hypothetically be achieved in different ways: high taxes on air travel, issuing a system of personal travel budgets or capacity restrictions within the (air) transport sector. The question is whether tourists will simply cancel their long haul trips when confronted with such measures and spend their money outside the tourism industry or if they will choose a closer destination. Currently, no research can give a conclusive answer to this question, but the reaction to the recent economic crisis perhaps gives an indication. The OECD concludes that in times of economic uncertainty people tend to save on international and air travel and shift towards domestic and car travel [118]. This same effect has been found in a time series of travel by Dutch holidaymakers. The

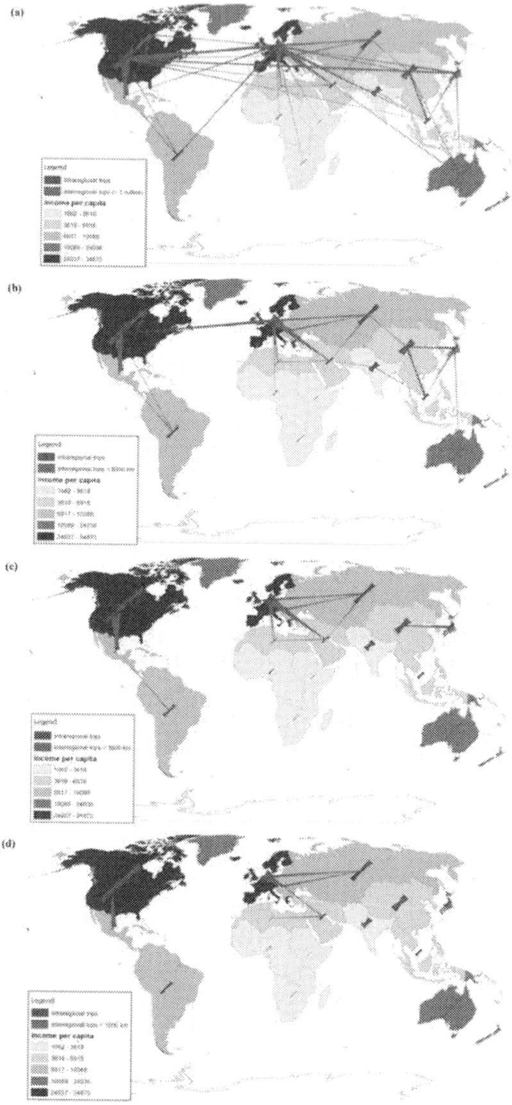

Figure 13. Impacts of distance cut-off on global tourism flows. Income per capita for 2003 in 2000 USD. (**a**) Unrestricted tourism flows, inter- and intraregional. Trips: 4.625×10^9, Distance: 11.360×10^{12} pkm; (**b**) Cut-off distance 6000 km, inter-and intraregional. Trips: 4.625×10^9, Distance: 9.482×10^{12} pkm, Distance reduction: 16.5%; (**c**) Cut-off distance 3000 km, inter- and intraregional. Trips: 4.625×10^9, Distance: 7.286×10^{12} pkm, Distance reduction: 35.9%; (**d**) Cut-off distance 1500 km, inter-and intraregional. Trips: 4.625×10^9, Distance: 5.370×10^{12} pkm, Distance reduction: 52.7%.

crisis in 2009 caused a reduction of the holidaymakers' carbon footprint by 3.8%, while in 2010, in the middle of an on-going crisis, the emissions increased again by 2.6%. Therefore, a model has been created based on the GTTD that first calculates the number of departures from each country that would cover a larger distance than the cut-off distance and then redistributes these trips proportionally over all of the remaining destinations for each country. Thus, the total number of trips remains constant while all of the trips above the cut-off distance are redistributed.

Figure 13 shows the results for cut-off distances of 6,000, 3,000 and 1,500 km (one-way) distance, in terms of distance reductions of the travel pattern on global maps. Distance reductions vary between 16.5% for a 6,000 km cut-off distance to and 52.7% for 1,500 km one-way cut-off distance. These cut-offs produce is substantial reductions of distances. However, the reduction of emissions may even be stronger as average emissions are some 20% lower for return distances below 2,500 km (see Figure 14). More importantly, the opportunities for emission reductions also improve greatly for the same short haul trips as the lowest attainable emission factors drop by 50–60% compared to return distances above 5,000 km. To achieve such larger reductions, strong transport mode shifts will be needed.

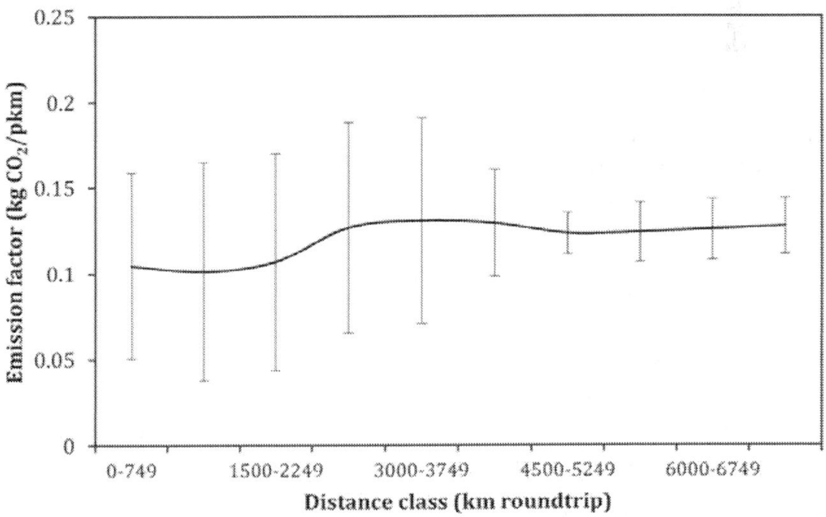

Figure 14. Average emission factors and standard deviation based scatter (source: [119]).

5. CONCLUSION AND DISCUSSION

This paper answered three questions: (1) is current tourism developing sustainably; (2) are current geographies/studies well equipped to assist policy makers in creating sustainable tourism development; and (3) how may the geography of tourism have to change to develop tourism sustainably. In short, the answers are: (1) tourism is not developing sustainably with respect to climate change; (2) the way in which current tourism geography is applied gives incomplete and sometimes misleading information to policy makers; and (3) sustainable development of tourism is impossible without substantial reductions of global travel distances in combination with a strong modal shift and additional improvements in energy efficiency in both transport and accommodations. These conclusions were reached by further developing older ideas from tourism geography and integrating them with some elements of transport geography; 'New' approaches include a focus on travel distance and travel time, the inclusion of all tourism (international and domestic) and a reconsideration of the nation as a basis for tourism flow data.

The first conclusion is founded in the literature [1,2,3,8,9]. Tourism related emissions are expected to continue to grow through the end of this century, and emission reductions of more than 80% are required to avoid 'dangerous' climate change. The main drivers for the environmentally unsustainable development of tourism are the increasing number of tourism trips and, on top of that, an increase of average travel speed caused by an increasing use of air transport, and resulting in an increase of average distances travelled. By 2050, tourism may cause all of the CO_2 emissions globally allowable within a scenario that assumes preventing 'dangerous climate change'. This projection includes assumptions about improved technology but excludes a significant role of biofuels because the prospects for biofuels are generally estimated to be higher than can be realized in the short to medium term. The role of transport speed has been shown in the boost of distances travelled, that grow faster than the number of tourist-nights. A sustainable tourism sector should somehow embrace the idea of decoupling its growth from the growth of transport. The sector must somehow get away from the highest speed transport modes or put more effort into selling other properties of travel than the shortest travel time.

To answer the second question it has been shown that a spatially-based theoretical approach to tourism geography had been founded already by the 1970s, but that a lack of operationalization at the global level and the use of flawed global tourism data [33], though being a useful starting point, led to knowledge gaps and misleading information. First, all domestic tourism, approximately 80% of all global trips, is not included. Second, the data takes the nation as the basis for measurement. This basis is unfortunate as nations vary by up to eight orders of magnitude in size (geographic, population, economic), which means that data points are not consistent (e.g., compare domestic travel within China with that within Monaco, both of which are an equal record in a nation based global database). Finally, almost all tourism databases fail to provide origin-destination data and thus exclude transport volumes. We developed the GTTD using the UNWTO data for international arrivals, additional EU and Dutch data, and some known and novel general relationships or 'rules of thumb'. A linear relationship between tourist trips and GDP per capita delivered the number of domestic trips, while domestic distances were extracted from the relation between the size of a country and the average distance of domestic tourism found for EU tourism. The possible existence of a power law describing the trip frequency as function of travel distance, related to the phenomenon of distance decay, allowed for redistribution of the largest domestic markets (e.g., China, the USA) over several distance classes. Although distance decay has been demonstrated by some authors for international tourism, detailed data for Dutch holidaymakers revealed that travel time (total in-transport time) follows the power law much better than travel distance. To find this result, the relation between travel distance and average travel speed for a range of transport modes has been found from another Dutch database that describes daily mobility. Figure 3 is the first published figure to use such data, although the idea for such relations stems at least from the 1970s [15]. Finally the problem of the nation as unit of measure, has been tackled by introducing the 18 SRES world regions, which are more even and differ by less than one order of magnitude in size. Ideally, a global database should be based on standardized areas (e.g., 100 × 100 km^2) containing data at the origin destination level for the number of trips, transport modes, distances, and spending habits. However, such a database may reach prohibitive sizes with current computer technology [120]. Interestingly, such tourism models might be coupled to down-scaled climate models, though some climatologists doubt the accuracy of these detailed models [120].

The GTTD is based on a large set of estimates and assumptions that create uncertainty about its validity. Therefore, we recommend the start of a new line of research on the intersection of tourism and transport geography to substantiate or reject the hypotheses about travel time and distance power laws, trip generation, and accessibility. However, the conclusion of this study, that sustainable tourism will not be achieved without somehow decoupling tourism growth from tourism transport volume growth has wide support [8,9,121,122,123,124,125]. A reduction of tourism trip one-way distances to below 1,500 km, affecting 20–25% of all trips, may reduce total travel distance by 57% and emissions by approximately 65%. Such reductions would also allow for a modal shift from air (and car) to rail and coach, enabling a total reduction of emissions of up to 90% with current technology and at current trip numbers. This two-pronged (combined transport volume reduction and a shift to more efficient transport modes) improvement of sustainability has thus far been ignored by most scholars and policy-makers but, as we have shown in the methodological sections of this paper, probably on flawed grounds.

A final methodological finding came from the analysis of the detailed Dutch holidaymaker data. It appeared that a 'Levy Flight' might be at work for travel *time* rather than travel *distance*. If that result is the case, this means that indeed travel speed will be the main or even the only parameter determining total distances travelled. Thus, effective mitigation policies should take reducing speed as a starting point, something also proposed by proponents of 'slow travel', although partly for different reasons [123,126,127]. It is recommended that further development of the approach to tourism geography as described above is necessary with the aims to better understand the mechanisms causing environmentally unsustainable development of tourism and to make the case for using reduced transport speeds as a policy to improve sustainable development.

In many discussions about reducing tourism transport because of its environmental impacts, there is reference to the detrimental effects of such measures on developing countries because the economy of such countries (e.g., small island nations) depends significantly on tourism. These ideas are based on a general notion that tourists always need to travel long distances to many of these developing countries and that reductions in long haul travel will always result in a reduction of the local tourism industry. However, these notions are based on an erroneous view of global tourism, which ignores most tourism trips (domestic) and assumes that a trip not taken for one destination will be lost for the whole tourism sector. Exclusion of domestic tourism causes an exaggeration of the impacts on tourism flows as including domestic tourism would dilute the impact on trips by a factor of up to 5. The idea that trips made impossible will be lost for the whole tourism sector is typical for destination-based thinking. Indeed it is correct for a specific destination. However, the incentive to travel does not depend on one specific destination, but on availability of time and income of the tourist. This means that such trips will redistribute to other destination, which thus will benefit from the measure.

Finally, our work has importance for the other, non-environmental, aspects of sustainability, such as the economic and social elements. Although we concentrated on environmental sustainability, it seems obvious that discussions about economic distribution effects and cross-cultural impacts would also likely benefit from a clear understanding of global tourism flows and the drivers causing them. Our recommendation is to do more research in establishing the economic and socio-cultural flows caused by tourism and then proceed to assess the positive and negative impacts.

ACKNOWLEDGMENTS

First, we thank Jeroen Nawijn and Nina Delbressine for their advice and work on some statistics used for this paper. Also, we are indebted to Michael Hall and Stefan Gössling for asking us to write this paper and for their very constructive comments on a draft version of this paper and assistance with writing proper English. Finally we thank the two anonymous reviewers for their constructive comments.

REFERENCES AND NOTES

1. Scott, D.; Peeters, P.; Gössling, S. Can tourism deliver its 'aspirational' greenhouse gas emission reduction targets? *J. Sustain. Tour.* **2010**, *18*, 393–408.
2. UNWTO-UNEP-WMO, *Climate Change and Tourism: Responding to Global Challenges*; UNWTO: Madrid, Spain, 2008.
3. Bows, A.; Anderson, B.; Peeters, P.M. Air transport, climate change and tourism. *Tour. Hosp. Plan. Dev.* **2009**, *6*, 7–20.
4. World Travel&Tourism Council (WTTC), *Climate Change. A Joint Approach to Addressing the Challenge*; WTTC: London, UK, 2010.
5. Chiesa, T.; Gautam, A. *Towards a Low Carbon Travel & Tourism Sector*; World Economic Forum: Geneva, Switzerland, 2009.
6. Cabrini, L.; Simpson, M.; Scott, D. *From Davos to Copenhagen and Beyond: Advancing Tourism's Response to Climate Change; Unwto Background Paper*; UNWTO: Madrid, Spain, 2009.
7. Gössling, S.; Hall, C.M.; Peeters, P.; Scott, D. The future of tourism: Can tourism growth and climate policy be reconciled? A climate change mitigation perspective. *Tour. Recreat. Res.* **2010**, *35*, 119–130.
8. Peeters, P.M.; Dubois, G. Tourism travel under climate change mitigation constraints. *J. Transp. Geogr.* **2010**, *18*, 447–457.
9. Dubois, G.; Ceron, J.P.; Peeters, P.; Gössling, S. The future tourism mobility of the world population: Emission growth*versus* climate policy. *Transp. Res. A* **2011**, *45*, 1031–1042.
10. Geography World What is Geography? Definition of Geography on the Web. *Geography World*, 2011. Available online:http://geographyworldonline.com/definition.html (accessed on 20 June 2011).

11. *Geography*; Answers.com: Jerusalem, Israel, 2011. Available online: http://www.answers.com/geography (accessed on 20 June 2011).
12. Williams, S. *Tourism Geography: A New Synthesis*, 2nd ed; Taylor & Francis: London, UK, 2009.
13. Mitchell, L.S.; Murphy, P.E. Geography and tourism. *Ann. Tour. Res.* **1991**, *18*, 57–70.
14. Pearce, D. *Tourism Today. A Geographical Analysis*, 2nd ed; Longman Scientific & Technical: Essex, UK, 1995; p. 202.
15. Miossec, J.-M. Elements pour une theorie de l'espace touristique. *Les Cahiers du Tourisme* **1976**, *Serie C*, 1–63.
16. Williams, A.V.; Zelinsky, W. On some patterns in international tourist flows. *Econ. Geogr.* **1970**, *46*, 549–567.
17. Butler, R.W. The concept of a tourist area cycle of evolution: Implications for management of resources. *Can. Geogr.***1980**, *24*, 5–12.
18. Hall, C.M. Space-time accessibility and talc: The role of geographies of spatial interaction and mobility in contributing to an improved understanding of tourism. In *The Tourism Area Life Cycle, Conceptual and Theoretical Issues*; Butler, R.W., Ed.; Channel View Press: Clevedon, UK, 2006; Volume 2, pp. 83–100.
19. Hall, C.; Page, S. *The geography of Tourism and Recreation: Environment, Place and Space*, 2nd ed; Routeledge: London, UK, 2002; p. 399.
20. Gössling, S.; Hall, C.M. Conclusion. Wake up ... This is serious. In *Tourism & Global Environmental Change. Ecological, Social, Economic and Political Interrelationships*; Gössling, S., Hall, C.M., Eds.; Routledge: London, UK, 2006; Volume 4, pp. 305–320.
21. Cooper, C. *Tourism: Principles and Practice*; Pearson Education Limited: Harlow, UK, 2008.
22. Budd, L.C.S. On being aeromobile: Airline passengers and the affective experiences of flight. *J. Transp. Geogr.* **2011**, *19*, 1010–1016.
23. Hall, C.M. *Tourism: Rethinking the Social Science of Mobility*; Pearson Education Limited: London, UK, 2005.
24. Mayor, K.; Tol, R.S.J. The impact of european climate change regulations on international tourist markets. *Transp. Res. Part D* **2010**, *15*, 26–36.

25. Veryard, D. *Study of International Visitor Flows and Greenhouse Gas Emissions for a Template to Examine the Impact on Apec Economies of Future Market-Based Measures Applying to International Transport*; GHD Meyrick: Canberra, Australia, 2009.
26. Mayor, K.; Tol, R.S.J. The impact of the UK aviation tax on carbon dioxide emissions and visitor numbers. *Transp. Policy* **2007**, *14*, 507–513.
27. Gössling, S.; Peeters, P.; Scott, D. Consequences of climate policy for international tourist arrivals in developing countries. *Third World Q.* **2008**, *29*, 873–901.
28. Becken, S.; Lennox, J. Implications of a long-term increase in oil prices for tourism. *Tour. Manag.* **2012**, *33*, 133–142.
29. Anger, A. Including aviation in the european emissions trading scheme: Impacts on the industry, CO_2 emissions and macroeconomic activity in the EU. *J. Air Transp. Manag* **2010**, *16*, 100–105.
30. UNWTO, *2007 Edition Tourism Market Trends. World Overview*; World Tourism Organisation: Madrid, Spain, 2008; 2006 Ed.
31. UNWTO, *UNWTO World Tourism Barometer*; UNWTO: Madrid, Spain, 2009; Volume 7, nr. 2.
32. *China Statistical Yearbook-2008*; National Bureau of Statistics of China: Beijing, China, 2008. Available online: http://www.stats.gov.cn/tjsj/ndsj/2008/indexee.htm (access ed on 12 Septermber 2009).
33. *UNWTO World Statistics Day*; UNWTO World Tourism Organization: Madrid, Spain. Available online: http://statistics.unwto.org/en/content/world-statistics-day (accessed on 22 October 2011).
34. UNWTO. *Compendium of tourism statistics (2005–2009)*; UNWTO: Madrid, Spain, 2011. Available online: http://www.wtoelibrary.org/content/v486k6/?v=search (acc essed on 28 July 2011).
35. Xu, J.; Yao, L.; Mo, L. Simulation of low-carbon tourism in world natural and cultural heritage areas: An application to Shizhong district of Leshan city in China. *Energy Policy* **2011**, *39*, 4298–4307.
36. Dwyer, L.; Forsyth, P.; Dwyer, W. *Tourism Economics and Policy*; Channel View Publications: Bristol, UK, 2010.
37. Rodrigue, J.P.; Comtois, C.; Slack, B. *The Geography of Transport Systems*; Routledge: London, UK, 2006.

38. Bigano, A.; Hamilton, J.M.; Lau, M.; Tol, R.S.J.; Zhou, Y. *A Global Database of Domestic and International Tourist Numbers at National and Subnational Level; Working Paper FNU-54*; Research Unit Sustainability and Global Change, Hamburg University and Centre for Marine and Atmospheric Science: Hamburg, Germany, 2004.
39. UNWTO, *Compendium of Tourism Statistics (2007 Edition)*; World Tourism Organisation: Madrid, Spain, 2007.
40. UNWTO, *UNWTO World Tourism Barometer (English Version)*; UNWTO: Madrid, Spain, 2011; Volume 9, pp. 1–57.
41. IMAGE-team, *The IMAGE 2.2 Implementation of the SRES Scenarios. A Comprehensive Analysis of Emissions, Climate Change and Impacts in the 21st Century; CD-ROM 500110001 (former 481508018)*; National Institute for Public Health and the Environment: Bilthoven, The Netherlands, 2006.
42. Swartz, K.L. *Great Circle Mapper*, 2010. Available online: http://gc.kls2.com/ (accessed on 22 January 2010).
43. Central Intelligence Agency (CIA), *The World Factbook 2005*; CIA: Washington, DC, USA, 2005; ISSN 1553-8133.
44. Central Intelligence Agency (CIA). *The World Factbook 2008*; CIA: Washington, DC, USA, 2008; ISSN 1553-8133.
45. Zahavi, Y.; Talvitie, A. Regularities in travel time and money expenditures. *Transp. Res. Record* **1980**, *750*, 13–19.
46. Zahavi, Y.; Lang, P.J. *Traveltime Budgets and Mobility in Urban Areas*; US Department of Transportation: Washington, DC, USA, 1974.
47. Zahavi, Y. *The Unified Mechanism of Travel (UMOT) Model*, 1976. Available online: http://www.surveyarchive.org/Zahavi/TheUMOTModel.pdf (accessed on 9 November 2011).
48. Hägerstrand, T. What about people in regional science? *Papers Reg. Sci.* **1970**, *24*, 6–21.
49. Levinson, D.; Kumar, A. Activity, travel, and the allocation of time. *J. Am. Plan. Assoc.* **1995**, *61*, 458–470.
50. Hupkes, G. The law of constant travel time and trip-rates. *Futures* **1982**, *14*, 38–46.
51. Arentze, T.A.; Timmermans, H.J.P. A dynamic model of time-budget and activity generation: Development and empirical derivation. *Transp. Res. Part C* **2011**, *19*, 242–253.

52. Schäfer, A. Regularities in travel demand: An international perspective. *J. Transp. Stat.* **2000**, *3*, 1–31.
53. Schäfer, A. The global demand for motorized mobility. *Transp. Res. A* **1998**, *32*, 445–477.
54. Schäfer, A.; Victor, D.G. The future mobility of the world population. *Transp. Res. A* **2000**, *34*, 171–205.
55. Shu-wei, Z.; Ke-jun, J.; De-shun, L. Macro-level transport modal split model: Theory and application in China. In Proceedings of International Conference on Management Science and EngineeringICMSE'06, Lille, France, 5–7 October 2006; pp. 2033–2038.
56. Metz, D. *The Limits to Travel. How Far will We Go?* Earthscan: London, UK, 2008; p. 164.
57. Banister, D. The trilogy of distance, speed and time. *J. Transp. Geogr.* **2011**, *19*, 950–959.
58. Mokhtarian, P.L.; Chen, C. TTB or not TTB, that is the question: A review and analysis of the empirical literature on travel time (and money) budgets. *Transp. Res. Part A* **2004**, *38*, 643–675.
59. Kölbl, R.; Helbing, D. Energy and scaling laws in human travel behaviour. *New J. Phys.* **2003**, *5*, 48:1–48:12.
60. Lin, C.-H.; Morais, D.B. The spatial clustering effect of destination distribution on cognitive distance estimates and its impact on tourists' destination choices. *Travel Tour. Mark.* **2009**, *25*, 382–397.
61. Projectteam MON, *Mobiliteitsonderzoek Nederland 2005. Het Onderzoek*; Ministerie van Verkeer en Waterstaat, Rijkswaterstaat Adviesdienst Verkeer en Vervoer: Den Haag, The Netherlands, 2006.
62. Mitchell, L.S. Tourism research in the united states: A geographic perspective. *GeoJournal* **1984**, *9*, 5–15.
63. Fotheringham, A.S. Spatial structure and distance-decay parameters. *Ann. Assoc. Am. Geogr.* **1981**, *71*, 425–436.
64. Wilson, A.G. The use of entropy maximising models, in the theory of trip distribution, mode split and route split. *J. Transp. Econ. Policy* **1969**, *111*, 108–126.
65. Ravenstein, E.G. The laws of migration. Second paper. *J. R. Stat. Soc.* **1889**, *52*, 241–305.
66. Ravenstein, E.G. The laws of migration. *J. Stat. Soc. Lond.* **1885**, *48*, 167–235.
67. Eldridge, J.D.; Jones, J.P. Warped space: A geography of distance decay. *Prof. Geogr.* **1991**, *43*, 500–511.

68. Tobler, W.R. A computer movie simulating urban growth in the Detroit region. *Econ. Geogr.* **1970**, *46*, 234–240.
69. Tobler, W. Thirty five years of computer cartograms. *Ann. Assoc. Am. Geogr.* **2004**, *94*, 58–73.
70. Mckercher, B.; Lew, A. Distance decay and the impact of effective tourism exclusion zones on international travel flows. *J. Travel Res.* **2003**, *42*, 159–165.
71. McKercher, B.; Chan, A.; Lam, C. The impact of distance on international tourist movements. *J. Travel Res.* **2008**, *47*, 208–224.
72. Smith, S. US vacation travel patterns: Correlates of distance decay and the willingness to travel. *Leis. Sci.* **1985**, *7*, 151–174.
73. Christaller, W. Some considerations of tourism location in Europe: The peripheral regions-underdeveloped countries-recreation areas. *Papers Reg. Sci.* **1964**, *12*, 95–105.
74. Hall, C. Time, space, tourism and social physic. *Tour. Recreat. Res.* **2005**, *30*, 93–98.
75. Ball, P. *Critical Mass*; Farrar, Strauss and Giroux: New York, NY, USA, 2004.
76. Banerjee, A.; Yakovenko, V.M. Universal patterns of inequality. *New J. Phys.* **2010**, *12*, 075032.
77. Clauset, A.; Shalizi, C.R.; Newman, M.E.J. Power-law distributions in empirical data. *SIAM Rev.* **2009**, *51*, 661–703.
78. James, A.; Plank, M.J. On fitting power laws to ecological data. *Arxiv Prepr.* **2007**. arXiv:0712.0613v1..
79. Newman, M. Power laws, Pareto distributions and Zipf's law. *Contemp. Phys.* **2005**, *46*, 323–351.
80. Sornette, D. *Critical Phenomena in Natural Sciences: Chaos, Fractals, Selforganization, and Disorder: Concepts and Tools*, 2nd ed; Springer: Berlin, Germany, 2006.
81. Stanley, H.E.; Amaral, L.A.N.; Buldyrev, S.V.; Gopikrishnan, P.; Plerou, V.; Salinger, M.A. Self-organized complexity in economics and finance. *Prod. Natl. Acad. Sci. USA* **2002**, *99*, 2561–2565.
82. Coelho, R.; Richmond, P.; Barry, J.; Hutzler, S. Double power laws in income and wealth distributions. *Phys. A* **2008**,*387*, 3847–3851.
83. Mandelbrot, B.B. *Fractals and Scaling in Finance: Discontinuity, Concentration, Risk*; Springer Verlag: New York, NY, USA, 1997.
84. Mandelbrot, B. The variation of certain speculative prices. *J. Bus.* **1963**, *36*, 394–419.

85. Brockmann, D.; Hufnagel, L.; Geisel, T. The scaling laws of human travel. *Nature* **2006**, *439*, 462–465.

86. Gonzalez, M.C.; Hidalgo, C.A.; Barabasi, A.-L. Understanding individual human mobility patterns. *Nature* **2008**, *453*, 779–782.

87. Humphries, N.E.; Queiroz, N.; Dyer, J.R.M.; Pade, N.G.; Musyl, M.K.; Schaefer, K.M.; Fuller, D.W.; Brunnschweiler, J.M.; Doyle, T.K.; Houghton, J.D.R.; *et al.* Environmental context explains Lévy and Brownian movement patterns of marine predators. *Nature* **2010**, *465*, 1066–1069.

88. Viswanathan, G.M. Ecology: Fish in Lévy-flight foraging. *Nature* **2010**, *465*, 1018–1019.

89. James, A.; Plank, M.J.; Edwards, A.M. Assessing Lévy walks as models of animal foraging. *J. R. Soc. Interface* **2011**. in press..

90. Peeters, P.M.; van Egmond, T.; Visser, N. *European Tourism, Transport and Environment. Final version*; NHTV CSTT: Breda, The Netherlands, 2004.

91. NBTC-NIPO. *Continu Vakantie Onderzoek (continuous holiday survey)*; NBTC-NIPO: Amsterdam, The Netherlands, 2011. Available online: http://www.nbtcniporesearch.nl/nl/Home/Producten-en-diensten/cvo.htm (accessed on 5 July 2011).

92. De Bruijn, K.; Dirven, R.; Eijgelaar, E.; Peeters, P. *Travelling Large in 2009. The Carbon Footprint of Dutch Holidaymakers in 2009 and the Development Since 2002*; NHTV University for Applied Sciences in collaboration with NBTC-NIPO Research: Breda, The Netherlands, 2010.

93. Nicolau, J.L.; Más, F.J. The influence of distance and prices on the choice of tourist destinations: The moderating role of motivations. *Tour. Manag.* **2006**, *27*, 982–996.

94. Daan, S.; Deerenberg, C.; Dijkstra, C. Increased daily work precipitates natural death in the Kestrel. *J. Anim. Ecol.* **1996**, 539–544.

95. Masman, D.; Daan, S.; Dijkstra, C. Time allocation in the Kestrel (falco tinnunculus), and the principle of energy minimization. *J. Anim. Ecol.* **1988**, 411–432.

96. Masman, D.; Dijkstra, C.; Daan, S.; Bult, A.; Reyer, U. Energetic limitation of avian parental effort: Field experiments in the Kestrel (falco tinnunculus). *J. Evol. Biol.* **1989**, *2*, 435–455.

97. Peeters, P.M.; Middel, J. Historical and Future Development of Air Transport Fuel Efficiency. In Proceedings of an International

Conference on Transport, Atmosphere and Climate (tac), Oxford, UK, 26–29 June 2006; Sausen, R., Blum, A., Lee, D.S., Brüning, C., Eds.; DLR Institut für Physic der Atmosphäre: Oberpfaffenhoven, Germany, 2007; pp. 42–47.

98. *CBS Statline*; CBS: The Hague, The Netherlands, 2011. Available online: http://statline.cbs.nl/StatWeb/start.asp?LA=nl&DM=SLNL&lp=Search/Search (accessed on 1 Septermber 2011).

99. Mayor, K.; Tol, R.S.J. Scenarios of carbon dioxide emissions from aviation. *Glob. Environ. Change* **2010**, *20*, 65–73.

100. Lee, D.; Pitari, G.; Grewe, V.; Gierens, K.; Penner, J.; Petzold, A.; Prather, M.; Schumann, U.; Bais, A.; Berntsen, T. Transport impacts on atmosphere and climate: Aviation. *Atmos. Environ.* **2010**, *44*, 4678–4734.

101. kerman, J. Sustainable air transport—On track in 2050. *Transp. Res. D* **2005**, *10*, 111–126.

102. Peeters, P.; Williams, V.; de Haan, A. Technical and management reduction potentials. In *Climate Change and Aviation: Issues, Challenges and Solutions*; Gössling, S., Upham, P., Eds.; Earthscan: London, UK, 2009; pp. 293–307.

103. Air Transport Action Group (ATAG), *Powering the Future of Flight. The Six Easy Steps to Growing a Viable Aviation Biofuels Industry; web version*; ATAG: Geneva, Switzerland, 2011.

104. The Air Transport Association (IATA), *The IATA Technology Roadmap Report*; IATA: Montreal, Canada, 2009.

105. *Aviation Global Deal Group: A Sectoral Approach to Addressing International Aviation Emissions; Discussion note 2.0*; AGD: London, UK, 9 June 2009. Available online: http://www.agdgroup.org/pdfs/090609_AGD_Discussion_Note_2.0.pdf(accessed on 22 December 2009).

106. ICAO. Comparison of life cycle GHG emissions from select alternative jet fuels, caaf/09-ip/6. In Conference on Aviation and Alternative Fuels; ICAO: Rio de Janeiro, Brazil, 2009.

107. Bailis, R.E.; Baka, J.E. Greenhouse gas emissions and land use change from Jatropha curcas-based jet fuel in Brazil.*Environ. Sci. Technol.* **2010**, *44*, 8684–8691.

108. Ariza-Montobbio, P.; Lele, S. Jatropha plantations for biodiesel in Tamil Nadu, India: Viability, livelihood trade-offs, and latent conflict. *Ecol. Econ.* **2010**, *70*, 189–195.

109. Findlater, K.M.; Kandlikar, M. Land use and second-generation biofuel feedstocks: The unconsidered impacts of Jatropha biodiesel in Rajasthan, India. *Energy Policy* **2011**, *39*, 3404–3413.
110. Gerbens-Leenes, W.; Hoekstra, A.Y.; van der Meer, T.H. The water footprint of bioenergy. *Proc. Natl. Acade. Sci. USA***2009**, *106*, 10219–10223.
111. Kant, P.; Wu, S. The extraordinary collapse of Jatropha as a global biofuel. *Environ. Sci. Technol.* **2011**, *45*, 7114–7115.
112. Sanderson, K. Wonder weed plans fail to flourish. *Nature* **2009**, *461*, 328–329.
113. Achten, W.; Verchot, L.; Franken, Y.; Mathijs, E.; Singh, V.; Aerts, R.; Muys, B. Jatropha bio-diesel production and use.*Biomass Bioenergy* **2008**, *32*, 1063–1084.
114. Edgerton, D. *The Shock of the Old*; Profile Books Ltd.: London, UK, 2007.
115. Edgerton, D. The charge of technology. *Nature* **2008**, *455*, 1030–1031.
116. Hall, C.M. Degrowing tourism: Décroissance, sustainable consumption and steady-state tourism. *Anatolia* **2009**, *20*, 46–61.
117. Hall, C.M. Changing paradigms and global change: From sustainable to steady-state tourism. *Tour. Recreat. Res.* **2010**,*35*, 131–143.
118. OECD, *Oecd Tourism Trends and Policies 2010*; OECD: Paris, France, 2010; ISBN 978-92-64-07742-3.
119. De Bruijn, K.; Dirven, R.; Eijgelaar, E.; Peeters, P. *Travelling Large in 2010. The Carbon Footprint of Dutch Tourism in 2010 and Developments Since 2002*; NHTV University for Applied Sciences in collaboration with NBTC-NIPO Research: Breda, The Netherlands, 2011; in press.
120. Kerr, R.A. Vital details of global warming are eluding forecasters. *Science* **2011**, *334*, 173–174.
121. Kallis, G. In defence of degrowth. *Ecol. Econ.* **2011**, *70*, 873–880.
122. Peeters, P. Transport technology, tourism and climate change. In *Tourism and the Implications of Climate Change: Issues and Actions*; Schott, C., Ed.; Emerald: Bingley, UK, 2010; Volume 3, pp. 67–90.
123. Dickinson, J.E. Adapting tourism for a lower carbon future: A slow travel approach. In Transport and Tourism. Challenges, Issues and Conflicts. Proceedings of the Travel and Tourism Research Association Europe, Annual Conference, Rotterdam; Breda, The

Netherlands, 22–24 April 2009; Landré, M., Ed.; TTRA: Breda, The Netherlands, 2009; pp. 32–45.

124. Peeters, P.; Szimba, E.; Duijnisveld, M. Major environmental impacts of european tourist transport. *J. Transp. Geogr.* **2007**, *15*, 83–93.

125. Hall, C.M. Policy learning and policy failure in sustainable tourism governance: From first- and second-order to third-order change? *J. Sustain. Tour.* **2011**, *19*, 649–671.

126. Lumsdon, L.M.; McGrath, P. Developing a conceptual framework for slow travel: A grounded theory approach. *J. Sustain. Tour.* **2010**, *19*, 265–279.

127. Dickinson, J.; Lumsdon, L. *Slow Travel and Tourism*; Earthscan: London, UK, 2010.

Chapter 11

A Sustainable Tourism Paradigm: Opportunities and Limits for Forest Landscape Planning

Dina Rizio [1,2] and Geremia Gios [2]

[1]*School of Social Sciences, University of Trento, 26, via Verdi, Trento 38122, Italy*

[2]*Department of Economics and Management, University of Trento, 5, via Inama, Trento 38122, Italy*

ABSTRACT

The promotion of sustainable tourism models has been widely debated; many pages have been devoted to the attempt to provide the subject with a strong theoretical base and coherent structure. This said, it is still the case that, although such frameworks are crucial for the development of appropriate planning and policy instruments, their actual implementation continue to be fraught with difficulties. These problems are exacerbated when sustainable tourism entails development opportunities which require the support of the local community and the management of natural resources which are typically common goods. Under these circumstances, new management structures, which can both satisfy the needs of the local community and ensure the appropriate stewardship of the natural resources, must be created. Management solutions are not always easy to define and often need to be considered within a general framework, based on which individual cases are then formulated. This study analyses the connections between models of sustainable tourism and natural resource management considering the forest landscape case. This relationship is first examined from a theoretical perspective and then within a case study, in order to highlight the dual approach—both general and within a specific context.

Keywords: sustainable tourism; natural resource management; forest landscape; local communities

1. INTRODUCTION

The tourism sector is experiencing strong growth despite the recent economic and financial crises [1]. Developing countries perceive tourism as an economic opportunity and the sector is also performing very well in industrialized countries [2,3]. In recent decades a significant body of academic research has focused on the need for a sustainable paradigm of tourism activities. However, starting from the Brundtland report definition of sustainable development [4], now so well known, further definitions and tools, the use of which is not fully consistent, particularly in the tourism sector, have been proposed.

This lack of consistency can be attributed not only to the heterogeneity and pervasiveness of tourism activities, but also to the evolution of the concept of sustainability. "Our common future" definition highlights the need for development, in order to be sustainable, to satisfy human needs without compromising future generations' possibility to satisfy those same needs [5]. Further definitions of sustainable development emphasize the role play by the environmental component such as that contained in the Rio Declaration in 1992 [6] and the transfer of this concept to tourism acknowledges, as that contained in the Lanzarote Charter, 18 principles which strongly underline the need for a sustainable approach to tourism [7]. Thus, the first definition of sustainable development has been expanded and modified over the years in an attempt to define the concept more concretely, thereby enabling its extension into a range of development areas. Moreover, the initial anthropocentric understanding of the term has been gradually shifting towards a more ecocentric approach which grants natural resources a value not just in relation to their utility to humans but also to their indispensable role in the production of ecosystem services vital for the whole terrestrial system. From an anthropocentric point of view, natural resources are inputs into the economy and their availability is a determining factor within the whole economic system. The nonrenewable and non-substitutability nature of these natural resources, however, clearly are constraints on their exploitation and necessitates that their use be scheduled as efficiently as possible [8]. Underlying this kind of planning is an acceptance of the concept of an intergenerational pact regarding the use of natural resources which is clearly based on human needs. From an ecocentric perspective, natural resources provide a vast range of ecological services which don't just function as direct inputs into the economic system but are also involved in the regulation, support, and coordination of the Earth's ecosystem [9].

Climate change, an awareness of the ecological limits dictated by finite natural resources, the ever increasing consumption of nonrenewable energy sources, and the exploitation of natural resources considerably in excess of their carrying capacity and ability to regenerate, are all powerful drivers for such a change.

Indeed, the strong connections between tourism and global dynamics, the sector's wide ranging economic effects and the potential danger it poses to the environment, have all contributed to a push towards a sustainable paradigm in tourism development. However, both the intrinsic nature of the activities associated with tourism and the factors necessary for sustainability have made it very difficult to transfer the sustainable development paradigm to the tourist industry, and this process is by no means complete [10].

More recent versions of the concept of sustainable tourism development underline the extent to which, although it has been widely debated and, indeed, promoted in numerous studies, the concept is still far from having been fully articulated [11,12]. In practice, the application of the conceptual framework of sustainable tourist development is not straightforward, particularly where natural resources, like forests, are concerned.

Within this context, the objectives of this study are: to explore a forest ecosystem and identify its potential flows of utility, addressing those which best satisfy tourism activities and recreational purposes; to identify the most appropriate tools to manage the flows of utility based on sustainable principles which integrate tourism activities. This analysis is carried out using a case study—Paneveggio forest—and is then related to a more general framework.

The paper is organized as follows: Section 2 contains a review of the goods and services provided by Alpine forests and focuses on the tourist/recreational use of these resources. Section 3 contains an analysis of the sustainable management of a forest landscape. Section 4 analyses a case study of forest management. Section 5 is the conclusion.

2. THE GOODS AND SERVICES OF ALPINE FORESTS

A forest is an extremely complex ecosystem with a multiplicity of functions and uses, and of stakeholders and consumers—both actual and potential [13]. Using the widely accepted definition that the services provided by ecosystems can be defined as "the benefits which people obtain from ecosystems" [14] allows us to understand not only how a forest ecosystem contains all kinds of services but also how the utility flows it generates have both local and global effects. Similar results are obtained starting from an alternative definition that distinguishes between environmental services and goods: the former being the availability of services provided by a natural resource, and the latter the flow or stock of capital that it represents [15]. In the particular case of forests, both the environmental services they provide and their products are numerous and can be classified as timber based, non-timber based, or intangible [16]. Although timber has traditionally been considered the main good provided by forests, it is by no means the only forest product, and sometimes not even the most important benefit generated by a forest ecosystem [13].

The general classifications of ecosystems can also be applied to forests, distinguishing four different categories: carbon sequestration, the conservation of biodiversity, the protection of aquifers, and hydrogeological functions, the beauty of the natural landscape, enjoyed through forms of contingent use like tourism and ecotourism or the contemplation and awareness of spiritual values [17]. This last function entails a strong relationship between at least three elements: nature, culture, and local community. This triad highlights the role play by local culture and communities in natural resource management and use [18]. In a forest management context, this concept emphasis even more the multifunctional dimension of forests.

A growing awareness of the multifunctionality of forests has led to their widespread management from this perspective: recognizing both the above-mentioned functions and the numerous other services and goods provided by the forest, this approach emphasizes the need to plan forest management in two stages. The first is the identification of the significant functions performed by the forest for the ecosystem; the second is the identification of the type of forest management best suited to both performing these functions and obtaining these products and services. A forest system is no longer seen only as a provider of timber products, but also of diverse goods and services, satisfying many different social needs and providing new opportunities for use [19].

The demand for more varied forest goods and services which are capable to satisfy such multiple needs is, thus, both changing and increasing. This transformation arises compatibility and coexistence issues for both forest management per se and in respect to the satisfaction of local communities' claims. The fulfillment of forest functions can be understood at two levels: the global and the local. The former refers to the general functions obtained by particular forest management patterns; the latter entails other functions related to goods and service production which can both satisfy the needs of local communities and generate additional sources of income for them.

Forest management which can achieve both these ends is necessary for long term sustainability and operating and management methods which simultaneously guarantee long term sustainability and allow products and services to be translated into sources of income for local populations are therefore necessary [20].

This production represents potential utility streams able to satisfy a variety of consumer demands, although their appropriation and management are not always conflict-free. From this perspective, the landscape is one of the most controversial utility streams involved in forest management. The most critical issues concern not only the definition of landscape and the establishing of its value, but also determining access to, and the conservation and management of, the resources that can be understood as the several flows of utility generated by this ecosystem.

Forest Landscapes from the Perspective of Tourist Use

The important role of landscape is widely recognized and includes various meanings and definitions [21]. The definition of landscape as the result of continuous changes caused by the interaction between human activities and the environment implies two considerations. The first regards the analysis of the factors which determine such transformations and the other regards the valuation of the landscape which, given the constant changes to which it is subject, must be dynamic and appropriate to the particular historical context of the area [22]. From the point of view of tourism, landscape (including that of forests) is considered to be a crucial input for recreational activities [23]. In this context, landscape stands for all elements, both natural and built, thus including socio-cultural features and dynamics [24]. Considering the value of landscapes from the demand perspective, we can distinguish three elements: scenic value, recreational value and evocative value [25].

Briassoulis [23], in his analysis of the relationship between common natural resources and sustainable tourism development, devotes considerable space to landscape, and also refers to Jafari's definition of landscape as providing "background tourism elements" [24,26].

Indeed, the forest landscape can serve a variety of functions within the ambit of tourism. It can be instrumental to tourism activities, whether as a "resource container" or simply as a background. In the first case, the forest landscape is seen as a clearly defined resource whose use by the consumer is either partial or full. Adventure parks, and acrobatic parks in general, are examples of this type of function. Only a limited area of forest is used but the access to it is regulated in order to guarantee exclusive and specific use of the resource. In the second case, the forest landscape constitutes a general resource, with no single specified recreational use. The area involved is very large but clearly defined. It is not, however, possible to ensure exclusive use. In the third case, the forest is not seen as crucial to the carrying out of the particular activity. The area involved is very large, indeed from the perspective of the user the boundaries will not be very clear, and it is not possible to guarantee exclusive use [22].

This initial description of the value and function of forest landscapes as tourism activity inputs has revealed two important characteristics which determine consumers' opportunities to access and use forest landscapes: excludability and rivalry. The economic utility of all, including forest, landscapes can be usefully framed by means of these two characteristics.

Different levels of excludability and rivalry define public goods, private goods, and common goods (or club goods). The operating and management models are different for each of these categories, thus entailing different pathways to sustainability. However, ensuring the appropriate and sustainable management of a single forest function or good, e.g., timber, does not necessarily mean that the remaining functions will be sustainably managed. Moreover, since new forest functions may be recognized at any time, with a subsequent alteration of the relative importance of the utility flows guaranteed by forests, it is necessary to update operating and management methods constantly in order to ensure efficiency and sustainability. Given the need to plan management frameworks over the long term, it is very difficult in practice to select the most appropriate tools or mechanisms. And so it is not by chance that since the late 1980s the need to move away from old models of economic development and uses of environmental goods and services has become increasingly clear. World population growth, while feeding a growing demand for these goods, also risks compromising their production due to the inability of the current economic system to work within a framework of long term sustainability.

Environmental resources in general and forest resources in particular are becoming increasingly sought after and subject to competing claims for their use. Consequently, it is necessary to devise new models to manage and assign economic value to these resources in order to create effective tools for their appropriate maintenance and use.

A wide range of intervention tools and mechanisms for the management of natural resources exists [27]. Payment systems for environmental services (PESs) have recently been introduced: financial mechanisms to commercialize natural goods and services through the introduction of incentives which encourage local actors to produce the relevant goods and services.

Other mechanisms include integrated conservation and development projects and sustainable forest management practices aimed not only at encouraging management strategies which ensure the availability and quality of these goods over time, but also sensitize local populations to the presence of these environmental services and guarantee them a supplementary source of income.

3. SUSTAINABLE FOREST LANDSCAPE MANAGEMENT

Based on current evidence, for forest management to be sustainable in the long term it must both be appropriate and deliver a sufficient income for the local population. Appropriate forest management requires that the production of those goods and services that form the flows of utility for an ecosystem and its local population be guaranteed. Such management must be able to provide a supplementary income to local populations; it cannot only represent the costs inevitably incurred for the conservation and production of the goods and services provided by the forest. In the specific case of forest landscapes the link between local populations and the forest involves a process of continual transformation. This process is a result not only of physical interventions, but also of the socio-cultural background of the local populations, who, through their interpretation and appropriation of the traditions and customs handed down from generation to generation, have shaped the forest landscape. This connection between local populations and the forest landscape and environment is very important to Alpine communities.

The conservation and management of the forest landscape in a given community is often an expression of the connection between that community and its surroundings. Traditionally, the Alpine environment was seen as hostile to human settlement; in order to survive and carve out a life in the mountains, it was always felt, people were forced to adapt and make numerous compromises. Multifunctional forest management takes into account the multiple potential products and related sources of income.

In fact, as has already been mentioned, one of the most important functions of appropriate forest management in the Alpine region, alongside the obtaining of raw materials, is the conservation of the landscape and natural resources as a background for recreational activities and the preservation of the local social network [28].

Forest management in the Alps has to be multifunctional, has to take the forest's status as a public good into consideration and has to be able to provide local populations with sufficient products and services for them to feel that the forest represents a source of income. Given these objectives, the appropriate tools to promote such management involve a mix of self-government of public goods and the introduction of payment mechanisms.

4. CASE STUDY: THE FOREST OF PANEVEGGIO

The forest of Paneveggio is situated in the autonomous Province of Trento, in the north eastern Italian Alps. Covering over 4300 hectares, it is the largest state owned forest in the area and extends within the administration of four different municipalities; the area includes a mix of meadow, forest and unproductive zones (see Table 1).

Table 1. Subdivision of the forest by land use.

Land Use	Area in hectares	As a percentage of the total area
Wood	2787.14	64.16%
Meadow	1263.52	29.08%
Unproductive	293.63	6.76%
Total	4344.29	100%

Source: own elaboration from [29].

The forest as it is now constituted dates back to at least the 19th century; it was included in economic plans drawn up towards the end of that century. It is home to a rich and varied fauna and is famous for the production of so called "resonant wood", obtained from the red spruce, which is much prized by luthiers for the high quality of the sound produced by string instruments whose bellies are made from it. Alongside the recognized value of the resonant wood, this fame generates considerable visitor flows and income for the whole zone [30]. In fact, there are several flows of utility that this forest generates both at a global and local level.

Indeed, as well as its importance in carbon sequestration, biodiversity conservation, the protection of aquifers and its hydrogeological functions, this forest area is an extremely precious landscape resource, imbued with, and modified by, the constant presence of local communities which have, through their labors, defined its shapes and uses (see Table 2).

From an administrative point of view, Paneveggio is owned by the Autonomous Province of Trento, its public ownership is, in fact, inalienable; the Autonomous Province of Trento is responsible for its management and has appointed an agency, APROFOD, to carry out this task. As has been mentioned, the forest falls under the jurisdiction of four different municipal administrations, and also of the National Park of Paneveggio and Pale di San Martino. Moreover, precisely because it is a public good, the forest also has a large number of stakeholders, both local and not: associations, environmental groups, tourists, public bodies, and local communities... (See part a, Figure 1).

Table 2. Flows of utility generated by Paneveggio forest.

Goods	Non-rival	Congestible	Rival
Non-excludable	Entirely public goods: absorption of carbon dioxide; non-use values of landscape	Open access resources: landscape as resource-based input for recreation; cultural and spiritual values	
Excludable to non-community members	Local public good: biodiversity; protection of aquifers; hydrogeological function. Landscape as scenic component	Common property resources: hunting, fishing, gathering	
Excludable	Toll goods: landscape as an option value	Club goods: trails	Private goods: timber, landscape as resource user-oriented for recreation; handicraft

Source: own elaboration from [22].

The large number of organizations and stakeholders involved in the management of the Paneveggio forest reflects the multiple interests that the opportunities and potential uses of the resource represent (see part b, Figure 1).

Tourism development plays a primary role, both in the use of the resource and in according it appropriate value; it represents a pull factor and connotes the territory as suitable for recreation, thereby offering economic opportunities through the creation of valid alternative methods by which to generate supplementary incomes. Tourism development is only one of the potential uses of this resource and the economic advantages for local populations are, in turn, only one of the needs which the resource is called upon to satisfy. The other utility flows, provided by the forest landscape, and the forest as a whole, cannot be ignored, nor can the need to conserve and protect the natural environment. Nevertheless, in order to ensure the long term sustainable management of the forest, all these interests and functions must be considered together. The critical issues for such management arise, on the one hand, from the need to coordinate all the local stakeholders and organizations, and, on the other, from the evaluation of all the diverse interests and functions involved, and the contribution which each of these elements brings to the overall, long-term management of the forest system, understood as the interaction between natural resources and the local community, at both the local and the global level.

The management tools for congestible goods also depend on the level of excludability. On the one hand, appropriate management tools for congestible and non-excludable goods can range from public compensation mechanisms to common property regimes; on the other hand, congestible goods—that are excludable to non-community members—can be managed by either public or common property regimes. Finally, compensation and payment mechanisms can be implemented for goods which are both congestible and excludable.

The implementation of management tools for Paneveggio forest, which, due to the specific characteristics of its utility flows, must necessarily be complex, also faces further challenges because of the large number of stakeholders involved in the governance system. It is therefore imperative that the implementation of these management tools takes the particular features of the local system into account.

A combination of management tools based on stakeholder involvement is potentially beneficial and, in certain cases, mandatory, in order not only to protect but also to maximize the utility flows generated by these natural resources.

Consequently, the utility flows generated by Paneveggio forest require a mix of management tools. The forest management should take a multifunctional approach in order to maximize utility flows. This approach must acknowledge those goods which require a public management regime—such as the conservation of biodiversity, the protection of aquifers, and the improvement of hydrogeological functions—and combine this regime with appropriate payment mechanisms for tourism activities. These tourism activities can be based on forest direct use (use value), e.g., adventure parks, and/or indirect or background use (scenic or option value) e.g., theme trails. Activities such as hunting, fishing or the gathering of soft fruit, truffles and mushrooms should follow common property management regimes for community members and payment mechanisms for non-community members. The limiting of permit availability allows forest management systems which husband and increase these resources to be rewarded and fully regulates their consumption, thus preventing overexploitation.

For utility flows with more clearly definable market values (such as timber), payment mechanisms are the most appropriate management tool.

Recognizing the difficulties involved in the coordination of such a complex system, denoted by multiple interests, stakeholders, and utility flows—the whole set of management tools and mechanisms needs an integrated and collaborative management approach in order to achieve long term sustainability. Nevertheless, the development of tourism activities based on the goods and services generated by forests certainly offers an economic opportunity for local communities, since such activities can create additional sources of income, and supplement these traditionally provided by forests. However, these opportunities are considerably limited by the need to control the exploitation of the resources used in tourism activities in order to provide a real opportunity to ensure the long term sustainability of these activities.

When the management of forest landscapes involves the development of tourism activities not only the two elements noted above (level and time frame of exploitation of resources) but also the other flows of utility which are provided by forest, need to be taken into account. The recognition and assessment of all flows of utilities is crucial, as are the definitions of interests and needs specified by stakeholders; this first step both defines priorities and identifies the most likely conflicts of interest. These interests need to be balanced and integrated to obtain a sustainable management plan. The choice of these management tools requires a concerted approach in both the selection and the implementation phase.

The limits of certain tools are well known and generally relate to poor managerial design; tools must be sufficiently solid and appropriate, and incentives, to comply with them, must be strong enough to be effective. Other limits are connected to the potential conflicting in using certain flows of utility.

Following a sustainable tourism paradigm, appropriate forest management which includes recreational needs can both offer a number of opportunities and provide a more long term perspective.

Adopting a sustainable approach both to tourism and to forest planning, further integration of the management of both is necessary to address the above mentioned limits.

Generally, the management of a forest landscape which is appreciated as an enjoyable setting for tourism activities demands a public management component which can be integrated with both compensation mechanisms to those stakeholders who contribute to the maintenance of the forest landscape and, when viable, with mechanisms requiring payment from those tourist operators who gain indirect advantages from the resource. When direct use of this good is made, in situations involving excludability and rivalry, it is possible to introduce payment mechanisms.

To manage other forest flows of utility which have different degrees of excludability and rivalry, the tools discussed and their implementation—both singly and in combination—need to be case appropriate. The reference framework can certainly be global, but individual solutions will necessarily be strongly colored by their local context.

AUTHOR CONTRIBUTIONS

The authors equally contributed to this work.

REFERENCES AND NOTES

1. The economic impact of travel & tourism 2013. World Travel and Tourism Council (WTTC). Available online: http://www.wttc.org/site_media/uploads/downloads/world2013_1.pdf (accessed on 29 November 2013).
2. Blake, A.; Arbache, J.S.; Sinclair, M.T.; Teles, V. Tourism and poverty relief. *Ann. Tour. Res.* **2008**, *35*, 107–126.
3. del P. Pablo-Romero, M.; Molina, J.A. Tourism and economic growth: A review of empirical literature. *Tour. Manag. Perspect.* **2013**, *8*, 28–41.
4. World Commission on Environment and Development (WCED). *Our Common Future*; Oxford University Press: Oxford, UK, 1987.
5. Pearce, D.W.; Markandya, A.; Barbier, E. *Blueprint for a Green Economy*; Earthscan: London, UK, 1989.
6. Agenda 21: Programme of Action for Sustainable Development. Rio Declaration on Environment and Development. In Proceedings of United Nations Conference on Environment and Development (UNCED), Rio de Janeiro, Brazil, 3–14 June 1992.
7. Charter for Sustainable Tourism. In Proceedings of World Conference on Sustainable Tourism, Lanzarote, Canary Islands, Spain, 27–28 April 1995.
8. However, these constraints are not fully recognized by all the authors who analyze this problem. In fact, for this reason, it is worth noting the definitions of both strong and, in particular, weak sustainability. For more details see Neumayer (2003). Neumayer, E. *Weak versus Strong Sustainability: Exploring the Limits of Two Opposing Paradigms*, 2nd ed.; Edward Elgar Publishing Limited: Cheltenham, UK, 2003.
9. Norgaard, R.B. Ecosystem services: From eye-opening metaphor to complexity blinder. *Ecol. Econ.* **2010**, *69*, 1219–1227.
10. Sharpley, R. Tourism and sustainable development: Exploring the theoretical divide. *J. Sustain. Tour.* **2000**, *8*, 1–19.

11. Jamal, T.; Camargo, B.A.; Wilson, E. Critical omissions and new directions for sustainable tourism: A situated macro-micro approach. *Sustainability* **2013**, *5*, 4594–4613.
12. Martini, U. Approccio alla sostenibilità, governance e competitività delle destinazioni turistiche: Stato dell'arte e prospettive. In *Management per la Sostenibilità dello Sviluppo Turistico e la Competitività delle Destinazioni*; (In Italian). Franch, M., Martini, U., Eds.; Il Mulino: Bologna, Italy, 2013; pp. 17–61.
13. Cesaro, L.; Romano, R.; Zumpano, C. *Analisi e Approfondimenti sul Post 2013: Foreste e Politiche di Sviluppo Rurale*; (In Italian). INEA: Roma, Italy, 2013.
14. Millenium Ecosystem Assessment (MEA). *Millennium Ecosystem Assessment: Ecosystems and Human Well-Being—A Framework for Assessment*; Island Press: Washington, DC, USA, 2003; pp. 1–25.
15. Farley, J.; Costanza, R. Payments for ecosystem services: From local to global. *Ecol. Econ.* **2010**, *69*, 2060–2068.
16. Wibe, S. *Non Wood Benefits in Forestry: A Survey of Valuation Studies*; United Nation Economic Commission for Europe: New York, NY, USA, 1995.
17. Wunder, S. *Payments for Environmental Services: Some Nuts and Bolts*; Occasional paper 42; Center for International Forestry Research (CIFOR): Bagor, Indonesia, 2005.
18. Staiff, R. Cultural Inscriptions of Nature: Some Implications for Sustainability, Nature-based Tourism and National Parks. In *Tourism, Recreation, and Sustainability: Linking Culture and the Environment*, 2nd ed.; McCool, S.F., Moisey, R.N., Eds.; CABI: Wallingford, UK, 2008; pp. 220–235.
19. Reimoser, F. Il ruolo della selvicoltura nella gestione faunistica. *Sherwood* **2005**, *112*, 19–23. (In Italian).
20. Gios, G.; Rizio, D. Payment for forest environmental services: A meta-analysis of successful elements. *iForest* **2013**, *6*, 141–149.
21. Deffontaines, J.-P. L'objet dans l'espace agricole. Le regard d'un géoagronome. *Nat. Sci. Soc.* **2004**, *12*, 299–304. (In French).
22. Gios, G.; Clauser, O. Forest and tourism: Economic evaluation and management features under sustainable multifunctionality. *iForest* **2009**, *2*, 192–197.
23. Briassoulis, H. Sustainable tourism and the question of the commons. *Ann. Tour. Res.* **2002**, *29*, 1065–1085.

24. Healy, R.G. The "common pool" problem in tourism landscapes. *Ann. Tour. Res.* **1994**, *21*, 596–611.
25. Dillman, B.L.; Bergstrom, J.C. Measuring environmental amenity benefits of agricultural land. In *Farming and the Countryside: An Economic Analysis of External Costs and Benefits*; Hanley, N., Ed.; CABI: Wallingford, UK, 1991; pp. 250–271.
26. Jafari, J. The tourism market basket of goods and services. In *Studies in Tourism Wildlife Parks Conservation*; Singh, T., Kaur, J., Singh, P., Eds.; Metropolitan Book: New Delhi, India, 1982; pp. 1–12.
27. Many different tools of intervention have traditionally been used in the management of forest resources. Command and control tools are based on legally binding restrictions. These measures, which come into play at the end of the production cycle, establish a final "cure", but do not address the root of the problem. However it is worth noting that procedural standards have also been introduced, and in some cases certain products, or the use of certain raw materials, have been banned. Cap and trade mechanisms are based on emissions limits and allow two parties to sign an agreement under which the party emitting lower quantities of greenhouse gas cedes the rest of their quota to the party whose emissions exceed the maximum limit. These are voluntary agreements, usually subsidized by governments and aimed at lowering polluting emissions. There are numerous other tools in place for voluntary use: corporate social responsibility, environmental certification, various systems of environmental management
28. Galizzi, G. *Una Politica per l'Agricoltura della Montagna Alpina*; (In Italian). Franco Angeli: Milano, Italy, 1992.
29. Provincia Autonoma di Trento (PAT); Ufficio Foreste Demaniali. *Piano di Assestamento dei beni Silvopastorali di Paneveggio 1991–2000*; (In Italian). Provincia Autonoma di Trento: Trento, Italy, 2006.
30. Zugliani, G.; Dotta, L. Legno di risonanza Gestione, selezione e lavorazione nelle Foreste Demaniali di Paneveggio (TN). *Sherwood* **2009**, *155*, 14–18. (In Italian).
31. Hardin, G. The tragedy of the commons. *Science* **1968**, *162*, 1243–1248.
32. Ostrom, E. *Governing the Commons: The Evolution of Institutions for Collective Action*; Cambridge University Press: Cambridge, UK, 1990.

33. Lachapelle, P.R.; McCool, S.F. Exploring the concept of "Ownership" in natural resource planning. *Soc. Nat. Resour.* **2005**, *18*, 279–285.
34. Ostrom, E. Come Alcune Comunità Hanno Evitato la Tragedia delle Risorse Collettive. In Quarto Meeting Annuale del Centro Studi e Documentazione sulle Proprietà Collettive, Trento, Italy, 7–8 November 1998. (In Italian).

Chapter 12

Land Use and Natural Resources Planning for Sustainable Ecotourism Using GIS in Surat Thani, Thailand

Khwanruthai Bunruamkaew and Yuji Murayama

Division of Spatial Information Science, Graduate School of Life and Environmental Sciences, University of Tsukuba, 1-1-1 Tennodai, Tsukuba, Ibaraki 305-8572, Japan

ABSTRACT

The primary objective of this study is to evaluate the land use and natural resources for future sustainable ecotourism site planning using GIS as a tool. The study is based on 2007 land use land cover data and ecotourism suitability data which are then integrated with other GIS datasets to evaluate the land use and natural resources at a district level in Surat Thani province. The final step of this study was the prioritization of the area that is best suited for ecotourism in assessing ecotourism sustainability in Surat Thani province. The result is useful for tourism facilities development and ecotourism resource utilization where ecotourism could be more developed. Additionally, the results can be used for managers and planners working in local and central governments and other non-governmental organizations. These integrated approaches cover complex and universal issues such as sustainable development of ecotourism, biodiversity conservation and protected area management in a tropical and developing country such as Thailand. Moreover, it is believed that this study can be used as a basis for evaluating the suitability of other areas for ecotourism. In addition, it may also serve as a starting point for more complex studies in the future.

Keywords: GIS; ecotourism; land use; Surat Thani; Thailand

1. INTRODUCTION

Tourism is a highly complex activity and thus requires tools to assist in effective decision making to come to terms with the competing economic, social, and environmental demands of sustainable development [1]. Similarly, tourism activities generally can create various negative impacts on the surrounding environment. Increased human interference in ecologically fragile areas can cause irreversible change in the existing ecological processes. These problems can be reflected in degrading natural resources, vegetation structure and the size of the habitat patch, increasing deforestation and decreasing upstream water flows. Ecotourism encompasses a spectrum of nature-based activities that foster visitor appreciation and understanding of natural and cultural heritage and are managed to be ecologically, economically and socially sustainable [2]. Therefore, ecotourism is accepted as an alternative type of sustainable development [3]. Ecotourism has attracted increasing attention in recent years, not only as an alternative to mass tourism, but also as a means to promote a country's economic development and environmental conservation [4]. For example, in Costa Rica, Ecuador, Nepal, Kenya, Madagascar and Antarctica, ecotourism represents a significant portion of the gross domestic product and economic activity [5,6]. This necessitates a methodical management of ecotourism destinations, which can minimize the impacts from ecotourism activities while offering benefits to local communities. Ecotourism is based on the natural environment with a focus on natural and cultural heritage. Hence, establishing a set of warning indicators and criteria is necessary for ecotourism management [7]. Abidin (1995) identified a set of 15 criteria and 58 indicators for evaluating sustainable ecotourism development in Taman Negara National Park (TNNP), Malaysia, on the basis of Delphi method [8]. Boyd *et al.*(1994) and Boyd and Butler (1996) used a resource inventory to identify potential ecotourism sites in Northern Ontario in Canada. GIS techniques were used to measure the ranking of different sites according to the set criteria (naturalness, wildlife, cultural heritage, landscape and community) for ecotourism [9,10]. Bukenya (2000) employed six criteria—a high number of species, wildlife management potential, endangered species, potential to attract more tourists, less susceptibility to encroachment and degradation over long period—to prioritize the potential national parks in Uganda [11]. In Samoa, 20 critical sustainable indicators were developed by field testing and screening to assess the overall sustainability [12]. There is no doubt that ecotourism represents more friendly alternatives for economic use of natural resources, when compared to other consumptive economic uses, such as mining, logging, farming, and others [7]. Furthermore, Kumari (2010) integrated five indicator indices to identify and prioritize the potential ecotourism sites in West District of Sikkim state in India by using analytical hierarchical process (AHP) and GIS. The primary variables used for generating various indices were landform, elevation, land use/cover, vegetation diversity, density and endemism, wildlife, tourism attraction features and the infrastructure facilities [13]. Godratollah *et al.* (2011) identified 61 indicators on environmental (ecological), social, economic, cultural and institutional aspects affecting sustainability of ecotourism. The study shows that it is possible to define a set of indicators for monitoring ecotourism sustainability in Northern forest of Iran [3]. Use of ecotourism indicators will help to identify the potential ecotourism sites based on the physical environmental and socio-economic parameters [13].

Land use management practices have a major impact on natural resources including water, soil, fertility, plants and animals [14,15]. In Thailand, being a tropical country, one of the prominent characteristics of land use/cover change (LUCC) is the decline in forest and woodlands due to land conversion, and particularly agricultural expansion for cash crop production such as cassava, corn, rice, and sugarcane [16,17,18]. In order to determine the most desirable direction for future development, the suitability for various land uses should be carefully studied with the aim of directing growth of the land use to the most appropriate sites. Land suitability evaluation is the process of determining the fitness of a given tract of land for proper use [19]. Careful planning of the use of land resources is based on land evaluation, which is the process of assessing the suitability of land for alternative land uses [20]. Land use is characterized by combination of land cover and human activities which can change over time. In the definition of land use, a direct link is established between land cover and the actions of human beings in their environment [21]. The process of land resource plans and decision making aims to identify the most appropriate spatial pattern for future land uses [22]. Therefore, the monitoring of LUCC provides significant input information for effective natural resources rehabilitation, protection and conservation planning.

Among the most fundamental issues that need careful consideration when pursuing such a research endeavor, are the quality of the data and the adoption of appropriate techniques. Likewise, GIS has also been very helpful in facilitating LUCC studies [23]. These capabilities make them a promising tool, especially for land use planning and LUCC modeling [24,25]. Additionally, GIS technology can be used for suitability evaluation and management of land resources [26]. Site selection is also an important use of GIS in tourism planning. For instance, by using suitable location identification tools and topology it is possible to identify potential areas for further tourism development [27]. The land suitability classifications will be determined by overlaying thematic maps and by analyzing attribute data with the support of GIS [28]. Minagawa and Tanaka (1998) used GIS to locate areas suitable for tourism development at Lombok Island in Indonesia. Using

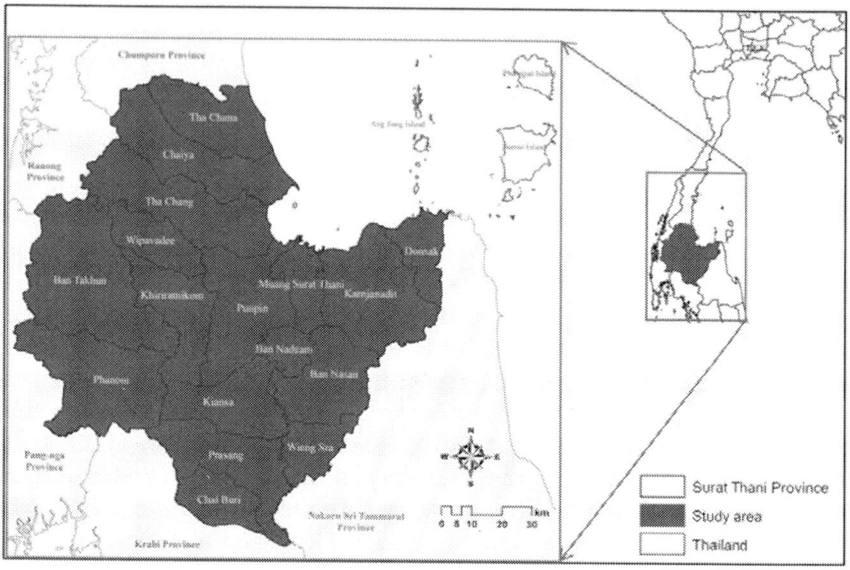

Figure 1. Location of the study area.

3. DATA AND METHODS

3.1. Data Acquisition and Pre-processing

The primary input data of the 2007 LUC and topographic map are derived from the Department of Land Development, Thailand at the scale of 1:50,000. The thematic maps are created and edited, overlaid and visualized on ArcGIS 9.3. There are 5 main classified LUC classes in Surat Thani province. These are urban and built-up land (U), agricultural land (A), forest (F), water body (W) and miscellaneous land (M). This study considered some important issues related to ecotourism potential resources. There were 10 LUC classes determined in this study. The definition and characteristics of each LUC class is described as follows:

1. Crop land (AC) includes paddy fields, field crops, truck crops and horticulture.
2. Farm land (AF) includes pasture and farm houses, poultry farm houses and aqua cultural land.
3. Orchard (AO) is an agricultural land which consists of mixed orchard and orchard (orange, durian, rambutan, coconut, and cashew).
4. Plantation (AP) is an agricultural land which consists of mixed perennials and perennials (oil palm, para-rubber, eucalyptus, teak, magosa, casuarinas, and coffee).

5. Dense forest (F) consists of the main natural forest including evergreen forest, swamp forest and mangrove forest.
6. Open forest (FO) is forest lands in which further the recreational lands are included in mixed forest plantations, forest plantations and agro-forestry that are caused by human activity, such as conversion of forest areas, vegetated areas for new cultivation.
7. Degraded forest (FD) includes disturbed evergreen forest, disturbed swamp forest, disturbed mangrove forest and disturbed forest plantations.
8. Miscellaneous land (M) includes rangelands (grass, scrub, and bamboo), marsh and swamp, mines and pits (laterite pits, sand pits, and soil pits) and other.
9. Urban and built-up land (U) is composed of cities, towns and commercial, residential houses, villages, institution land, industrial land, transportation, communications, utilities and so forth.
10. Water body (W) includes both natural water bodies (rivers, canals, and lakes) and man-made (reservoirs, dams, farm ponds, and irrigation canals).

3.2. Classified LUC Map

This part is to generate LUC map 2007 for ecotourism development focuses in land ecosystems of Surat Thani province. The various LUC types were identified according to the bio-physical vegetation characteristics for ecotourism potential resources and the background knowledge of the study area. Therefore, the LUC 2007 was reclassified into 10 LUC classes presented in Figure 2. The result of the area coverage of each LUC class in hectares shows in Table 1. It could be seen that, from the total area of 1,250,957.61 ha of the province being classified, about 60.12% was used for various agricultural uses, such as oil palm, para-rubber, orchard and other crops/farm land, while 31.27% was classified as forest area. Less than 3% was classified as urban and built-up area.

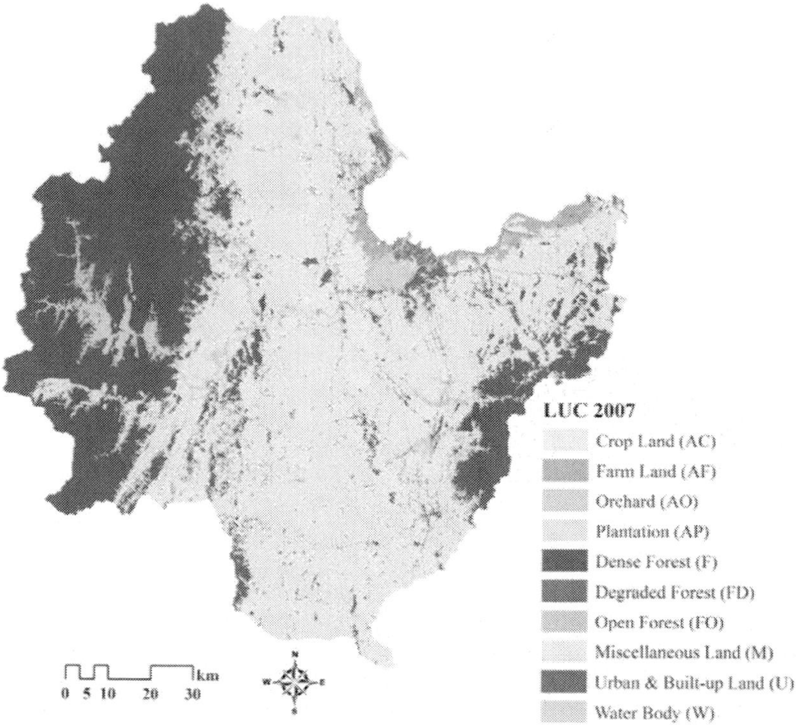

Figure 2. Classified land use/cover (LUC) map of Surat Thani province in 2007.

Table 1. Area coverage of land use/cover change (LUCC) classes in 2007.

LUC Type	Area (ha)	Proportion (%)
Crop Land (AC)	24,793.02	1.98
Farm Land (AF)	18,050.94	1.44
Orchard (AO)	43,688.62	3.49
Plantation (AP)	665,634.10	53.21
Dense Forest (F)	381,372.70	30.49
Degraded Forest (FD)	8,893.26	0.71
Open Forest (FO)	825.03	0.07
Miscellaneous Land (M)	45,822.42	3.66
Urban & Built-up Land (U)	33,064.38	2.64
Water Body (W)	28,813.14	2.30
Total area	**1,250,957.61**	**100.00**

3.3. Classification of Suitable Areas for Ecotourism

This study has been built upon the land suitability map for ecotourism to classify suitability by each LUC type at district level [34]. This study demonstrates a methodology for identifying potential ecotourism sites using GIS and AHP as case study of Surat Thani province, Thailand. The main steps to produce land suitability map for ecotourism are: finding suitable factors to be used in the analysis, assigning factor priority, weight and class weight (rating) to the parameters involved and generating land suitability map of ecotourism. In this process, experts' opinions were asked to calculate the relative importance factors involved. The AHP is one of the most extended Multi-Criteria Decision Making (MCDM) techniques. This method provides a structural basis for qualifying the comparison of decision elements and criteria in a pair wise technique. The decision criteria and factors were divided into two main categories according to the land ecosystems of Surat Thani: physical environmental and socio-economic parameters. This study selected five criteria and nine factors in the form of nine GIS-based layers in determining what areas are best suited for ecotourism development. These are landscape/naturalness (visibility, land use/cover), wildlife (reservation/protection, species diversity), topography (elevation, slope), accessibility (proximity to cultural sites, distance from roads) and community characteristics (settlement size). These criteria and factors were chosen based on the opinion experience and expertise of experts, surveying of authenticated literatures and analysis of historical data. The flow chart of the suitability analysis methodology is shown in Figure 3.

First, the AHP method was applied to determine the relative importance of all selected criteria and factors. Each criterion and factor received a weight and a score which represented its relative importance in the suitability evaluation. The total score for suitability is achieved by multiplying criterion score with its appropriate weight and adding all weighted scores. At the next stage, spatial data of the 9 factors as a set of GIS layers were overlaid together for final suitability classification for ecotourism. Further analysis was done in raster-based format and was then reclassified into 4 classes using the 'reclassify' function. Finally, the suitability map for ecotourism was produced using GIS overlay. At the end of this process, the suitability map for ecotourism development is generated. The degree of suitability of each factor classified as highly suitable (S1), moderately suitable (S2), marginally suitable (S3) and not suitable (N) is displayed in Figure 4.

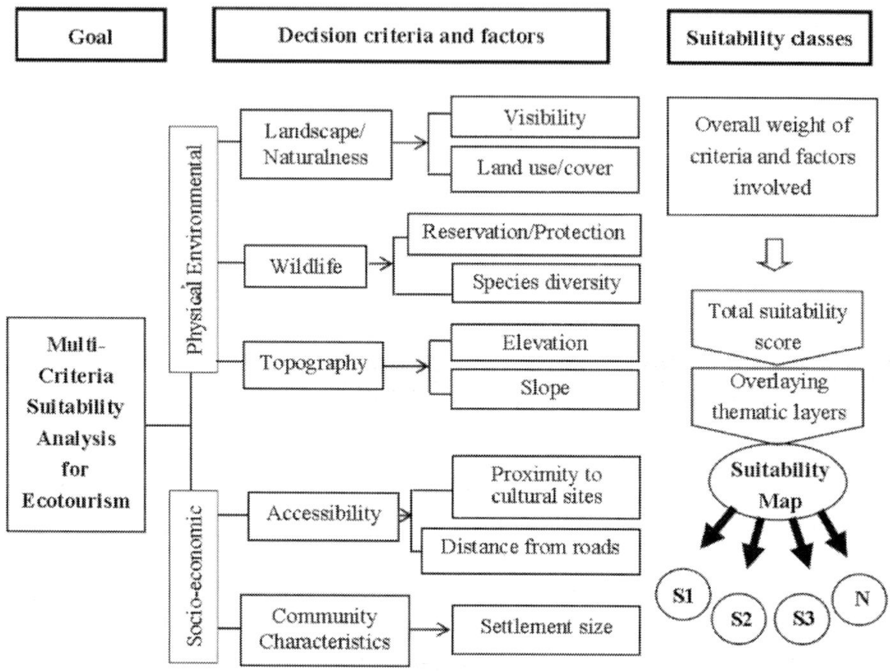

Figure 3. Flow chart of the suitability analysis.

Result of the area coverage in the classified suitability map for ecotourism is shown in Table 2. The map contains pixels with varying degrees of suitability from 0 to 1. A higher pixel score shows a higher suitability level. The map was reclassified, for easier representation, into four classes based on the structure of the FAO suitability classification [35]: the highly suitable (0.75–1.00), the moderately suitable (0.50–0.75), the marginally suitable (0.25–0.50), and the least suitable (0–0.25). The highly suitable is the suitable capacity of locations with high and satisfies all criteria set up. The moderately suitable is the suitable capacity of locations with medium and satisfies most of the criteria set up, but some criteria are not satisfied. The marginally suitable is the suitable capacity of locations with low and satisfies some of the criteria set up, but most of the criteria are not satisfied. The not suitable assumes that all of criteria are not satisfied. The extent of each class is summarized in Table 2.

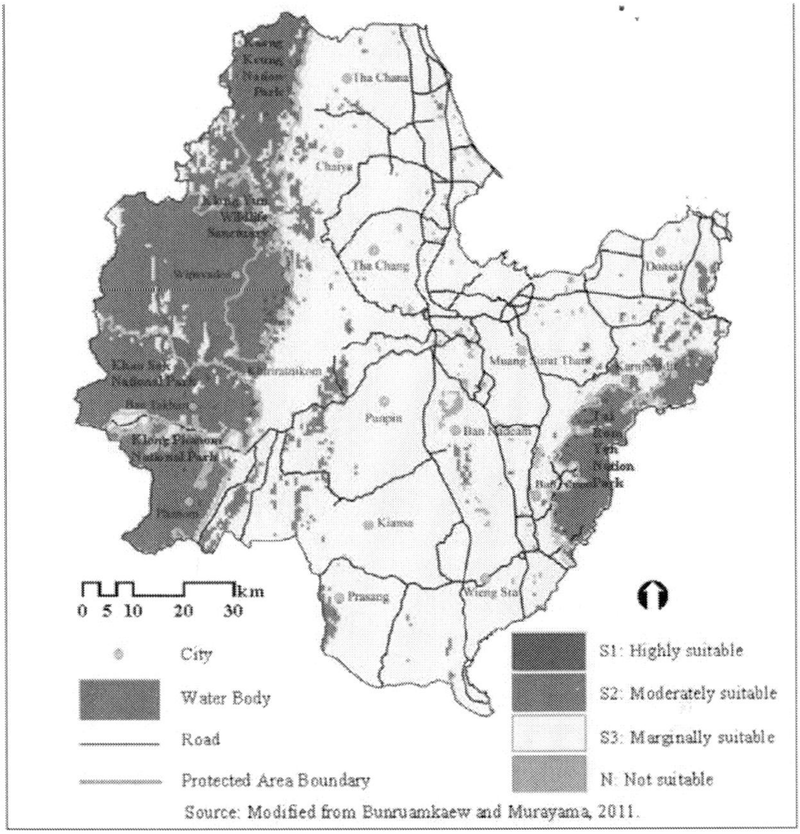

Figure 4. Land suitability map for ecotourism in Surat Thani province [34].

Table 2. Area coverage for ecotourism suitability classes.

Suitability Class	Score Range	Area Coverage	
		Hectares (ha)	%
Highly suitable (S1)	0.75–1.00	4,995.43	0.40
Moderately suitable (S2)	0.50–0.75	361,525.77	28.90
Marginally suitable (S3)	0.25–0.50	873,507.55	69.83
Not suitable (N)	0.00–0.25	10,928.86	0.87
Total area		**1,250,957.61**	**100.00**

The result indicates that 0.4% (4,995.43 ha) of the total study area was found to be the highly suitable class. These areas are mainly located in the sensitive areas that characterized with very rich biological diversity, endemic and endangered species. However, these areas can be considered as the most ecotourism attractions but under the controllable and limitations of visitor, in order to protect and preserve the most of the biodiversity value and their environmental condition. If people are attracted to the most valuable areas too much, this could again harm the quality of the areas. In addition, the moderately suitable class was found to be 28.9% (361,525.77 ha) of the territory that can be considered as moderately attractive for ecotourism. These areas are mainly characterized as very attractive and have recreational potential for ecotourism, such as beautiful scenery, spectacular landscapes, wildlife, abundant and different plant communities. Both of the highly and moderately suitable classes were 29.3% (366,520.20 ha) of the total area. This result highlights that the highly and the moderately suitable areas have been found in the protected areas (as seen in Figure 4) that are suitable for ecotourism requirement. Reasonably, the number of eco-destinations expands with the increases in park numbers [36]. The moderately suitable class was found to be 69.83% (873,507.55 ha) of the area with low sensitivity. These areas can provide eco-tourist facilities by facilitating proper ecotourism infrastructures and services under the policy guidelines. For example [34], the areas could accommodate physical structure to support ecotourism activities, such as green hotels, eco-lodge, restaurants and public convenience facilities. Specifically, infrastructure should be developed in harmony with local identity and with nature. Ecotourism infrastructure development and maintenance should be reflected best practice minimal impact design and procedures appropriate to the setting. Provide safe, reliable, sustainable and appropriate access to ecotourism attractions in and nearby natural areas. While, only a few percentage of the area is about 0.87% (10,928.86 ha) which was classified as not suitable for ecotourism refers to the areas with several effects of development from human activity and degraded forest. There is decreasing of forest because of deforestation for agriculture areas, the cultivation of steep slopes, trespassing upon public area to be dwelling of local people and illegal wood processing. Such areas are in a deteriorated condition or have been destroyed.

3.4. Comparison between LUC Map and Suitability Map

The spatial matching offered valuable information to identify whether the land was optimally utilized in the region [37]. The result of overlaying the suitability map (Figure 4) with the LUC map of 2007 (Figure 2) is presented in Table 3. This table displays the suitability of each LUC type in the study area.

Table 3. Classification of suitability by each LUC type.

LUC Type	Highly Suitable (S1)		Moderately Suitable (S2)		Marginally Suitable (S3)		Not Suitable (N)		Total Area
	ha	%	ha	%	ha	%	ha	%	ha
AC	0.00	0.00	38.61	0.01	24,229.35	2.77	525.06	4.80	24,793.02
AF	0.00	0.00	13.77	0.00	17,717.67	2.03	319.50	2.92	18,050.94
AO	3.6	0.07	2,119.2	0.59	41,025.37	4.70	540.45	4.95	43,688.62
AP	202.77	4.06	19,952.97	5.52	643,562.73	73.68	1,915.63	17.53	665,634.10
F	4782.04	95.73	331,527.84	91.70	44,918.73	5.14	144.09	1.32	381,372.70
FD	5.22	0.10	400.86	0.11	8,076.87	0.92	410.31	3.75	8,893.26
FO	0.00	0.00	183.96	0.05	641.07	0.07	0.00	0.00	825.03
M	0.00	0.00	723.96	0.20	39,911.76	4.57	5,186.70	47.46	45,822.42
U	1.80	0.04	330.75	0.09	31,185.09	3.57	1,546.74	14.15	33,064.38
W	0.00	0.00	6,233.85	1.72	22,238.91	2.55	340.38	3.11	28,813.14
Total	**4,995.43**	**100.00**	**361,525.77**	**100.00**	**873,507.55**	**100.00**	**10,928.86**	**100.00**	**1,250,957.61**

The LUC map indicates that the four major land uses are plantation (655,634.10 ha), dense forest (381,372.70 ha), miscellaneous land (45,822.42 ha) and orchard (43,688.62 ha), which account for 53.21%, 30.49%, 3.66% and 3.49%, respectively, of the total study area. Dense forest (F) that is mainly dry evergreen forest and mixed deciduous forest is mainly located in the protected areas. Considering plantation is mainly characterized by para-rubber and oil palm that is the most important cash crop in Southern Thailand. Miscellaneous land includes rangelands, marsh and swamp. The orchard consists of mixed orchard (such as rambutan, durian, mangosteen, orange, *etc*,) and coconut.

As expected, the highly suitable and moderately suitable areas for ecotourism were found in the dense forest but these utilized lands should be controlled under the use of certain limitations and guidelines. If too many people claim the most valuable areas, this could again harm the quality of the areas. The result indicates that 95.73% of the highly suitable class was distributed over the dense forest, while only 4.06% of this classification was located in the plantations. With respect to the moderately suitable class, 91.70% was also found in the dense forest, whereas only 5.52% of the class was located in the plantation. However for the marginally suitable area, 73.68% was found in the plantations. This is because plantation areas should be properly monitored and protected from any encroachment. In addition, the unsuitable class was distributed over the miscellaneous land (47.46%). This class was also found in the plantations (17.53%) and the urban and built-up land (14.14%) that is generally suitable for eco-tourist infrastructure development.

3.5. Classification of Suitable Areas for Ecotourism at District Level

The zonal analysis method was used to determine the average suitability for each district. This method is the creation of an output raster (or statistics table) in which the desired function is computed on the cell values from the input value raster that intersect or fall within each zone of a specified input zone dataset. The ecotourism land suitability analysis at district level was also assessed and the results are shown in Table 4.

Table 4. Classification of suitable areas for ecotourism at district level.

District	Highly Suitable (S1)		Moderately Suitable (S2)		Marginally Suitable (S3)		Not Suitable (N)	
	ha	%	ha	%	ha	%	ha	%
Tha Chana	315.45	7.34	6,338.70	1.75	59,013.36	6.78	1,126.26	10.06
Chaiya	446.94	10.41	29,117.70	8.06	61,376.67	7.05	1,489.50	13.30
Tha Chang	87.12	2.03	32,950.89	9.12	76,571.64	8.79	986.04	8.81
Ban Takhun	290.70	6.77	123,723.63	34.25	34,942.05	4.01	87.12	0.78
Wipavadee	526.86	12.27	22,603.68	6.26	20,031.84	2.30	0.00	0.00
Donsak	0.00	0.00	4,085.01	1.13	29,469.06	3.38	0.00	0.00
Karnjanadit	248.40	5.78	22,564.62	6.25	83,084.40	9.54	648.36	5.79
Punpin	0.00	0.00	313.92	0.09	93,665.52	10.75	2,973.24	26.55
Muang Surat Thani	0.00	0.00	43.56	0.01	33,273.90	3.82	78.03	0.70
Khiriratnikom	1,329.03	30.94	14,945.04	4.14	51,322.05	5.89	75.24	0.67
Phanom	306.90	7.14	69,532.47	19.25	49,318.11	5.66	132.66	1.18
Ban Nasan	692.82	16.13	25,640.73	7.10	47,035.71	5.40	946.89	8.46
Ban Nadeam	0.00	0.00	11.61	0.00	20,121.21	2.31	1,453.68	12.98
Kiansa	0.00	0.00	1,885.41	0.52	69,457.68	7.97	468.72	4.19
Wieng Sra	7.65	0.18	3,680.28	1.02	31,178.61	3.58	601.83	5.37
Prasang	42.57	0.99	2,022.21	0.56	75,213.90	8.64	130.68	1.17
Chai Buri	0.99	0.02	1,739.97	0.48	35,900.55	4.12	0.00	0.00
Total	**4,295.43**	**100.00**	**361,199.43**	**100.00**	**870,976.26**	**100.00**	**11,198.25**	**100.00**

Note: District locations are shown in Figure 1.

The result indicates that the highly suitable areas (S1) are mostly located in the districts of Khiriratnikom (30.94%) and Ban Nasan (16.13%). These districts are characteristically endowed with lush green forests, wildlife sanctuaries and a rich cultural heritage and they are mainly located in protected areas. Therefore, the management of these areas is comprehensive and addresses issues of resource conservation, environmental management, and the control of tourism development. For instance, these areas can be used for education as well as conservation of biodiversity and maintenance of the ecosystem. Then, the moderately suitable areas (S2) are mainly located in the districts of Ban Takhun (34.25%) and Phanom (19.25%). Most of these are largely free from urban settlements, with a unique and outstanding natural beauty, diverse attractions and great tourism potential. Most of them are located in the protected areas, where they are attractive and have recreational potential for ecotourism. Subsequently, the three areas with the highest percentages of marginally suitable land (S3) are located in Punpin (10.75%), Karnjanadit (9.54%) and Tha Chang (8.79%) districts. The most appropriate areas are mainly located in urban area. The areas are appropriate for facilities development because there are no conflicts or any restrictions on usage areas. These areas can provide ecotourism services which take into account the condition of the natural environment, local society and culture. Finally, the percentages of unsuitable areas (N) for ecotourism are often located in urban areas of Punpin district (26.55%) and Chiya district (13.30%).

More specifically, ecotourism development needs to consider additional environmental problems which can be controlled using sustainable ecotourism site planning where the areas are suitable for ecotourism. This classification of suitable areas for ecotourism at district level also encouraged the development of a database and inventory of natural areas with ecotourism potential to minimize fragmentation and duplication in planning.

4. RESULTS AND DISCUSSION

This study is at the preliminary stage of assessing whether land is likely to be practical and successful for future sustainable ecotourism sites. The study was conducted from the perspective of natural resource management and future sustainable ecotourism development using GIS. The GIS data offered LUC information that was crucial to examining the spatial matching between the potential suitability areas and the current LUC patterns. This information helped to identify priority areas for ecotourism and whether future land uses can be modified for future development within the province. This study further classified suitable areas for ecotourism at district level of Surat Thani where ecotourism sites should be promoted. The application of this paper can be useful for the managers and planners working in local and central governments and other non-governmental organizations. Since tourism in Surat Thani is largely dependent on a major use of natural resources and biodiversity, it is recommended that tourism be specifically addressed by regional policies that deal with biodiversity and conservation. More specifically to this study, it should be noted that the land suitability map for ecotourism is intended to guide regional land use decisions. It can be used for a decision making process that allocates land to the uses that provide the greatest benefits in terms of conservation of biodiversity and other ecosystem services in this province. Therefore, an ecotourism development plan should be incorporated into the development plans at district, provincial and regional levels to link resources which can complement each other. This will make it easier to redirect tourists away from tourism resources whose capacity is already exceeded.

In addition, this study result appears practically useful for the development of tourism facilities and ecotourism resource utilization: First, for the highly suitable areas (4,995.43 ha), most of which are located in the protected areas (Figure 4) characterized by dense forest and need to be conserved. Its high value of natural resources is suitable for research and education because some are involved in the sensitive areas. To maintain ecotourism, development must control and manage the resources in order to retain their original conditions as far as possible, and to avoid or to abstain from travelling in sensitive areas [38]. Thus, this area can still be considered for ecotourism attraction, but with the use of certain limitations and guidelines. Hence, the example of guidelines to be used to limit the number of tourists and duration of access to the areas is the necessary code of conduct. Activities suggested for these areas include those for purposes of research and education [39]. With regards to preserving most of the biodiversity value, ecosystem protection is the first issue for ecotourism development in environmentally sensitive areas. Moreover, effective and responsible marketing of ecotourism opportunities can be used as a form of visitor management, creating realistic expectations and an understanding of allowable and acceptable behavior while visiting sensitive natural areas. However, reducing the amount of visitor pressure on these sites is recommended.

Secondly, the moderately suitable class was found to consist of 361,525.77 ha of the total area that could serve as major attractions for ecotourism. Most of them are located in the protected areas. The majority of the marginally suitable land (873,507.55 ha of the total area) can be considered for ecotourism by incorporating appropriate infrastructures and services under the policy guidelines. Finally, about 10,928.86 ha of the total area was classified as unsuitable for ecotourism because some are degraded forest areas and destroyed public areas that would cause severe environmental degradation from an effective use of land for human requirements. Thus, it should be strictly restricted from tourism and any activities. It is very important that more people, especially Surat Thani visitors, are educated in the concept of sustainable development and the principles of sustainable living [40]. Therefore, issues of ecosystem management and conservation of biodiversity should be addressed with local communities and stakeholders in protecting ecosystem and biodiversity [41,42,43]. These will increase opportunities for local people and communities to participate in ecotourism and will help in distributing the income. More importantly, apart from local communities, the government and international agencies play a very important role for promotion as well as demotion of any initiative. Civil societies need to be more responsible for sustainable ecotourism to safeguard public lands, protected forests, water bodies, and violation of local and indigenous customary rights [44]. This consideration is a very important aspect for high-quality tourism in this province.

However, the GIS tool is only a preliminary way to assess appropriate land areas which may be suitable for ecotourism. This study is just a beginning, its regular monitoring and needs-based implementation of activities is crucial for this work. Carrying capacity should be dealt with higher up where the marketing element is discussed. Local communities can play a crucial role in planning, organizing community and stakeholder engagement, finding other forms of land use and environmental impact assessments, as well as performing social impact assessments. Provincial and federal governments also should recognize the contribution of the local people. Likewise, ecotourism management must take the character and potential of existing resources from the suitability map for ecotourism in order to arrange appropriate activities and to ensure compatibility between ecotourism and the original activities carried out in the area. This should include the avoidance of any serious conflict, especially in areas that are highly suitable for ecotourism, with other forms of tourism. For example, zoning, *i.e.*, establishing a tourism zone, a conservation zone, a buffer zone, and a rehabilitation zone, which should be done not only in protected areas but also in areas which are currently not suitable for ecotourism, the land having severe limitations that preclude the given type of use, but can be improved by specific management. We believe that if such tourism strategies are introduced according to these zones, they can prevent further deforestation and improve the appropriate use of land in the buffer zone. In addition, ecotourism marketing will have to wait until the resources are more ready, but in the meantime existing tourism activities in potential ecotourism areas are in need of proper control and guidelines if they are not to be subject to environmental deterioration and destructive competition in order to attract ever greater numbers of visitors which would have a greater impact on the goal of promoting sustainability. This is a good form of alternative tourism but numbers and land use should be carefully monitored. Therefore, the ecotourism market is expected to grow somewhat slowly until many of the issues mentioned here are addressed.

Additionally, ecotourism development must promote educational development and create awareness in people of the need to jointly maintain the ecosystem of the area. There is a need to implement development plans and manage natural resources in a way that ensures ecological and environmental integrity. Environmental education and interpretation is the key to creating an enjoyable and meaningful ecotourism experience, and is one of the key points of differentiation between ecotourism and other tourism products. Successful interpretive components of ecotourism products will foster appreciation and support for conservation efforts, local communities and culture [2]. Moreover, a further study should be done implementing other sub-related models such as the carrying capacity model in establishing a comprehensive ecotourism resource management plan. Furthermore, this can be done by allowing local people, communities and organizations to participate in environmental and resource management [45] and by creating legal and economic measures to control environmental and resource use, including setting up a management system that can foster economic development and improvements in people's quality of life. In addition, villagers should be empowered to manage natural resources within the boundary of their own villages. This study recommended that local people should be involved in the planning process under participatory approaches, *i.e.*, when decision makers prepare a plan, they must consider local people's opinions and their interest [27].

Furthermore, this study might be useful to those who are interested in the GIS technique for assessment and monitoring of natural resources, especially ecotourism sites and mapping. GIS plays a key role in documenting natural conditions, developments and documenting the suitability of resources for tourism, exposing conflicts, and revealing cause-effect relationships [46,47]. As previously mentioned, the use of GIS is not only ideal for reducing the time and cost of site selection, but also provides a digital data bank for long-term and beneficial monitoring of sites [48]. The study shows that GIS can evaluate dynamic patterns of LUCC as well as providing a preliminary study for identifying priority areas for ecotourism and thus, it contributes to improving the efficiency of conservation and of sustainable land management in the Surat Thani province. This approach can be used as a basis for evaluating the suitability of other areas for ecotourism and it may also serve as a starting point for more complex studies in the future taking into consideration those limitations encountered in this study.

Finally, land use decisions are based not only on such information but also on other assessments, such as economic analysis of land uses and environmental impact assessments. Since only a small number of socio-economic factors have been considered for the evaluation of land suitability, it is recommended to incorporate more factors to represent a holistic view of the actual process, such as distance from local community or from protected areas. Moreover, research can be extended to explore the application of other information sources such as remote sensing images, Global Positioning System (GPS), *etc.*which should be encouraged as it will help bring real changes to land use and management strategies.

ACKNOWLEDGMENTS

The authors gratefully thank the Royal Thai Government Scholarship for financial support. Likewise, I would also like to express thanks to Ko Ko Lwin for his valuable comments and suggestions. Finally, special thanks are given to the anonymous referees and the editor for their useful suggestions in the revision of this manuscript.

REFERENCES AND NOTES

1. Fadahunsi, J.T. Application of Geographical Information System (GIS) technology to tourism management in Ile-IFe, Osun State, Nogeria. *Pac. J. Sci. Technol.* **2011**, *2(13)*, 274–283.
2. Tourism Queenland, *Queensland Ecotourism Plan 2003–2008*; Tourism Queensland: Brisbane, Australia, 2002.
3. Godratollah, B.; Azlizam, A.; Manohar, M.; Mohd, H.I.; Syed, M.H. Delphi technique for generating criteria and indicators in monitoring ecotourism sustainability in Northern forests of Iran; case study on Dohezar and Sehezar Watersheds. *Folia Forestalia Polonica Series A* **2011**, *53(2)*, 130–141.
4. Tewodros, K.T. Geospatial Approach for Ecotourism Development: A Case of Bale Mountains National Park, Ethiopia. Master Thesis, Faculty of Natural Science, Department of Earth Sciences, Addis Ababa University, Addis Ababa, Ethiopia, 2010.
5. Isaacs, J.C. The limited potential of ecotourism to contribute to wildlife conservation. *The Ecologist* **2002**, *28(1)*, 61–69.
6. Eadington, W.R.; Smith, V.L. *The Emergence of Alternative Forms of Tourism, in Tourism Alternatives: Potentials and Problems in the Development of Tourism*; University of Pennsylvania Press: Pennsylvania, PA, USA, 1992.
7. Li, W. Environmental Management indicators for ecotourism in China's nature reserve: A case study in Tianmushan nature reserve. *Tourism Manage.* **2004**, *25(5)*, 559–564.
8. Abidin, Z.Z. The Identification of Criteria and Indicators for the Sustainable Management of Ecotourism in Taman Negara National Park, Malaysia: A Delphi Consensus. Ph.D. Dissertation, West Virginia University, Morgantown, WV, USA, 1995.
9. Boyd, S.; Butler, R.; Haider, W.; Perera, A. Identifying areas for ecotourism in Northern Ontario: Application of a Geographical Information System methodology. *J. Appl. Recreat. Res.* **1994**, *19(1)*, 41–46.
10. Boyd, S.W.; Butler, R.W. Seeing the forest through the trees: using GIS to identify potential ecotourism sites in Northern Ontario. In *Practicing Responsible Tourism: International: Case Studies in*

Tourism Planning, Policy and Development; Harrison, L.C., Husbands, W., Eds.; J. Wiley & Sons: New York, NY, USA, 1996; pp. 380–403.

11. Bukenya, J.O. Application of GIS in Ecotourism Development Decisions: Evidence from the Pearl of Africa; Research Paper; Natural Resource Economics Program. West Virginia University: Morgantown, WV, USA, 2000. Available online: http://www.rri.wvu.edu/pdffiles/bukenya2012.pdf (accesse d on 4 April 2010).
12. Twining-ward, L. Monitoring sustainable tourism in Samoa. *Industry Environ.* **2002**, *24*, 46–47.
13. Kumari, S.; Behera, M.D.; Tewari, H.R. Identification of potential ecotourism sites in West District, Sikkim using geospatial tools. *Trop. Ecology* **2010**, *51(1)*, 75–85.
14. Di Gregorio, A.; Jansen, L.J.M. A new concept for a land cover classification system. *The Land* **1998**, *2*, 55–65.
15. Baniya, N. Land Suitability Evaluation Using GIS for Vegetable Crops in Kathmandu Valley, Nepal. Ph.D. Dissertation, Institute of Horticulture Science, Faculty of Agriculture, Humboldt University of Berlin, Berlin, Germany, 2008.
16. FAO. *State of the World's Forests 1997*; Food and Agriculture Organization of the United Nations (FAO): Rome, Italy, 1997.
17. FAO. *State of the World's Forests 2003*; Food and Agriculture Organization of the United Nations (FAO): Rome, Italy, 2003.
18. Barbier, E.B.; Burgess, J.C. The economics of tropical deforestation. *J. Economic Surveys* **2001**, *15(3)*, 413–421.
19. Marsh, S.P.; MacAulay, T.G. Land reform and the development of commercial agriculture in Vietnam: Policy and issues. *Agribus. Rev.* **2002**, *10(1)*, 1–19.
20. Fresco, L.O.; Huizing, H.; Keulen, H.V.; Luning, H.A.; Schipper, R.A. *Land Evaluation and Farming Systems Analysis for Land Use Planning, FAO Working Document*; FAO: Rome, Italy, 1994.
21. Jamal, B.N. Land Evaluation for Land Use Planning with Especial Attention to Sustainable Fodder Production in the Rouzeh Chai Catchment of Orumiyeh Area, Iran. Master Thesis, International Institute for Geo-Information Science and Earth Observation, Enschede, The Netherland, 2003.
22. Arabinda, L. Integrating GIS and Multi-Criteria Decision Making Techniques for Land Resource Planning. Master Thesis,

International Institute for Geo-Information Science and Earth Observation, Enschede, The Netherland, 2003.

23. Estoque, R.C.; Murayama, Y. Spatio-temporal urban land use/cover change analysis in a hill station: the case of Baguio city, Philippines. *Procedia. Soc. Behav. Sci.* **2011**, *21*, 326–335.

24. Serneels, S.; Lambin, E.F. Proximate causes of land use change in Narok District Kenya: A case spatial statistical model. *Agr. Ecosyst. Environ.* **2001**, *85(1–3)*, 65–81.

25. Wu, Q.; Li, H.; Wang, R.; Paulussen, J.; He, Y.; Wang, M.; Wang, B.; Wang, Z. Monitoring and predicting land use change in Beijing using remote sensing and GIS. *Landscape Urban Plan.* **2006**, *78(4)*, 322–333.

26. Obi Reddy, B.P.; Maji, A.K.; Srinivas, C.V.; Kamble, K.H. and Velayutham, M. GIS-based basin morphometric information system for terrain and resources analysis. In Proceedings of First National Conference on Agro-Informatics, Dharwad, India, June 2001; pp. 37–42.

27. Azizur, R.M. Application of GIS in Ecotourism Development: A Case Study in Sundarbans, Bangladesh. Master Thesis, Department of Social Science, Mid-Sweden University, Ostersund, Sweden, 2010.

28. Chuong, M. Multi-Criteria Land Suitability Evaluation for Selected Fruit Crops in Hilly Region of Central Vietnam. Ph.D. Dissertation, Humboldt University of Berlin, Berlin, Germany, 2007.

29. Minagawa, Y.; Tanaka, N. Application of Geographic Information Systems to tourism development planning: A case study of Lombok, Indonesia. UNCRD Research Report Series No. 27 . United Nations Centre for Regional Development: Nagoya, Japan, 1998.

30. Bahaire, M.; Elliot-White, M. The application of Geographical Information Systems (GIS) in sustainable tourism planning: A review. *J. Sustain. Tourism* **1999**, *7(2)*, 159–174.

31. Bauer, T.; Dowing, R.K. Ecotourism policies and issues in Antarctica. In *Ecotourism Policy and Planning*; Fennell, D.A., Dowind, R.K., Eds.; CAB International: Wallingford, UK, 2003; pp. 309–330.

32. Anomasiri, W. Eastern Philosophy of Ecotourism Management Model in Mae Hong Son Province, Thailand. Master Thesis, Technology of Environmental Management, Faculty of Graduate Studies, Mahidol University, Bangkok, Thailand, 2004.

33. Mieczkowski, Z. *Environmental Issues of Tourism and Recreation*; University press of America: Maryland, MD, USA, 1995.

34. Bunruamkaew, K.; Murayama, Y. Site suitability evaluation for ecotourism using GIS & AHP: A case study of Surat Thani province, Thailand. *Procedia Soc. Behav. Sci.* **2011**, *21*, 269–278.
35. FAO. *A Framework for Land Evaluation*; Soils Bulletin No. 32; Food and Agriculture Organization of the United Nations (FAO): Rome, Italy, 1976.
36. Zurick, D.N. Adventure travel and sustainable tourism in the peripheral economy of Nepal. *Ann. Assoc. Am. Geogr.* **1992**, *82(4)*, 608–628.
37. Khoi, D.D.; Murayama, Y. Delineation of suitable cropland areas using a GIS based multi-criteria evaluation approach in the Tam Dao National Park region, Vietnam. *Sustainability* **2010**, *2*, 2024–2043.
38. Eurblarp, S. Ecotourism policy in Thailand. In *Linking Green Productivity to Ecotourism: Experiences in the Asia-Pacific Region*; University of Queensland: Brisbane, Australia, 2002; pp. 236–241.
39. Yaakup, A.; Che'Man, N.; Hosni, N.; Haron, H.W.; Sulaiman, S. A GIS approach in evaluation of metropolitan green area: A case of Sungai Pulai wetland. In proceeding of Urban Forestry Conference: Managing Urban Green for Sustainable Cities, Kuala Lumpur, Malaysia, 6–7 June, 2006.
40. IUCN/UNEP/WWF. *Caring for the Earth: A Strategy for Sustainable Living*; International Union for Conservation of Nature and Natural Resources: Gland, Switzerland, 1991; Second Report on World Conservation and Development .
41. WCMC. *Global Biodiversity: Status of the Earth's Living Resources*; Compiled by the World Conservation and Monitoring Center, Chapman and Hall: London, UK, 1992.
42. Maser, C. *Sustainable Forestry: Philosophy, Science, and Economics*; St. Lucie Press: St. Lucie, FL, USA, 1994.
43. Ceballos-Lascurain, H. *Tourism, Ecotourism, and Protected Areas*; IUCN Protected Areas Program: Gland, Switzerland, 1996.
44. Bhattacharya, P.; Kumari, S. The Commons in an age of global transition: Challenges, risk and opportunities. In Proceeding of IASCP Bi-Annual Conference, Oaxaca, Mexico, 9–14 August 2004.
45. Shackleton, S.; Bruce, C.; Wollenberg, E.; Edmunds, D. Devolution and community-based natural resource management: Creating space for local people to participate and benefit? Revenue or conservation interests rather than local livelihood. *Nat. Resour. Per.* **2002**, *76*, 1–6.

46. Dye, A.S.; Shaw, S.L. A GIS-based spatial decision support system for tourists of the Great Smoky Mountains National Park. *J. Retailing Consum. Serv.* **2007**, *14(4)*, 269–278.
47. Rannveig, O.; Micael, C. A GIS approach to evaluating ecological sensitivity for tourism development in fragile environments: A case study from SE Iceland. *Scand. J. Hosp. Tour.* **2009**, *9*, 22–38.
48. Mazaher, M. Siting MSW landfill using weighted linear combination and analytical hierarchy process (AHP) methodology in GIS environment (case study: Karaj). *Waste Manag.* **2010**, *30*, 912–920.

Chapter 13

Young Tourists and Sustainability. Profiles, Attitudes, and Implications for Destination Strategies

Federica Buffa

Department of Economics and Management, University of Trento, Via Inama 5, Trento 38122, Italy

ABSTRACT

Global trends highlight the growing tourist interest in authentic and sustainable holiday experiences. Designing strategies that enable destinations to catch this tourist segment is, therefore, becoming more and more important for competitiveness. A long-term outlook calls into question the "next generation" of actual and potential tourists, *i.e.*, young people: understanding their attitude towards sustainability is paramount to drive tourism development in a direction which is coherent to the forthcoming demand. Drawing from an *ad hoc* survey of 1156 members of the largest Italian association of student and youth tourism, this contribution (a) discusses youth attitudes towards sustainability, their travel motivations and behavior (b) identifies different profiles of young tourists with reference to sustainability. By using characteristics and dimensions linked with "harder ecotourists" we identify hard path young tourists (HPYT) and soft path young tourists (SPYT). The findings confirm young people's interest in certain dimensions of sustainability and the influence this interest has on their decision-making processes, motivations, and behaviors. HPYT and SPYT are profiles which should be considered in destination strategies: the strong sensitivity of HPYT to sustainability suggests the possibility of creating offers that optimize the unique features of a territory.

Keywords: sustainable tourism; destination strategy; young tourists; youth attitudes; youth behaviors; tourist profile; sustainability

1. INTRODUCTION

The influence of megatrends on the tourism sector highlights how important it is that destinations and enterprises understand these changes and the consequences that they can have on both tourist values and attitudes and on tourism forecasts [1]. The data on certain trends related to social drivers are particularly meaningful. Among them, individualism, the search for a variety of experiences, self-improvement, experimental, and increased social and environmental awareness are trends which show a growing demand for authentic holiday experiences [1]. Tourists seek involvement and interaction with local people and the opportunity to discover a destination's unique features and to acquire new knowledge. Increased authenticity [2,3,4,5,6,7] can boost tourist satisfaction and improve the tourism experience [8,9,10,11,12] and attract new demand segments, as has been shown in studies carried out in mature destinations [13].

From a managerial point of view, destinations evaluating their own competitiveness need to take into account the existence and effects of both megatrends and tourists' motivations and behavior. In accordance with the main principles of destination management (see the review in [14]) and with the current debate on the role of DMOs in the definition of tourist offers [15], it is clear that destinations should follow proactive strategies that create new tourist products adapted to current trends. It should, thus, be possible to overcome the limitations of the "impersonal, non-interactive system of mass tourism" [1] (p. 67). The implications of these megatrends for tourist demand also indicate the extent to which strategies for sustainability can become drivers for destination competitiveness [16,17,18,19], lending weight to the thesis of the inseparability of sustainability and competitiveness which is amply illustrated in Ritchie and Crouch [20] and subsequent studies.

Sustainable tourism and related issues are vastly debated and sometimes questioned: the actual possibility to promote sustainable local development and improve local quality of life through tourism is alternatively supported or doubted [21,22], while feasible strategies and best practices are still under discussion [22,23]. Researchers call for noteworthy efforts to promote key elements of sustainability and educate actual and future generations [24], the latter being crucial for the destination's development in the long run [25]. Understanding holiday motivations and behaviors of this segment are, therefore, key factors to design proper, effective, and long-term destination strategies.

In the light of these considerations, this paper focuses on tourist demand, identifying the travel behaviors and motivations of a specific segment: young tourists. The aim of the research is two-fold. First, to analyze the sensitivity of young tourists to sustainability. The behavior and motivations of young tourists is a topic of current interest for both international organizations and the scientific community. Young people are one of the main targets of the EU "Europe 2020 strategy", aimed at increasing their opportunities to study and/or work abroad (see the initiative Youth on the move). Internationally, young travellers' mobility is encouraged by the World Youth Student and Educational (WYSE) Travel Confederation, an affiliate of UNWTO (www.wysetc.org/unwto). Data from the UNWTO [26], the UNWTO and WYSE Travel Confederation [27], and the WYSE Travel Confederation [28,29] highlight the current and future importance of young tourists—both in quantitative terms (the size of the segment and the growth of tourist flows) and with regard to spending power. The growing importance of this market is also underlined by the strategic partnership formed in June 2014 between the UNWTO and the WYSE Travel Confederation, intended to enable collaboration on projects that will increase understanding of the youth, student, and educational travel sector. Notwithstanding the importance of this segment, in fact, a knowledge gap does exist, with regard to travel motivations and behaviors [30]. In addition, little research has been done to date into young tourists' interest in, and sensitivity to, sustainability and sustainable tourism offers [30,31,32,33,34,35,36].

The second aim of the research is to examine whether the young tourist segment includes profiles with different degrees of sensitivity to sustainability. Based on previous research on the links between sustainable tourism, tourist demand, and young tourists, our study identifies two different profiles. The characteristics and dimensions linked to "harder ecotourists" described by Weaver and Lawton [37] allow for the identification of "hard path young tourists" and "soft path young tourists" and highlight differences in their organizational, motivational, and behavioral profiles. From a strategic point of view these results allow us to reflect on the relevance of values and needs of this tourists segment for destinations in the (re)definition of tourism products.

2. LITERATURE REVIEW

2.1. Sustainable Tourism and Tourist Demand

Following the publication of "The Brundtland Report" [38] and the debate it ignited on the subject of sustainable development, since the late 1980s academics in the field of tourism have also been considering the necessity/opportunity of striking the right balance between the consumption of resources involved in tourism and local development within a territory [39,40]. Inskeep [41] and Swarbrooke [42] welcome such stimuli and draw attention to the links between the use of resources for tourism and the sustainability of tourist development, highlighting the need to find a balance between resource use and the three pillars of sustainability: the economic, the environmental, and the social. Starting from these considerations, many scholars have investigated the main topics of sustainable tourism, focusing on its particular characteristics, criticalities, and implications in relation to:

- the planning, governance, management, and competitiveness of tourism destinations [20,43,44,45,46,47,48,49,50,51,52].
- the analysis of the behavior and motivations of the segments of demand that are interested in sustainable tourism offers [37,53,54,55,56,57,58,59,60].

Some authors are of the opinion that the sheer quantity of research that has been done in this area has led to confusion around the terminology used in discussions of sustainable tourism [39,61,62,63,64,65]. This phenomenon is especially evident in the research on tourist demand, and particularly that into "new travellers" interested in new/alternative forms of tourism [66]. In fact, there has been considerable interest in the study of these segments, most notably in relation to ecotourism. Numerous studies have analyzed the motivational and behavioral factors involved and have identified four main segments:

- ecotourists [67,68,69,70,71,72,73,74,75,76,77]
- nature-based tourists [78,79,80,81,82,83,84,85]
- responsible tourists [86,87,88,89,90]
- green tourists [91,92]

One of the first points revealed by these researches is that there is now widespread concern about the issues of sustainability, across different segments of demand [93], and, furthermore, a growing trend in this direction [66,77] which, it has been predicted, may even result in its becoming a megatrend in the tourism sector [1].

In contrast to the consensus on a growing interest in sustainable tourism, it is much more difficult to identify the actual motivations and behaviors that distinguish these segments. As has been noted in the various definitions of sustainable tourism and ecotourism (see in particular [22,23,75,94]), even analyses of segments of demand and their related characteristics are not easily made. These difficulties are increased by the fact that there are socio-demographic, motivational, attitudinal, and behavioral perspectives that cross segments [71,76,77,91,95] making the identification of the "so-called true ecotourist", and distinguishing from other segments, difficult [96].

The general observation can be made, however, that these segments (in particular ecotourists and nature-based tourists) are older, better educated, and better off than general travellers and mass tourists [71,72,74]. Various studies have produced conflicting data on tourist gender: some show a prevalence of male/female tourists, in others there is no significant difference between the genders (see particularly the review in [72] (p. 260)).

Regarding motivation and behavior, on the other hand, the main similarity between the segments is tourist interest in products and holiday experiences consistent with sustainable principles. What differentiates them is their levels of sensitivity to various aspects of sustainability. In fact, although it differs in intensity, a focus on the environment and natural resources is the main characteristic shared by ecotourists, nature-based tourists, and green tourists. However, the first group also expresses particular interest in the educational and learning aspects of tourism, while the second focuses on the environment, identifying the contemplation of nature as their main motivation [95,97,98,99,100]. Lastly, green tourists are often identified as being interested in rural destinations [91,92]. The social dimension of sustainability, on the other hand, is more significant for responsible tourists, who are particularly sensitive to the relations between tourists and host communities and the socio-cultural impact of tourism on a destination.

Due to the difficulty of discriminating between these segments, Weaver and Lawton's reflections [75] on Laarman and Durst [101] acquire great significance. Indeed, through an understanding of the varying levels of interest in the ecotourism offer expressed by different sectors of the demand, Laarman and Durst highlight its "soft" and "hard" dimensions: segments are not only differentiated by their motivations but also by the strength of these motivations. Following this approach, other researchers have highlighted this aspect and defined "weaker (lighter green) and stronger (darker green) interpretations of sustainable tourism" [102] (p. 10); the "soft" and "hard" behavior of the ecotourist demand [68,103,104]; "hard path" and "soft path" ecotourism [105], "hard", "medium" and "soft" ecotourism [80]; rather than the "harder", "structured" and "softer" ecotourists as ideal types positioned on an ecotourism spectrum [37]. Weaver and Lawton [37] describe the behaviors, motivations, and opinions revealed in the statements of 1180 tourists at two eco-lodges in Lamington National Park (Australia). The research identifies and describes "harder" ecotourists as being strongly committed to the environment, interested in enhancing sustainability, physically active, expecting few or no services, placing emphasis on personal experience, preferring specialized trips, long trips, small groups and being willing "to donate extra money to support ecotourism sites" [37] (pp. 274, 279).

These researchers introduce elements of particular relevance to the study of tourist demand. The new tourists are distinguished by different levels of sensitivity; their behavior and motivation is not necessarily compatible with sustainable tourism. "Soft" tourists, in fact, although interested in nature and the beauty of landscapes, may require levels of service and facilities that mean they choose offers/destinations which do not necessarily focus on sustainability. As Weaver and Lawton [66] point out "soft ecotourism activity often occurs on a large scale within certain accessible protected areas, leading to the observation that ecotourism can and often does also occur as a form of mass tourism" (p. 363). Weaver [106] also points out that mass tourism does not have to be unsustainable; nevertheless it is clear that the management of "hard" or "soft" segments can require very different strategies on the part of destinations which do not necessarily imply the creation of sustainable tourism offers. Softer ecotourists favor multi-purpose trips, short trips, and larger groups, are moderately committed to the environment and to steady state sustainability, are physically passive, expect services to be available, and give emphasis on interpretation [37] (p. 279).

2.2. Sustainable Tourism and Young Tourists

The complexity of tourist behavior has been frequently stressed in numerous sociological, psychological, and managerial studies, dating back to the 1970s. A wide variety of analytical models and variables have been used (socio-economic, demographic, psychographic) to identify different tourist profiles and their particular choice mechanisms, motivations, and behaviors [107,108,109,110,111,112,113,114,115]. The complexity of this process is clearly revealed in the large body of work already mentioned on the analysis of sustainable tourism, in which the similarities and differences between various segments are not always easily understood.

In recent years, the scientific community has increasingly turned its attention to an examination of one particular segment of tourist demand—the young. Youth are "the new visitors in the tourism market" [35] (p. 5). "The importance of this market segment lies in the fact not only that it is becoming larger but also that it represents the market of the future" [116] (p. 1). The study of this section of the market requires, first of all, that the criteria and variables used to define an individual as a member of the "young" be identified so that their decisional processes, motivations, and behavior can subsequently be observed and described. The socio-demographic identification of this segment gives us objective information, but does not resolve the question of what age an individual ceases to be "young". As Pendergast [35] points out, generational theory, which takes into account the demographic aspects, makes an important contribution to population analysis through the identification of generations by linking each birth year to its relative age range. Such an analysis reveals that the generation which currently includes the young is Generation Y (Gen Y)—people born between 1982 and 2002. Although the actual age range and, consequently, also the socio-demographic characteristics, associated with Gen Y, are still debated [117], it is possible to connect this cohort with events (notably the birth of the new technologies and the 9/11 terrorist attacks) and with social phenomena (*i.e.*, greater independence, better education, changing roles within the family, and changing family structures, *etc.*) with which particular habits and attitudes are associated (*i.e.*, frequency of Internet use, access to social networks, concerns about security and crime, leaving home later, more working mothers). Certain inconsistencies emerge, however [30,35] (for in-depth analysis of the social characteristics and consumer patterns of Gen Y see the review in [33]; for detailed research on the online behavior of Gen Y see [34] and cited works).

Little research has been done to date on Gen Y and tourism [30], but it is a particularly interesting area of study given that Gen Y is "one of the largest consumer cohorts today with some of the highest levels of discretionary spend at their disposal" [33] (p. 5). Most of the data on the motivations and behaviors of the young in regard to holidays come from the studies done on a sub-cohort of Gen Y, youth, and student travellers. Chhabra [31] emphasizes the importance of understanding their consumer black box because "it is important for marketers to understand what happens inside the black box and how various stimuli are transformed into responses that reflect in the travel choices of tourists" (p. 345). The UNWTO [26] underlined the growing importance of the young tourist segment (compared to other market segments), in terms of both tourist flow and spending power. The WYSE Travel Confederation [28] underlines the fact that, in comparison with the past, this segment includes a wider population range: people aged 15–30+, rather than 18–24. From the behavioral point of view, it underlines the fact that youth and student travellers are keen to discover and learn about different cultures and to be able to interact with local populations. These factors open up the possibility of promoting, through this segment of tourist demand, behaviors that are more responsible and more aware of the various aspects of sustainability. Data from the UNTWO and WYSE Travel Confederation [27] supports these findings, describing youth and student travellers as "paving the way for responsible tourism". Other studies provide a detailed analysis of students' environmental attitudes and ecological beliefs, highlighting a general interest among the young in these questions (see particularly [118] and their bibliography).

The young traveller segment is also examined in studies of backpacker motivations and behaviors, but here too there are gaps in our knowledge of young peoples' sensitivity to issues of sustainability. This may be explained by the following factors:

- Demographics: many, but not all, backpackers are young. In Loker-Murphy and Pearce's [119] conceptual framework, age (along with budget and degree of organization) is one of the factors that distinguishes backpackers from other segments: the authors are referring to the youth tourist segment, but older tourists may also fall into this category. Cohen [120] makes it clear that while most backpackers are young (between 20 and 30), older backpackers certainly exist, and their motivations and journey lengths tend to differ from those of the young. Larsen, Øgaard, and Brun [121] point out that backpackers do not fall within a clearly defined age range; in support of their argument they cite the research carried out by Elsrud [122] on backpackers aged between 18 and 71.
- Territorial context: backpacking, although practiced in Europe (particularly in Scandinavia, the UK, and Germany), is much more common elsewhere (Australia, New Zealand, North America, and South East Asia in particular) [119,121,123,124,125].
- Motivations and behaviors: backpackers' motivations and behaviors are influenced by their socio-cultural backgrounds [126,127] which are also reflected during their holidays "in the scope of their interactions with other backpackers" [120] (p. 99). Analysis of their motivations and behaviors is often clouded by a stereotypical view of backpackers and by the motivations which originally gave rise to backpacking as a "rite of passage" between youth and adulthood [120,128], referring back to the educational meaning of travel typical of the days of the Grand Tour (for details on the origin of backpacking, see [119,125,126]).

Loker-Murphy and Pearce [119] define backpackers as "travellers who exhibit a preference for budget accommodation; an emphasis on meeting other people (locals and travellers); an independently organized and flexible travel schedule, longer rather than brief holidays; and an emphasis on informal and participatory recreation activities". An overall picture emerges of people who distance themselves from mass tourism, favor locally-produced goods and services [123], sporting activities and contact with nature, are interested in environmentally-oriented activities, the educational side of travelling and having authentic holiday experiences [119,129] and are less interested in luxury, comfort and relaxing [121].

Backpackers, however, are very heterogeneous: the above profile cannot be generalized to the whole segment, elements of which are strongly effected by their social and national backgrounds [127,130,131] and lifestyles [132]. Larsen *et al.* [121] also point out that some characteristics, a desire to discover new cultures, for example, are not peculiar to backpackers, but rather are shared by other segments of demand.

So, we see that this strand of research investigates the motivations and behaviors which also engage the young more generally, but some specific characteristics of backpackers (including their sensitivity to certain dimensions of sustainability) [129] cannot also be attributed to the wider segment of youth and student travellers.

In the light of these observations, we consider that the investigation of young tourists' motivations for travel and behavior while travelling and of their sensitivity to sustainability, thus, appears to be an interesting area of study which could contribute, on the one hand, to increasing our knowledge of this segment and, on the other, to suggesting destinations how to improve and/or redefine their tourist offer. This seems to be particularly the case given the limited attention paid, until now, by researchers in the area of sustainable tourism to this segment of demand [32].

3. METHODOLOGY

3.1. The Population

Our field research analyses the holiday behavior and motivations of young tourists in Italy. We have enjoyed the cooperation of the Centro Turistico Studentesco e Giovanile (student and youth tourism centre) (CTS). This association, founded in 1974, is the largest Italian organization involved in the promotion, development, and organization of youth tourism. CTS sees travel as a means of spreading knowledge, preserving natural habitats and historic, artistic and cultural heritage and enriching young people's leisure time. Among its objectives are the promotion and diffusion of forms of responsible tourism coherent with the principles of sustainable development and environmental protection, and respectful of the rights of local communities (http://associazione.cts.it/).

Given the above, we began with the hypothesis that CTS tourists were an interesting subject for a case-study investigating the motivations and behaviors of young tourists interested in the aspects of sustainable tourism.

3.2. The Questionnaire

Our collaboration with CTS gave us access to the mailing list of its 20,140 registered users. CTS sent an email to its registered members in which it explained the research objectives and gave a link to our online questionnaire. The questionnaire contains 28 questions in four sections.

These sections cover the social and personal data of tourists and go into greater detail on the following topics:

- Organizational factors: analysis of factors such as the size of the tourist group (also revealing connections with "do-it-yourself" rather than intermediated tourism), their destination, and type of accommodation. With regard to information sources (the Internet, word of mouth, tourist guides, documentaries and TV programs, specialized magazines/journals, travel agents, tour operators' catalogues, radio, advertising), particular attention is paid to how regularly specialized magazines, dealing with issues such as environment, ecology, culture, and social responsibility, are read.
- Factors influencing the decision-making process: analysis of how the environmental, economic, and social factors influenced choices of destination, accommodation, and tour operator. With regard to destinations, particular attention is paid to the analysis of how important tourists felt the discovery of new cultures and landscapes, the contemplation of a destination's artistic and natural heritage, and contact with nature and local community to be. Regarding choice of accommodation, our research reveals the importance of the outcome of tourism for the local community, through the use of local products and local staff. Lastly, in the choice of tour operator, we focus on the importance attributed to the ethical aspects of the journey and the company, and the extent to which tour operators supported a destination's development projects and environmental protection strategies and hired local guides. Our analysis of these factors used reply options which required respondents to give a score to each item, grading their attribution of importance on a four point scale (to avoid respondent's fatigue on a longer scale but at the same time have them take a position) from "very important" to "not at all important".
- Behavioral factors: analysis of statements about behaviors adopted during the holiday and factors which may have influenced the purchase of locally-produced products. Particular attention is paid to respondents' willingness to integrate with the local community and to be involved in events organized by it and to commitments to learning about the local environment, culture, and history in order to minimize the negative impact of tourism upon the host community. For the analysis of behaviors, we also asked definite questions, using a projective technique (the third-person technique) in order to reveal less biased opinions than those stated by the respondents. Projective techniques were introduced in psychology but are regularly deployed also in marketing research. One of the most common is the "third-person test". In this technique the respondent is required to describe how s/he would tackle a task or a situation by identifying him/herself in a hypothetical subject, *i.e.*, a third person. This allows researchers to

address potentially sensitive issues, and enable the respondent to analyze situations from a certain distance, thus expressing his/her opinions/evaluations on specific topics more objectively (for further information see [133]). The questionnaire developed for this research included the question "How much does Daisy look like you?" with reference to issues related to sustainable tourism behavior. Possible answers ranged from "a lot" to "not at all".

Our research also investigated people's willingness to pay a premium (on the next holiday) in order to protect the local environment, support the culture and the local community, limit the negative impact of tourism, and guarantee decent wages and good working conditions to local workers. The willingness to pay was evaluated by giving four reply options: paying a premium "of more than 20%", "between 10%–20%", "less than 10%" or "no premium at all".

3.3. Statistics and Data Analysis

The research is based on 2020 online replies to our questionnaire. Most respondents (57.2%) were between 19 and 29 years old, 25.9% were between 30 and 39, 9.4% between 40 and 49, and 7.5% were over 50. As the focus of research was young tourists, and in the light of the points highlighted in the literature review regarding socio-demographic characteristics, the decision making-process, motivations and behaviors of the 1156 respondents aged between 19 and 29 were analyzed. Although the exploratory nature of the research does not permit us to extend the results to the entire population observed, we consider that the high number of questionnaire responses, consistent with other international research (see [37]) makes it possible to identify characteristics which contribute to the analysis of the young tourist profile.

The elaboration of the data involves two steps. The first uses descriptive statistical tools and analyses aggregate data on young tourists to map their main organizational, motivational, and behavioral characteristics and their sensitivity to sustainability (see Section 4.1). This step allows us to achieve the first research goal.

The second step identifies "hard path young tourists" (HPYT) and "soft path young tourists" (SPYT), measures the composition of the two profiles and analyses the personal, organizational, motivational, and behavioral characteristics of the two groups. The identification of HPYT uses variables associated with harder ecotourists [37]. The results of this analysis are described in the Section 4.2. The analysis and comparison of the HPYT and the SPYT profiles use descriptive statistical tools and are supported by the results from non-parametric tests (two proportions test). The two proportions measured are HPYT and SPYT. The comparisons of these two proportions is performed with α = 0.01 and α = 0.05. This second step of the analysis allows us to describe the organizational, motivational, and behavioral profiles of HPYT and SPYT (see Section 4.3), thus fulfilling the second aim of the research.

4. RESULTS AND DISCUSSION

4.1. Young Tourists and Sensitivity to Sustainability

The young tourists are mainly students (with a university degree or a high school diploma), women, living in Northern Italy, unmarried, with no children (Table 1). The main characteristics linked to the organizational profile of the respondents are: over 70% prefer to organize their holidays independently; about 44% state that they never travel with organized groups and 51% prefer small groups (less than 10 people). The Internet is the main information source (94.3%), followed by word-of-mouth (81.9%) and the consultation of tourist guides (72.8%). About 58% are interested in reading supplements of magazines devoted to environmental, ecological, and cultural matters. Italy (85.3%) and other European countries (77.7%) are the most visited destinations. Hotels are the most frequently used accommodation (75.2%).

Table 1. Socio-demographic variables of young respondents (age: 19–29 years) (note: data expressed in the form of percentages).

Socio-Demographic Variables	Main Results (*n* = 1156)
Gender	women (73.0)–men (27.0)
Place of residence	northern Italy (50.6)–central Italy (30.6)–southern Italy (18.8)
Marital status	single (94.4)–married/cohabiting (5.6)
Family unit	no children (92.9)–independent children (5.4)–small children (1.6)
Job	student (55.6)–employee (23.7)–self-employed (8.9)–unemployed (4.8)–others (6.9)
Education	university degree (53.8)–high school diploma (44.7)

The findings confirm young people's interest in certain dimensions of sustainability and the influence this interest has on their decision-making process, motivations and behaviors (see Table 2).

Table 2. Dimensions of sustainability influencing the decision-making process of young tourists in the choice of the destination, accommodation, and tour operator (*n* = 1156) (note: data expressed in the form of percentages).

	Very Important	Quite Important	Not Very Important	Not Important at All
CHOICE OF DESTINATION				
discovery of new cultures	75.2	21.9	2.3	0.6
discovery of new landscapes	71.5	26.2	2.0	0.3
contemplation of natural/artistic heritage	66.1	28.3	4.8	0.9
contact with the local community	47.1	38.3	13.1	1.5
contact with nature	43.9	40.8	13.4	1.9
CHOICE OF ACCOMMODATION				
profits benefiting the local community	14.0	31.4	36.0	18.6
CHOICE OF TOUR OPERATOR				
ethicality of the journey and of the organization	41.3	43.1	12.1	3.5
support for local development projects and environmental protection	28.0	40.9	24.7	6.4
involvement of local guides	23.2	41.6	27.8	7.4

With regard to the decision-making process we see that:

- in the *choice of destination* the motivations considered most important are the discovery of new cultures, discovery of new landscapes, contemplation of natural and artistic heritage, contact with local community, contact with nature;
- in the *choice of accommodation* price is the most influential factor (over 65% consider it very important and more than 30% quite important); factors more relevant to sustainability, like the actual return and benefit to the local community from tourism, are considered of little importance by the majority of respondents;
- in the *choice of a tour operator* price is the most influential factor (over 67% consider it very important); the ethicality of the journey and the organization are also very important; most respondents consider support for local development projects and environmental protection and the involvement of local guides, to be quite important.

Their statements about behavior show that about 40% of respondents are willing to pay a premium of up to 10% on their next holiday in order to support local cultures and communities, protect the environment, limit the negative effects of tourism, and ensure that local workers have good wages and working conditions.

Most young tourists say they prefer local food, adapt as much as they can to the traditions and customs of the place in which they are holidaying, try to learn about their destination before travelling, would be willing to be involved in events organized by the local community and to interact with it, demonstrate interest in the protection of the authenticity of the destination, even if this means going without certain comforts, find out how to protect the local environment and reduce waste, and are concerned to ensure that their spending benefits the local population.

4.2. Identification of Hard Path Young Tourists (HPYT) and Soft Path Young Tourists (SPYT)

Acknowledging the general interest of young tourists in sustainability, this research examines in detail the characteristics of those young people who express interest in all the dimensions of sustainability covered in this research, identifying hard path young tourists (HPYT) and soft path young tourists (SPYT) and analyzing their personal, organizational, motivational, and behavioral characteristics. To this end, variables and criteria, by which each interviewee is associated with one profile or the other, have been defined.

The choice of variables is based on the main characteristics to emerge from the literature of the "hard" profile, referring in particular to those linked with harder ecotourism described by Weaver and Lawton [37]: small groups, enhancing sustainability, strong environmental commitment, and emphasis on personal experience. These characteristics are traced back to motivational and behavioral variables which illustrate an interest in sustainability (environmental, economic, and social). For example, the willingness to pay a premium in order to protect the local environment and limit the negative impact of tourism is a variable which illustrates a behavior linked to the characteristics of strong environmental commitment and a desire to enhance sustainability (environmental and social dimensions) as described by Weaver and Lawton [37].

With regard to the exploratory nature of the research, particularly restrictive criteria have been adopted in the association of each respondent to the HPYT profile. This allows us to clearly identify the two profiles and to record their organizational, motivational, and behavioral characteristics which will be more closely investigated through further research.

The variables and criteria used to identify the HPYT profile are shown in Table 3. This profile exhibits a high level of sensitivity to all the observed dimensions: it includes 198 respondents; that is about 17% of the young tourists. The other respondents that do not satisfy all the criteria are classified as SPYT.

An analysis and comparison of the personal, organizational, motivational, and behavioral characteristics of HPYT and SPYT follow below. Unless otherwise indicated, the results of the tests refer to values of α to 0.01.

Table 3. Aspects, questions, variables, and criteria used to identify the profile of hard path young tourists (HPYT).

Aspects	Questions	Variables	Criterion	Harder Ecotourists [37]
Organization of holiday	Composition of the tourist group	Composition of the tourist group	Never travel with organized groups or small groups (<10 people)	Small groups
Decision-making process and motivations	Choice of destination	Discovery of new cultures	All variables evaluated as "very important" or "quite important"	Emphasis on personal experience
		Discovery of landscapes		Strong environmental commitment Emphasis on personal experience
		Contemplation of natural/artistic heritage		Strong environmental commitment Emphasis on personal experience
		Contact with the local community		Emphasis on personal experience
		Contact with nature		Strong environmental commitment Emphasis on personal experience
	Choice of accommodation	Benefits accruing to the local community from tourism	Variable evaluated as "very important" or "quite important"	Enhancing sustainability
	Choice of tour operator	Ethicality of the journey and the tour operator	All variables evaluated as "very important" or "quite important"	Enhancing sustainability
		Tour operator's declared support for development projects and environmental protection in the local area		Enhancing sustainability
		Involvement of local guides		Enhancing sustainability

Behavior	Willingness to pay a premium (for next holiday)	To support the culture and the local community	Declaration of willingness to pay a premium	Enhancing sustainability
		To preserve the local environment and limit the negative impact of tourism		Enhancing sustainability Strong environmental commitment
		To guarantee decent wages to local workers and good working conditions		Enhancing sustainability
	Importance or accordance declared in relation to specific statements	Preference for local food	All variables evaluated as "very" or "quite"	Emphasis on personal experience
		Adaptation to local habits		Emphasis on personal experience
		Willingness to be involved in events organized by local communities		Emphasis on personal experience
		Commitment to learn about the local environment, culture and history		Emphasis on personal experience
		Protection of the authenticity of the destination		Strong environmental commitment Enhancing sustainability
		Interaction with local community		Emphasis on personal experience
		Commitment to reduce waste and limit the impact of tourism on the local community		Enhancing sustainability
		Spending benefits local population		Enhancing sustainability
		Tourism effects on local community		Enhancing sustainability

4.3. Hard Path Young Tourists vs. Soft Path Young Tourists

4.3.1. Personal Profile

The comparison between gender, family composition, and professions of HPYT and SPYT confirms, as described for the whole sample, the prevalence of women, family without children, students (over 50%) and employees (about 24%).

Statistically significant differences emerge in marital status, education, and participation in groups or associations. Although the majority of respondents in both profiles are single, this is more significant among the SPYT; most of the young have degrees or qualifications, but graduates are more prevalent among the HTPT; there are more students with high school diplomas in the SPYT profile ($\alpha = 0.05$). Although membership of groups and associations is not very widespread, the findings show that more HPYT participate in cultural and voluntary work; fewer than 20% of the HPYT are involved in environmental and political associations, but this still differentiates them from the SPYT ($\alpha = 0.05$). About a third of respondents in both profiles are members of sports associations.

4.3.2. Organizational Profile

The organizational profiles of HPYT and SPYT differ in relation to four specific factors: composition of tourist groups, information sources used, destination, types of accommodation.

Most respondents organize their holidays independently (HPYT 86.9% *vs.* 67.2% SPYT), but this is more prevalent in the HPYT profile; SPYT are more likely to look for travel agents. Tourist groups in both profiles are usually composed of friends and relatives, but almost half the HPYT say they also travel alone. In sharp contrast, fewer than 30% of the SPYT would travel alone.

Both profiles use the Internet and word of mouth as their main sources of information. HPYT use the Internet more and are also more willing to spend time collecting information, are more autonomous in seeking this information and use multiple sources ranging from tourist guides and television documentaries to specialized journals and radio programs. SPYT, on the other hand, tend to rely on travel agencies and tour operators for information.

The HPYT profile is characterized by more regular reading of magazines dedicated to environmental, ecological, geographical and cultural topics, civil society, social responsibility, and critical consumption.

Italy and Europe are the favorite destinations for both profiles, but about 15% of the HPYT had visited North or South America in the previous year (α =0.05).

Friends' or relatives' houses and hotels are young tourists' preferred accommodation. HPYTs, however, are more likely than SPYTs to choose private homes (67.2% *vs.* 57.8% SPYT; α = 0.05) and *vice versa* for hotels (77.2% SPYT *vs.* 65.2% HPYT). Campsites and agritourisms are also more popular among HPYT, residences and tourist villages are more often the choice of SPYT (in both cases results with α = 0.05).

4.3.3. Motivational Profile

The following motivations influence HPYTs' and SPYTs' choice of destination, accommodation and, if relevant, tour operator:

- *Destination choice*: adventure is the most important factor for most HPYT; relaxing and/or entertainment and a search for comfort are more typical of the SPYT. Both profiles attribute little importance to sport.
- *Accommodation choice*: price is the most important variable for both profiles. It is, however, consistent with the HPYT profile that motivations linked to availability of local produce and locally-employed staff follow that of price for HPYT, unlike SPYT who value comfort more highly.
- *Choice of tour operator*: the motivations of both profiles are equally influenced by price, followed by the quality standards of the organization.

4.3.4. Behavioral Profile

Over 90% of HPYT and SPYT have been to restaurants, pubs or local lodgings. Both profiles are interested in visiting museums or places linked to local folklore and traditions, and buying local and DOC products, but this is more markedly the case for HPYT.

Buying fairly-traded and organic products is clearly more important to HPYT, most of whom would be willing to pay premiums of as much as 20%. Most SPYT would be willing to accept premiums up to 10% or more, while one in six tourists is not willing to accept increased prices; which differentiates them from the HPYT.

The SPYT differ from the HPYT in that they are willing to pay a premium (10%–20%) to ensure that their holiday will go according to plan, and to increase the comfort and the quality of their holiday.

HPYT and SPYT differ in their descriptions of their behavior while on holiday. For HPYT the importance of protected areas, typical local products, and organic products is greater; they also worry more about pollution, are more aware of the impacts of tourism on the climate and the environment, more willing to forgo skiing if the only available snow is artificial. Moreover, they judge many tourists not to be sufficiently respectful of the environment or local customs.

5. DISCUSSION AND IMPLICATIONS

The study analyzed a select population of young tourists (members of CTS) and focuses on their personal, organizational, motivational, and behavioral characteristics. The first step of the analysis confirms some of the trends highlighted in the UNWTO report [26] and UNWTO and WYSE Travel Confederation [27] regarding youth and student travellers' interest in several dimensions of sustainability. An interest in environmental attitudes and ecological belief as highlighted in Benckendorff *et al.* [118] is also confirmed, as is the importance for the young of the Internet as a source of information.

Of course, young tourists vary in the intensity of their interest in sustainability. The adoption of very strict criteria for segmentation has allowed us to identify and distinguish HPYT and SPYT and analyze their decision-making processes, motivations, and behaviors. The greater independence and autonomy shown by HPYTs in the organization of their holidays and their willingness to pay premiums find confirmation in the harder ecotourist profile described by Weaver and Lawton [37]. In contrast, the SPYT tendency to rely on agencies to make their travel arrangements, the importance they place on comfort and relaxation, and their willingness to spend more in order to ensure they can relax in comfort, correlates the SPYT profile with that of Weaver and Lawton's softer ecotourists.

The analysis of HPYT and SPYT reveals two profiles which could be considered by destinations in the planning on their strategic and operational choices. Examples of this might be the means of communication (supported by travel agencies or tour operator in the case of SPYT; specialized journals for HPYT *vs.* general publications for SPYT) and the promotion of offers in various accommodation facilities (campsites and agritourisms for HPYT *vs.* tourist villages and residences for SPYT) or the development of particular local attractions and products. Both profiles are interested in local products, but HPYT are more willing to accept limits to their consumption and are more aware of the impact of tourism on places and communities. In addition, the strong sensitivity of HPYT to sustainable offers suggests that destinations could create tourist offers that optimize the unique features of the territory. The optimization of a territory's distinctive features (natural resources, historical, artistic, and cultural heritage) alongside the limiting of access and resource use can allow destinations to develop new products and to differentiate the offer in order to better respond to the specific needs of the young profiles, and at the same time preserve the natural, economic, and socio-cultural capital of the territory and increase the authenticity of the tourism experience [3,9,12].

Our research has drawn attention to the motivations and behaviors of a share of tourist demand which shows a steady interest in sustainable tourism and analyzes a subject on which little work has yet been done. The exploratory nature of the research does not allow us to extend the results to the whole population of young tourists and therefore further investigation is required. The intention is to follow two analytical paths. The first provides for the extension of this analysis to other populations potentially interested in sustainable tourism offers (*i.e.*, members of the WWF or other environmental associations). The second examines sensitivity to sustainability by investigating the motivations and behaviors of generic populations of young tourists (such as university students).

The analysis and comparison of the results from these two research paths allows us:

- To verify the significance of the profiles which we have established and to compare the composition (weight) of HPYT and SPYT. Our research identifies about 17% of respondents as HPYT. This figure is lower than the cluster of harder ecotourists picked out by Weaver and Lawton [37], but the differences between the contexts and research methods prevent any direct comparison between the two profiles. We believe it necessary, however, to examine this issue, using the research methodology which we have adopted (variables and analytical criteria) to analyze populations of young tourists with similar/different characteristics to those of the young CTS travellers.
- To examine the relationship between young tourists and the price variable. The research reveals the strong influence this variable has on choices of accommodation and, when relevant, tour operator. Given the relatively high spending power of this tourist segment (as shown in international report and in the literature) (see [26,33]) it is worth examining this result more closely, both with regard to young tourists in general and to HPYT and SPYT.

We intend to conduct the next stages of the research using methods which integrate (a) online surveys of wide cross-sections of young tourists; (b) laboratory tests based on the methods of experimental economics; and (c) choice experiments on willingness to pay. The methodology thus adopted (bringing together research tools and techniques which have already been used by the author's research group) will allow us to verify the validity and wider applicability of the data collected in the course of this project, overcoming the limits inherent in an analysis of motivations and behavior based on personal statements.

ACKNOWLEDGMENTS

The author wish to thank the Italian Student and Youth Tourism Centre (CTS) for collaboration. The author is grateful to prof. Mariangela Franch and prof. Umberto Martini of the Department of Economics and Management (University of Trento, Italy) for their comments and suggestions.

REFERENCES

1. Dwyer, L.; Edwards, D.; Mistilis, N.; Roman, C.; Scott, N. Destination and enterprise management for a tourism future. *Tour. Manag.* **2009**, *30*, 63–74.
2. Chhabra, D. Understanding authenticity and its determinants. *J. Travel Res.* **2005**, *44*, 64–73.
3. Cohen, E. Authenticity and commoditization in tourism. *Ann. Tour. Res.* **1988**, *15*, 371–386.
4. Ferrari, S.; Adamo, G.E. Autenticità e risorse locali come attrattive turistiche. *Sinergie* **2005**, *66*, 80–112. (In Italian).
5. Hughes, G. Authenticity in tourism. *Ann. Tour. Res.* **1995**, *22*, 781–803.
6. Yeoman, I.; Brass, D.; McMahon-Beattie, U. Current issue in tourism. *Tour. Manag.* **2007**, *28*, 1128–1138.
7. Taylor, J.P. Authenticity and Sincerity in Tourism. *Ann. Tour. Res.* **2001**, *28*, 7–26.
8. Coghlan, A. Linking natural resource management to tourist satisfaction. *J. Sustain. Tour.* **2012**, *20*, 41–58.
9. McIntosh, A.J.; Prentice, R.C. Affirming authenticity—Consuming cultural heritage. *Ann. Tour. Res.* **1999**, *26*, 589–612.
10. Noy, C. This trip really changed me. *Ann. Tour. Res.* **2004**, *31*, 78–102.
11. Walker, K.; Moscardo, G. Encouraging sustainability beyond the tourist experience. *J. Sustain. Tour.* **2014**, *22*, 1175–1196.

12. Wang, N. Rethinking Authenticity in Tourism Experience. *Ann. Tour. Res.* **1999**, *26*, 349–370.
13. Sedmak, G.; Mihalič, T. Authenticity in mature seaside resorts. *Ann. Tour. Res.* **2008**, *35*, 1007–1031.
14. Pearce, D.G.; Schänzel, H.A. Destination management: The tourists' perspective. *J. Destin. Mark. Manag.* **2013**, *2*, 137–145.
15. Beritelli, P.; Laesser, C. Getting the cash-cow directors on board—An alternative view on financing DMOs. *J. Destin. Mark. Manag.* **2014**, *2*, 213–220.
16. Chen, C.-M.; Chen, S.H.; Lee, H.T. The destination competitiveness of Kinmen's tourism industry. *J. Sustain. Tour.* **2011**, *19*, 247–264.
17. Franch, M., Martini, U., Eds.; *Management per la Sostenibilità dello Sviluppo Turistico e la Competitività Delle Destinazioni*; Il Mulino: Bologna, Italy, 2013. (In Italian)
18. Ritchie, J.R.B.; Crouch, G.I. *The Competitive Destination*; CABI: Wallingford, UK, 2003.
19. Weaver, D.B. Organic, incremental and induced paths to sustainable mass tourism convergence. *Tour. Manag.* **2012**,*33*, 1030–1037.
20. Ritchie, J.R.B.; Crouch, G.I. Editorial—The Competitive Destination. *Tour. Manag.* **2000**, *21*, 1–7.
21. Moscardo, G.; Murphy, L. There is No Such Thing as Sustainable Tourism: Re-Conceptualizing Tourism as a Tool for Sustainability. *Sustainability* **2014**, *6*, 2538–2561.
22. Jamal, T.; Camargo, B.A.; Wilson, E. Critical Omissions and New Directions for Sustainable Tourism: A Situated Macro–Micro Approach. *Sustainability* **2013**, *5*, 4594–4613.
23. Aall, C. Sustainable Tourism in Practice: Promoting or Perverting the Quest for a Sustainable Development?*Sustainability* **2014**, *6*, 2562–2583.
24. Bowser, G.; Gretzel, U.; Davis, E.; Brown, M. Educating the Future of Sustainability. *Sustainability* **2014**, *6*, 692–701.
25. Benckendorff, P., Moscardo, G., Pendergast, D., Eds.; *Tourism and Generation Y*; CABI: Wallingford, UK, 2010.
26. UNQTO. *Youth Travel Matters*; Unwto: Madrid, Spain, 2008.
27. UNWTO & WYSE Travel Confederation. The Power of Youth Travel. 2011. Available online: http://www2.unwto.org/publication/am-reports-volume-2-power-youth-travel (accessed on 20 August 2015).

28. WYSE Travel Confederation. Youth and Student Travel Market—Data, Statistics and Trends. 2012. Available online: http://staywyse.org/research-2/industry-review-no-1/ (accessed on 30 August 2015).
29. WYSE Travel Confederation. Youth and Student Travel Market—Global Data, Statistics and Trends II. 2014. Available online: https://www.wysetc.org/wp-content/uploads/2015/01/Industry-Review-No.-5-Executive-Summary.pdf(accessed on 31 August 2015).
30. Moscardo, G.; Benckendorff, P. Mythbusting: Generation Y and travel. In *Tourism and Generation Y*; Benckendorff, P., Moscardo, G., Pendergast, D., Eds.; CAB International: Wallingford, UK, 2010; pp. 16–26.
31. Chhabra, D. What do student travelers really want? Revisiting the Buyer's Black Box. *J. Hosp. Mark. Manag.* **2011**, *21*, 344–355.
32. Cini, F.; Leone, L.; Passafaro, P. Promoting Ecotourism among Young People. *Environ. Behav.* **2012**, *44*, 87–106.
33. Leask, A.; Fyall, A.; Barron, P. Generation Y: Opportunity or challenge. *Curr. Issues Tour.* **2011**.
34. Nusair, K.K.; Parsa, H.G.; Cobanoglu, C. Building a model of commitment for Generation Y. *Tour. Manag.* **2011**, *32*, 833–843.
35. Pendergast, D. Getting to Know the Y Generation. In *Tourism and Generation Y*; Benckendorff, P., Moscardo, G., Pendergast, D., Eds.; CAB International: Wallingford, UK, 2010; pp. 1–15.
36. Scarpato, D.; Annunziata, A.; Ardeleanu, M.P. Sustainable tourism and the perception of the young generation: An analysis of the South of Italy. *Qual. Access Success* **2014**, *15*, 254–259.
37. Weaver, D.B.; Lawton, L.J. Overnight ecotourism market segmentation in the Gold Coast hinterland of Australia. *J. Travel Res.* **2002**, *40*, 270–280.
38. WCED. *Our Common Future*; Oxford University Press: Oxford, UK, 1987.
39. Butler, R.W. Sustainable Tourism: A state-of-the-art review. *Tour. Geogr.* **1999**, *1*, 7–25.
40. McKercher, B. Some fundamental truths about tourism. *J. Sustain. Tour.* **1993**, *1*, 6–16.
41. Inskeep, E. *Tourism Planning*; Van Nostrand Reinhold: New York, NY, USA, 1997.

42. Swarbrooke, J. *Sustainable Tourism Management*; CABI: Wallingford, UK, 1999.
43. Bahaire, T.; Elliott-White, M. Community Participation in Tourism Planning and Development in the Historic City of York, England. *Curr. Issues Tour.* **1999**, *2*, 243–276.
44. Bramwell, B. Governance, the state and sustainable tourism. *J. Sustain. Tour.* **2011**, *19*, 459–477.
45. Bramwell, B.; Lane, B. Sustainable tourism and the evolving roles of government planning. *J. Sustain. Tour.* **2010**, *18*, 1–5.
46. Jamal, T.; Getz, D. Collaboration theory and community tourism planning. *Ann. Tour. Res.* **1995**, *22*, 186–204.
47. Nelson, J.G., Butler, R.W., Wall, G., Eds.; *Tourism and Sustainable Development*; University of Waterloo: Waterloo, ON, Canada, 1993.
48. Plummer, R.; Fennell, D.A. Managing protected areas for sustainable tourism. *J. Sustain. Tour.* **2009**, *17*, 149–168.
49. Robinson, M. Collaboration and cultural consent. *J. Sustain. Tour.* **1999**, *7*, 379–397.
50. Selin, S. Developing a typology of sustainable tourism partnerships. *J. Sustain. Tour.* **1999**, *7*, 260–273.
51. Spenceley, A.; Meyer, D. Tourism and poverty reduction. *J. Sustain. Tour.* **2012**, *20*, 297–317.
52. Tosun, C. Roots of unsustainable tourism development at the local level. *Tour. Manag.* **1998**, *19*, 595–610.
53. Dolnicar, S. Insights into sustainable tourists in Austria. *J. Sustain. Tour.* **2004**, *12*, 209–218.
54. Dolnicar, S. Identifying tourists with smaller environmental footprints. *J. Sustain. Tour.* **2010**, *18*, 717–734.
55. Fiorello, A.; Bo, D. Community-based ecotourism to meet the new tourist's expectations. *J. Hosp. Mark. Manag.* **2012**,*21*, 758–778.
56. González, R.; Otero, A. Alternative tourism activities management in the Argentinean-Chilean Great Lakes Corridor.*Curr. Issues Tour.* **2002**, *5*, 193–207.
57. Hunter, C. Sustainable tourism as an adaptive paradigm. *Ann. Tour. Res.* **1997**, *24*, 850–867.
58. Kline, J.D. Tourism and Natural Resource Management: A General Overview of Research and Issues. 2001. Available online: http://www.fsl.orst.edu/lulcd/Publicationsalpha_files/Kline_2001_GTR506.pdf (accessed on 15 October 2015).

59. Middleton, V.T.C.; Hawkins, R. *Sustainable Tourism*; Butterworth-Heinemann: Oxford, UK, 1998.
60. Zografos, C.; Allcroft, D. The environmental values of potential ecotourists. *J. Sustain. Tour.* **2007**, *15*, 44–66.
61. Brown, F. Alternative tourism Insights. *Magazine of the English Tourist Board/British Tourist Authority* **1991**.
62. Clarke, J. A Framework of Approaches to Sustainable Tourism. *J. Sustain. Tour.* **1997**, *5*, 224–233.
63. Liu, Z. Sustainable tourism development. *J. Sustain. Tour.* **2003**, *11*, 459–475.
64. Harris, R., Leiper, N., Eds.; *Sustainable Tourism*; Butterworth-Heinemann: Oxford, UK, 1995.
65. Page, S.J.; Dowling, R.K. *Ecotourism*; Prentice Hall: Harlow, UK, 2002.
66. Weaver, D.B.; Lawton, L.J. *Tourism Management*; Wiley: Milton, Australia, 2002.
67. Ballantine, J.K.; Eagles, P.F. Defining Canadian ecotourists. *J. Sustain. Tour.* **1994**, *2*, 210–214.
68. Blamey, R.; Braithwaite, V. A social values segmentation of the potential ecotourism market. *J. Sustain. Tour.* **1997**, *5*, 29–45.
69. Buckley, R. *Case Studies in Ecotourism*; CABI: Wallingford, UK, 2003.
70. Fennel, D.A. *Ecotourism Programme Planning*; CABI: Wallingford, UK, 2002.
71. Kerstetter, D.L.; Hou, J.S.; Lin, C.H. Profiling Taiwanese ecotourists using a behavioral approach. *Tour. Manag.* **2004**,*25*, 491–498.
72. Juric, B.; Cornwell, T.B.; Mather, D. Exploring the usefulness of an ecotourism interest. *J. Travel Res.* **2002**, *40*, 259–269.
73. Lawton, L. A Profile of Older Adult Ecotourists in Australia. *J. Hosp. Mark. Manag.* **2002**, *9*, 113–132.
74. Meric, H.C.; Hunt, J. Ecotourist motivational and demographic characteristics. *J. Travel Res.* **1998**, *36*, 57–61.
75. Weaver, D.B.; Lawton, L.J. Twenty years on: The state of contemporary ecotourism research. *Tour. Manag.* **2007**, *28*, 1168–1179.
76. Wight, P. North American ecotourism markets. *J. Travel Res.* **1996**, *34*, 3–10.
77. Wurzinger, S.; Johansson, M. Environmental Concern and Knowledge of Ecotourism among Three Groups of Swedish Tourists. *J. Travel Res.* **2006**, *45*, 217–226.

78. Buckley, R.; Pickering, C.; Weaver, D.B. *Nature-Based Tourism, Environment and Land Management*; CABI: Wallingford, UK, 2003.
79. Dolnicar, S.; Leisch, F. Selective marketing for environmentally sustainable tourism. *Tour. Manag.* **2008**, *29*, 672–680.
80. Hassan, S.; Mohd, A.; Mariapan, M.; Husin, M.Z. Segmentation of nature tourists through motivation of travel to Kuala Gandah national elephant conservation centre. *Malays. Forest.* **2010**, *73*, 7–15.
81. Hvengaard, G.T.; Dearden, P. Ecotourism versus tourism in a Thai national park. *Ann. Tour. Res.* **1998**, *25*, 700–720.
82. Luzar, E.J.; Diagne, A.; Ecgan, C.; Henning, B.R. Profiling the Nature-based Tourist. *J. Trav. Res.* **1998**, *37*, 48–55.
83. Mehmetoglu, M. Segmenting the Nature-Based Tourists Based on Travel Mode Choice. *J. Hosp. Leis. Mark.* **2006**, *14*, 47–67.
84. Priskin, J. Assessment of natural resources for nature-based tourism. *Tour. Manag.* **2001**, *22*, 637–648.
85. Rinne, P.; Saastamoinen, O. Local Economic Role of Nature-based Tourism in Kuhmo Municipality, Eastern Finland.*Scand. J. Hosp. Tour.* **2005**, *5*, 89–101.
86. Colombo, L. *Il Turismo Responsabile*; Xenia: Milano, Italy, 2005. (In Italian)
87. Frey, N.; George, R. Responsible tourism management. *Tour. Manag.* **2010**, *31*, 621–628.
88. Goodwin, H.; Francis, J. Ethical and responsible tourism. *J. Vacat. Mark.* **2003**, *9*, 271–284.
89. Pennington-Gray, L.; Reisinger, Y.; Kim, J.E.; Thapa, B. Do U.S. tour operators' brochures educate the tourist on culturally responsible behaviours? *J. Vacat. Mark.* **2005**, *11*, 265–284.
90. Sambri, C.; Pegan, G. La domanda di turismo responsabile. *Mercat. Compet.* **2008**, *1*, 65–92. (In Italian).
91. Dolnicar, S.; Matus, K. Are Green Tourists a Managerially Useful Target Segment? *J. Hosp. Leis. Mark.* **2008**, *17*, 314–334.
92. Hong, S.; Kim, S.; Kim, J. Implications of potential green tourism development. *Ann. Tour. Res.* **2003**, *30*, 323–341.
93. Holtz, C.; Edwards, S. Linking Biodiversity and Tourism Policy. In *Ecotourism, Policy and Planning*; Fennel, D.A., Dowling, R.K., Eds.; CAB International: Oxon, UK, 2003; pp. 39–54.
94. Fennel, D.A. A content analysis of ecotourism definitions. *Curr. Issues Tour.* **2001**, *4*, 403–421.

95. Luo, Y.; Deng, J. The new Environmental Paradigm and Nature-Based Tourism Motivation. *J. Travel Res.* **2008**, *46*, 392–402.

96. Perkins, H.E.; Brown, P.R. Environmental Values and the So-Called True Ecotourist. *J. Travel Res.* **2012**, *51*, 793–803.

97. Blamey, R. Ecotourism: The search for an operational definition. *J. Sustain. Tour.* **1997**, *5*, 109–130.

98. Blamey, R. Principles of ecotourism. In *The Encyclopedia of Ecotourism*; Weaver, D.B., Ed.; CABI: Wallingford, UK, 2001; pp. 5–22.

99. Goodwin, H. In Pursuit of Ecotoruism. *Biodivers. Conserv.* **1996**, *5*, 277–291.

100. UNWTO. *The Italian Ecotourism Market*; Unwto: Madrid, Spain, 2002.

101. Laarman, J.; Durst, P. *Nature Travel and Tropical Forests (FREI Working Paper Series)*; Southeastern Center for Forest Economics Research: Raleigh, NC, USA, 1987.

102. Hunter, C. Sustainable Tourism and the Touristic Ecological Footprint. *Environ. Dev. Sustain.* **2002**, *4*, 7–20.

103. Diamantis, D. The characteristics of UK's ecotourists. *Tour. Recreat. Res.* **1999**, *24*, 99–102.

104. Palacio, V.; McCool, S. Identifying ecotourists in Belize through benefit segmentation. *J. Sustain. Tour.* **1997**, *5*, 234–243.

105. Fennel, D.A. The Canadian ecotourist in Costa Rica. *Int. J. Sustain. Dev.* **2002**, *5*, 282–299.

106. Weaver, D.B. A Broad Context Model of Destination Development Scenarios. *Tour. Manag.* **2000**, *21*, 217–224.

107. Crompton, J.L. Motivations for pleasure vacation. *Ann. Tour. Res.* **1979**, *6*, 408–424.

108. Dann, G.M.S. Anomie, ego-enhancement and tourism. *Ann. Tour. Res.* **1977**, *4*, 184–194.

109. McIntosh, R.W.; Goeldner, C.R. *Tourism. Principles, Practices, Philosophies*; Wiley: New York, NY, USA, 1990.

110. Moscardo, G.; Morrison, A.M.; Pearce, P.L.; Lang, C.T.; O'Leary, J. Understanding vacation destination choice through travel motivation and activities. *J. Vacat. Mark.* **1996**, *2*, 109–122.

111. Plog, S.C. Why destination areas rise and fall in popularity. *Cornell Hotel Rest. Q.* **1974**, *14*, 55–58.

112. Plog, S.C. Why destination areas rise and fall in popularity. An Update of a Cornell Quarterly Classic. *Cornell Hotel Rest. Admin. Quart.* **2001**, *42*, 13–24.
113. Ozdemir, B.; Aksu, A.; Ehtiyar, R.; Çizel, B.; Çizel, R.B.; İçigen, E.T. Relationships Among Tourist Profile, Satisfaction and Destination Loyalty. *J. Hosp. Mark. Manag.* **2012**, *21*, 506–540.
114. Schmidhauser, H. Tourist needs and motivations. In *The Tourism Marketing and Management Handbook*; Witt, S., Moutinho, L., Eds.; Prentice Hall: London, UK, 1989; pp. 569–572.
115. Schmoll, G.A. *Tourism Promotion*; Tourism International Press: London, UK, 1977.
116. Vukic, M.; Kuzmanovic, M.; Stankovic, M.K. Understanding the Heterogeneity of Generation Y's Preferences for Travelling: A Conjoint Analysis Approach. *Int. J. Tour. Adm. Res.* **2014**.
117. Huntley, R. *The World According to Y*; Allen and Unwin: Crows Nest, Australia, 2006.
118. Benckendorff, P.; Moscardo, G.; Murphy, L. Environmental Attitudes of Generation Y Students. *J. Teach. Travel. Tour.* **2012**, *12*, 44–69.
119. Loker-Murphy, L.; Pearce, P.L. Young budget travelers: Backpackers in Australia. *Ann. Tour. Res.* **1995**, *22*, 819–843.
120. Cohen, E. Backpacking: Diversity and Change. *J. Tour. Cult. Ch.* **2003**, *1*, 95–110.
121. Larsen, S.; Øgaard, T.; Brun, W. Backpackers and Mainstreamers. *Ann. Tour. Res.* **2001**, *38*, 690–707.
122. Elsrud, T. Risk creation in traveling: Backpacker adventure narration. *Ann. Tour. Res.* **2001**, *28*, 597–617.
123. Hampton, M.P. Backpacker tourism and economic development. *Ann. Tour. Res.* **1998**, *25*, 639–660.
124. Jennings, G.; Cater, C.; Lee, Y.-S.; Ollenburg, C.; Ayling, A. Generation Y: Perspectives of quality in youth adventure travel experiences in an Australian backpacker context. In *Tourism and Generation Y*; Benckendorff, P.J., Moscardo, G., Pendergast, D., Eds.; CABI: Oxfordshire, UK, 2010; pp. 58–72.
125. O'Reilly, C.C. From drifter to gap year tourist. *Ann. Tour. Res.* **2006**, *33*, 998–1017.
126. Mohsin, A.; Ryan, C. Backpakers in the Northern territory of Australia. *Int. J. Tour. Res.* **2003**, *5*, 113–131.
127. Maoz, D. Backpackers' motivations the role of culture and nationality. *Ann. Tour. Res.* **2007**, *34*, 122–140.

128. Noy, C., Cohen, E., Eds.; *Israeli Backpackers*; State University of New York Press: Albany, NY, USA, 2005.

129. Ooi, N.; Laing, J.H. Backpacker tourism: Sustainable and purposeful? *J. Sustain. Tour.* **2010**, *18*, 191–206.

130. Hannam, K.; Ateljevic, I. *Backpacker Tourism: Concepts and Profiles*; Channel View Pubblications: Clevedon, UK, 2007.

131. Uriely, N.; Yonay, Y.; Simchai, D. Backpacking experiences. *Ann. Tour. Res.* **2002**, *29*, 520–538.

132. Cohen, S.A. Lifestyle travellers. *Ann. Tour. Res.* **2011**, *38*, 1535–1555.

133. Chisnall, P.M. *Marketing Research*, 3rd ed.; McGraw-Hill: London, UK, 1986.

Chapter 14

Raising Competitiveness for Tourist Destinations through Information Technologies within the Newest Tourism Action Framework Proposed by the European Commission

Ray F. Iunius[1], Laura Cismaru[2,*] and Diana Foris[2]

[1]*École Hôtelière de Lausanne, Route de Cojonnex 18, Lausanne 1000, Switzerland*

[2]*Faculty of Food and Tourism, Transilvania University from Brașov, str. Castelului, nr. 148, 500014 Brașov, Romania*

ABSTRACT

Several challenges regarding the European tourism industry were recently identified at EU level which the experts of the European Commission tried to meet, thus formulating several priorities within the newest Tourism Action Framework: Stimulate long-term competitiveness in the European tourism sector, promote the development of sustainable and high-quality tourism, and consolidate the image and promotion of European tourist destinations. Due to the new generational context, information and communications technology ICT and innovation became keywords within the most recent European Tourism Policy. Considering the symbiotic relationship that exists at the European tourism level between sustainable development, innovative ICT solutions, and long-term competitiveness, the decision-makers in European destinations should focus on identifying innovative ways to implement the new Tourism Action Framework adopted by the European Commission, through ICT applications, in order to support long-term competitiveness achievement. Two such authentic proposals are formulated within the present discussion paper: the creation of a decision support system for the management of sustainable European destinations and the development of a trip-planner for quality-sensitive tourists based on an umbrella European certification/labeling system for tourism quality. There is still a fertile field in these areas and, therefore, more innovative ICT tools to support the long-term competitiveness of European tourist destinations can be developed.

Keywords: European action framework for tourism; long-term competitiveness; sustainability; innovative ICT tools; tourist destinations

1. INTRODUCTION

The concept of "tourist destination" is a complex one, all tourist destinations actually being amalgams of products, services, and experiences, with many different stakeholders involved. Buhalis formulated the following definition of the tourist destination: "a defined geographical region which is understood by its visitors as a unique entity, with a political and legislative framework for tourism marketing and planning" [1]. According to Ritchie and Crouch, "the tourist destination is the fundamental unit on which all the many complex dimensions of tourism are based" [2]. For the first time, in 1979, tourism as a phenomenon has been officially defined through the notion of "tourist destination" by the British Tourism Society [3]. If during the 1990s few authors were writing about concepts like place/tourist destination branding, place/tourist destination marketing, or place/tourist destination management, after 2000, all these concepts started to bloom and nowadays there is truly fertile scientific literature in the field of tourist destination management and marketing [1,4,5]. This process of conceptual development took place due to a real need: For tourist destinations, being unique places with particular characteristics and specific needs, the traditional marketing and management approach was no longer enough to cover and treat all aspects related to such complex structures. Therefore, the process of strategic planning of tourist destinations always involves several dimensions and stakeholders. Managing tourist destinations means creating and developing places which satisfy the real needs of consumers/tourists, where the local population and businesses prosper, and where there is a proper balance between the impact of tourism (costs) and all the benefits it brings. In this context, developing tourism in a sustainable way appears to be the most appropriate solution for the management of tourist destinations [6].

If "tourist destination" is considered a complex concept, the "competitiveness of tourist destinations" is an even more complex one. As Pike suggested, the battle for gaining competitive advantage is no longer a battle of products and services, but a battle of the tourists' perceptions; therefore, it is a battle that takes place inside the tourists' mind [7]. From the beginning of the 1980s, tourism competitiveness started to be defined through two central concepts: profitability and sustainability. Therefore, one of the simplest definitions of tourist destination competitiveness has been formulated by Buhalis [1] as being the "long term success" of a tourism area. A successful destination is the tourism area which continuously attracts the desired tourists through quality and unique experiences and which continuously develops, ensuring benefits for all the stakeholders, for the entire local community, and for the place itself (resource preservation). The process of continuous development should follow the sustainable way in order to provide long-term success. This approach has been recently highlighted by the World Economic Forum, which, since 2007, has been releasing the Tourism and Travel Competitiveness Index, one of the most detailed and comprehensive studies regarding the tourism performances of world countries as macro-tourist destinations [8]. In the 2015 report, the experts from the World Economic Forum stated that the aim of the Competitiveness Index is "to provide a comprehensive strategic tool for measuring the set of factors and policies that enable *the sustainable development* of the Travel & Tourism sector, which in turn, contributes to the development and competitiveness of a country" [8]. In order to emphasize the relationship between sustainability and competitiveness, it was recently declared that a destination which does not develop in a sustainable way cannot be a competitive destination [2].

However, competitive advantage can only be obtained through a strategic planning approach [3]. Usually, Destination Management Organizations (DMOs) are responsible for strategically planning the development of tourist areas, but if no DMO exists, local authorities become the decision-makers. Within the process of strategically planning the way a tourist destination gains competitive advantage, evaluation plays a crucial role; the entire process is based on the correct, sincere, and comprehensive evaluation of the destination. After that, strategies can be formulated and action frameworks can be built and implemented periodically.

Also, in the same context, the correct evaluation of the external environment (social, political, economic and technological) is of vital importance and the focus always should be on consumers/tourists. For example, we are currently witnessing a major generational shift, with Generation Y entering on the scene [9,10,11]. The new generation of tourists and tourism employees is very different from the previous ones and the main generational gap is related to technology; starting with Generation Y, we are talking about "digital natives" [11] and "tech-savvy" people.

Technology is vital for new generations. In such a generational context, decision-makers in tourism destinations have to make constant and good use of new technologies in order to keep the destinations attractive and to gain a competitive advantage [9]. Destinations also live in this new social context, with new generations of consumers who are "digital natives" and "tech-savvy". Destinations have to be attractive for these new generations of consumers, they need to convince these new consumers to come and live the tourist experience they offer; therefore, tourist destinations have to "speak" the same language as their potential consumers, being forced to adapt to the profile of this new consumer. Information and communications technology ICT and innovation are a must in current society and, for these reasons, in order to be competitive, tourist destinations have to adhere to the set of values that characterize the new generations of consumers.

In the context presented above, the current paper discusses the efforts made by the decision-makers of Europe, as a macro-tourist destination, in order to keep European destinations competitive in this new social, political, economic and technological environment. The most recent Tourism Policy adopted by the European Commission in 2010 includes several actions which aim to support European tourist destinations in gaining competitive advantage through sustainable development and best use of ICT. Several ideas included within the Tourism Policy were connected and integrated into two authentic and concrete proposals formulated in the present paper, demonstrating how this new Tourism Action Framework adopted by the European Commission provides real support for the enhancement of the competitiveness of European destinations, through making best use of innovative ICT tools.

Taking into consideration the above-mentioned definitions of the "tourist destination", Europe can be considered a specific tourist destination, along with all the places/regions included in it. Europe, as a macro-tourist destination, is managed by the European Commission through tourism policies and specific programs and actions. The European Commission is the main decision-maker for Europe. Regarding the management of the composing European destinations, there is a joint management effort because, in order to properly implement the measures adopted by the European Commission, the national and/or local stakeholders need to be actively involved.

Developing European tourism in a sustainable way has always been a major concern for the European Commission and, within this context, the sustainable development of tourist destinations has become a focal matter. The reason why efforts have been made at the European level in order to help tourist destinations develop in a sustainable way is related to a key finding of the Commission's experts: the fact that unsustainable tourism patterns can be changed only by following two directions, the consumer behavior direction, and the production of tourism direction, which includes both tourism enterprises and destinations [6]. Because many efforts have been made at the European Union (EU) level for the sustainable development of tourism enterprises, and their impact was not the expected one, the European Commission understood that focusing on tourist destinations might be a more constructive approach. In recent years, the sustainable development of tourist destinations has become a central interest at the EU level.

In 2003, the Commission also stated its intention to "fully exploit the potential for better use of information and communication technologies (ICT)" in order to support European tourism development in a sustainable way [6] (p. 17). ICT are important tools for the promotion and distribution of tourist destinations and the experts of the European Commission proved to totally commit to this statement within the most recently adopted tourism policies. The Internet empowered both tourists and destinations. Tourists gained a significant "power of search and choice" and tourist destinations gained access to the global market and the power to "enhance the quality of the tourist experience" [2] (p. 6). However, these technologies can also become supporting tools for the development of European tourist destinations, helping them to gain a competitive advantage. At the European level, the concept of competitiveness, when it comes to tourist destinations, is based on the sustainable development of the tourism industry and the high quality of the tourism experience. The European competitive destination can therefore be defined mainly as the destination which develops in a sustainable way and offers high-quality experiences to tourists. These two important sources for the competitive advantage of European destinations need to be addressed from the ICT perspective, because "taking advantage of new technologies and the Internet can also enable destinations to enhance their competitiveness" [1] (p. 22). This perspective is embraced by the experts of the European Commission within the newest Action Framework for European tourism adopted in 2010.

2. A New Action Framework for European Tourism

In 2009, eight key megatrends were identified as strongly affecting the European tourism industry, including access to information, sustainability, and experience economy [12]. In order to remain competitive, Europe needs to find ways to face the challenges determined by these megatrends and:

- make the tourism industry part of the knowledge economy;
- develop EU tourism in a sustainable manner;
- better position the EU as the number one provider of authentic and quality tourist experiences.

The most recent European Tourism Policy, communicated by the European Commission in 2010, takes into consideration all these above-mentioned factors. The architecture of the policy is based on four priorities, all related to better promoting European tourism [6]. For each priority, several specific actions are proposed. Only actions of direct interest for the topic of the present paper will be shortly mentioned below.

The first priority within the policy is to stimulate competitiveness in the European tourism sector. Due to the fact that a close relationship has been identified between innovation and competitiveness, several concrete actions have been proposed in order to reach the goal of developing innovation in the European tourism industry. Action no. 3 is related to the creation and launching of an "ICT and tourism platform for stakeholders to facilitate the adaptation of the tourism sector and its businesses to market developments in new information technologies and improve their competitiveness by making the maximum use of possible synergies between the two sectors" [6] (p. 8). In order to encourage the extension of the tourism season, action no. 7 proposes to "develop a voluntary online information exchange mechanism to improve the coordination of school holidays in the Member States, without prejudice to their cultural traditions" [6] (p. 9). With the main aim of consolidating the socioeconomic knowledge base for tourism, the Commission intends to create a "virtual tourism observatory to support and coordinate research activities by the various national research institutes (research institutes, universities, public and private monitoring units, regional and national authorities and national tourism offices)" [6] (p. 10).

The second priority within the policy is to promote the development of sustainable, responsible and high-quality tourism. Action no. 11 presents the intention of the European Commission to develop a genuine set of indicators for evaluating the sustainability performances of European tourist destinations, followed by the creation of a label for promoting European sustainable destinations. Action no. 13 declares the intention of the Commission to develop a European "High Quality" label [6] (p. 11). Even if the two above-mentioned actions have no ICT elements, they have been noted within this paper because several proposals will be presented in the following section that introduce an ICT component.

The third priority within the policy is to consolidate the image and profile of Europe as a collection of sustainable and high-quality destinations. Action no. 19 states that efforts need to be made to promote the European tourism web-platform [13] "in order to increase the attractiveness of Europe as a collection of sustainable and high-quality tourist destinations, particularly among emerging countries" [6] (p. 10).

The fourth priority within the policy is to maximize the potential of EU financial policies and instruments for developing tourism but, unfortunately, it only expresses the intention of the European Commission to support innovative initiatives in the future, with no specific action or instrument.

Taking into consideration the above-mentioned priorities and selected actions which have been included within the most recent European Tourism Policy, adopted in 2010 by the European Commission, the present paper focuses on demonstrating how these ideas, formulated by European experts, can be implemented at the destination level in order to enhance competitiveness.

3. ICT TOOLS FOR COMPETITIVE EUROPEAN TOURIST DESTINATIONS

Tourist destination competitiveness was defined in the tourism literature as "its ability to increase tourism expenditure, to increasingly attract visitors while providing them with satisfying, memorable experiences, and to do so in a profitable way, while enhancing the well-being of destination residents and preserving the natural capital of the destination for future generations" [2] (p. 2).

As Ritchie and Crouch stated, tourist destination competitiveness is a multidimensional concept, including economic competitiveness, political competitiveness, sociocultural competitiveness, environmental competitiveness, and technological competitiveness [2]. As a consequence, in order to gain a competitive advantage, tourist destinations need to harmoniously combine all the above-mentioned dimensions of competitiveness.

The two great *sources for competitive advantage* identified by the European Commission within its most recent Tourism Policy are: *sustainability* and *high quality*. However, due to the large access to information and to the new generational context with more and more consumers who are "digital natives" and "tech-savvy", European decision-makers also need to find ways to "make the tourism industry part of the knowledge economy", as stated in the previous section. In such a context, several initiatives and projects which genuinely combine ICT and sustainability or high-quality promotion have been proposed and even totally or partially implemented at the EU level. Also, further possibilities for innovation can be identified and formulated in these areas.

One of the first projects that combined ICT and the promotion of European tourist destinations was the creation of the web platform VisitEurope [13] in 2006, where all European tourist destinations are being promoted. In 2009, the web platform was improved and, in 2011, its Chinese version was launched. Also, the Virtual Tourism Observatory (VTO) website [14] and the

Calypso web platform [15] have become functional. Several projects were co-financed by the European Commission, such as KNOWNET or FAST LANE [16]. All the above-mentioned projects mainly aim to promote and inform.

The TOURISMlink Project is a very important initiative funded by the European Commission. with its main target "to increase the competitiveness of small and medium enterprises (SMEs) across Europe in today's digital marketplace" [17]. The project is not completely finished, but the TOURISMlink platform was developed to offer three main elements: "a central reservation system (CRS) that centralizes all products and destinations into one database; a property management system (PMS); and channel management, allowing immediate and automatic integration and adaptation of all data" [17]. The TOURISMlink Project represents the implementation of action no. 3 from the newest European Tourism Policy, as it was cited above. The TOURISMlink Project is the perfect demonstration of how business intelligence and innovation can be used to offer real support to tourist destinations and businesses in order to enhance their competitive advantage. The entire project is based on the European principles of sustainable development, transforming the participating destinations and businesses into real partners.

Acknowledging the fact that tourism management is considered the key to gain and maintain competitive advantage [2], we affirm that more tools to support the management of tourist destinations should be encouraged, funded, and implemented at the EU level in order to help European tourist destinations gain a competitive advantage, following the successful example of the TOURISMlink web platform. Two such proposals will be introduced in the following paragraphs. Both proposals combine the concepts of competitiveness and ICT, being concrete examples of how innovative business intelligence tools can be developed in order to contribute to the performance management at the destination level, enhancing the competitiveness of European destinations. Also, both proposals are placed within the new generational context. The new generation of "digital natives" is entering the tourism market as well as the workforce. Tourists, employees, and managers need technology because it is a vital part of their lives. They make decisions based on and using new technologies and they completely understand the benefits of using innovative ICT tools. Therefore, both proposals respect this trend, which was also reinforced within the new Tourism Action Framework formulated in 2010 by the experts of the European Commission. Due to this new social context, innovation and new technologies became vital elements in order to gain a competitive advantage in tourism, as mentioned within the most recent European Tourism Policy [6].

3.1. Development of a Decision Support System for Managing European Tourist Destinations in a Sustainable Way

One of the priority areas identified by the experts of the European Commission in 2000 was to create tools for the benchmarking of European destinations in order to help them improve the quality of tourism. Also, for the first time, the need to create business intelligence tools like Destination Management Systems in order to integrate information and communication technologies in the tourism sector at the European destination level was mentioned [18] (p. 27). In 2001, the Commission stated the necessity to conceive and disseminate authentic tools and assessment methods for monitoring and benchmarking European tourist destinations [18] (p. 18). In 2013, the European Commission successfully implemented action no. 11 from the newest European Tourism Policy and delivered a genuine set of 67 indicators for the evaluation and benchmarking of sustainable European tourist destinations called the ETIS system (European Tourism Indicators System for Sustainable Destinations). As we have already proposed in a recent article [19], an innovative decision support system for tourist destinations management can be developed based on the ETIS indicators.

In 2014, based on the proposal presented above, the DIMAST Pilot Project (Destination Intelligent Management for Sustainable Tourism) [20] received financial support from the EU Structural Funds Program by the Executive Unit for Financing Higher Education, Research Development and Innovation in Romania. It is a collaborative applied research project, based on a consortium of seven partners (a university, a national research institute, a destination management organization, a non-governmental organization (NGO) for sustainable development, two tourism and hospitality enterprises, and a software firm), which is currently entering in its second phase, as shown in Figure 1. The DIMAST Project focuses on the design and development of an original and innovative technical process for implementing the European Tourism Indicators System for Sustainable Destinations (ETIS), through a group decision support system (GDSS). The project focuses on the county of Brasov, which is a top all-year-round destination in Romania. Due to the pronounced heterogeneous character of this tourist destination, the DIMAST project aims to create a flexible business intelligence solution that is highly adaptable to other European destinations willing to implement the ETIS System using ICT and develop in a sustainable way. One of its final goals is to create an innovative ICT tool which directly and actively supports the participatory management of sustainable tourism development at the destination level and is capable of offering real support to the achievement of long-term competitiveness for European destinations.

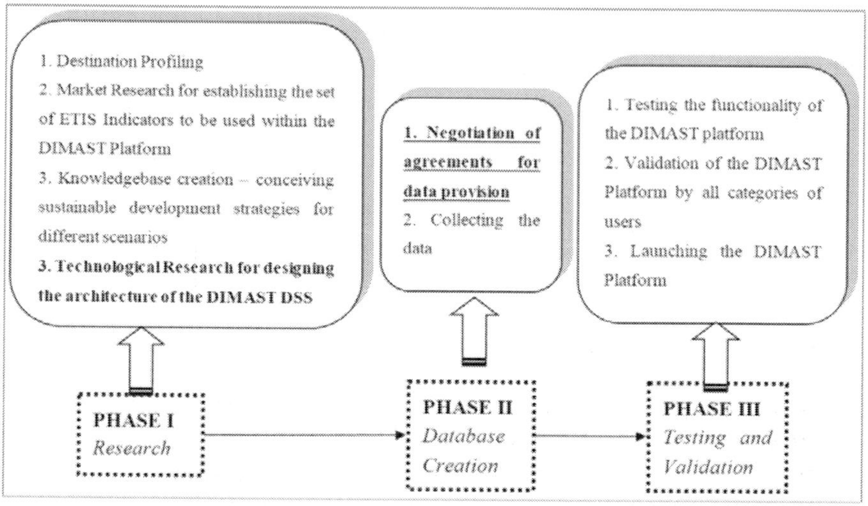

Figure 1. The three main phases of the DIMAST Project.

A tourist destination cannot gain competitive advantage by accident [3]; it is the result of a constant planning process. It was demonstrated that enterprises that use planning procedures have significantly higher performances compared to businesses that do not [3]. In such a context, more and more business intelligence solutions have been developed in order to support the planning process of enterprises, most of them being decision support systems (DSS). Due to the infinite possibilities of conceiving such systems, no broad and exact definition for a DSS can be formulated [21]. Nevertheless, taking into consideration the practical aims of a DSS, it can be concisely defined as an interactive software-type product that aims to assist the decision-makers within an organization to solve problems and/or take the best decisions in a given context through data interpretation and correlations [22]. An innovative computerized decision-making support system should support the decision-making process and not generate automated decisions [23]. The above-presented discussion perfectly fits the tourist destination management sector, too, but, as Buhalis and Costa [24] recently noticed, there are only a few projects approaching this issue and they are considered highly innovative; most decision-makers at the destination level do not know about the possibility of using such business intelligence tools in order to increase the competitiveness level of destinations. It has been recently stated that DSS systems can play a fundamental part within the decision-making process of tourist destinations, significantly contributing to formulating policies and strategies, to destination development, and to the implication of all the stakeholders [24]. As we have mentioned above, the European Commission has been supporting this perspective since 2000 [18]. Analyzing the recent tourism literature, a small number of studies which treat the idea of designing decision support systems for tourist destinations management was identified; some of these are just theoretical articles proposing the development of such systems [22,25] and only a few take into account the need for sustainability within European tourism destinations [26,27,28,29,30]. Nevertheless, most of these projects that have been implemented at the European level are not functional any longer. A successful and still-functional project is the TourMIS Platform [31,32], co-financed by the Austrian government, offering a decisional support system for tourist destination marketing. This system has been put together by the Modul University specialists in Vienna and was conceived to regroup different tourism stakeholders within a web platform upheld by a database, being one of the strengths of TourMIS through its user-friendly access. The platform not only addresses to tourist destinations' decision-makers, it also creates reports and useful information for a number of other organizations as well. The system also offers benchmarking features for tourist destinations willing to compare their performances. However, TourMIS does not clearly aim toward a sustainable development of tourism and this is the most significant difference in comparison with the DIMAST Project.

The DIMAST Project [20] is based on the ETIS tool created by the experts of the European Commission as a support tool for European tourist destinations willing to develop in a sustainable way and, accordingly, achieve a long-term competitive advantage. It is the first implementation project that includes the ICT component, aimed at transforming the ETIS tool into a more practical, useful and easy-to-implement instrument for European destinations interested in following the sustainable way, according to the European Union's principles. The innovative DIMAST DSS will be built on: (1) a database (primary and secondary data), mainly conceived of the selected ETIS indicators, (2) a specialized knowledgebase in the field of sustainable tourism strategies, (3) a model and algorithm base used for data processing, and (4) a geographic information sub-system (GIS) for the creation of thematic maps based on key indicators included in the project. The DIMAST system functionality will be tested through an integrated web platform. Being designed as a collaborative DSS (GDSS), the DIMAST Project includes several categories of users: the main user is the Destination Management Organization (DMO) and there are also three additional categories of (1) tourism and hospitality enterprises; (2) research organizations; and (3) other entities that might be interested in sustainable development. The basic architecture of the DIMAST GDSS is presented in Figure 2below.

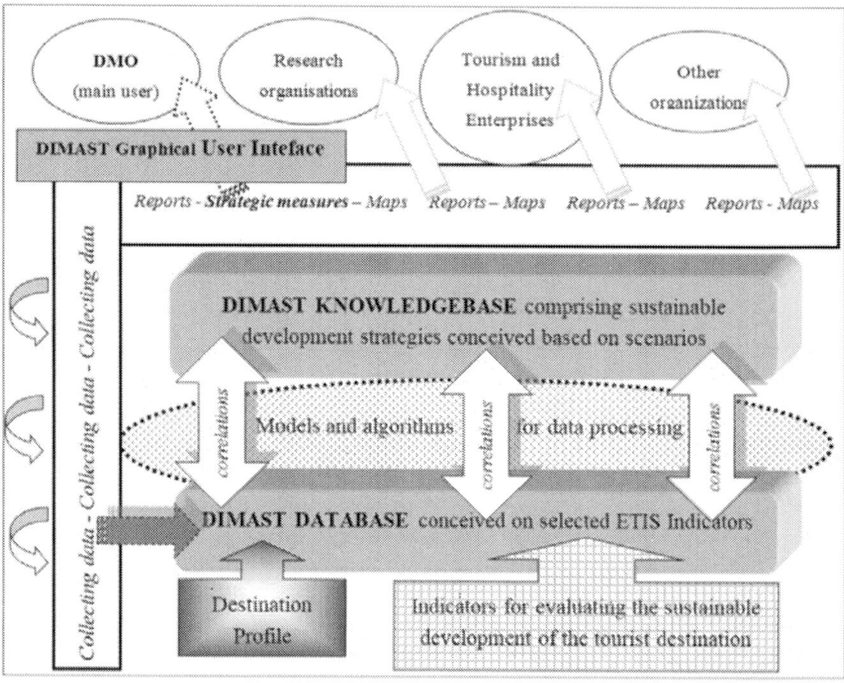

Figure 2. DIMAST DSS basic architecture.

The main objective of the DIMAST Project is to create an innovative ICT tool which can bring a real and useful contribution to the sustainable development of European destinations. If correctly implemented by several European destinations, it can develop into a European benchmarking tool for sustainable destinations [33,34]. By all means, the ultimate aim of the DIMAST DSS is to help European tourist destinations gain a long-term competitive advantage.

The proposal presented above, using the case study of the DIMAST Project, provides a concrete example of how the new Tourism Action Framework, proposed by the European Commission in 2010, can contribute to raising the competitiveness of European destinations through ICT innovative tools. The ETIS System of Indicators has been created based on this new Tourism Action Framework as an important management instrument for the decision-makers in European destinations who are willing to develop their destinations in a sustainable and, thus, competitive way. Including the technological dimension and creating a business intelligence instrument based on ETIS can provide significantly higher visibility and attractiveness for the ETIS tool and, therefore, exponential long-term benefits for European destinations. Using specialized software with a user-friendly interface which also offers strategic decision support is definitely much more attractive for decision-makers at the destination level who are willing to implement ETIS or who only want to manage their destinations toward gaining a long-term competitive advantage.

This proposal is directly addressed to decision-makers at destinations. The importance of conceiving such innovative business intelligence tools to support the management of tourist destinations has been acknowledged by the European Commission since 2000 in the context of a changing social, political, economic, and technical environment. These solutions proved to be beneficial for businesses and, at the same time, the new generations of employees and managers who have vital needs regarding the use of new technologies [11]. In such a context, proposing, designing and implementing innovative ICT solutions for the people who work and will work in destination management teams, such as the DIMAST Project, is the natural way of supporting the long-term competitiveness of European destinations. The proposal presented above also has an important asset—the European added value—due to the fact that it is based on ETIS, which is a European product issued by the European Commission within the process of implementing the new Tourism Action Framework.

3.2. Development of a Decision Support System Based on an Umbrella European Certification/Labeling System for Tourism Quality

In 2007, the European Parliament addressed the invitation to Member States and European tourism and hospitality enterprises to set up quality management schemes [35]. Nowadays, there are many tourism quality-labeling and standardization schemes which coexist at the European level, both private and public, local and national; however, as noticed by European experts [36,37], they lack consistency "when compared in respect of their sectorial scope, geographical coverage, governance, assessment".

The existing market is highly fragmented and the European added value cannot be found in any of these existing schemes. In such a context, the new Tourism Action Framework adopted by the European Commission in 2010 introduced the proposal to develop a European Tourism Quality Label, as mentioned above in this discussion paper. As noticed in this policy document, progress has been made in the direction of developing a European "Qualité Tourisme" brand, in order to "increase consumer security and confidence in tourism products and reward rigorous efforts by tourism professionals whose aim is quality of tourism service for customer satisfaction" [6] (p. 10). After the launching of the new Tourism Action Framework by the European Commission, an important study has been conducted on the estimated impacts, possible options, and legal instruments of the umbrella European tourism label for quality schemes [37]. In 2014, the European Commission conceived a formal legal proposal for a Council Recommendation on European Tourism Quality Principles [36]. The document explains in detail the necessity to introduce a set of common Tourism Quality Principles at the European level and proposes a minimal set of such principles "applicable to tourism services offered in the Union directly to consumers and (*which*) are recommended to be followed by public and private organizations providing services in the field of tourism" [36]. The formulation of a set of European Tourism Quality Principles was a necessary step on the way to adopting an umbrella European Quality Tourism Label in the future. The creation of such an instrument would help "quality-sensitive consumers to make better informed choices" [36], would support and reward the efforts of tourism and hospitality enterprises regarding the provision of higher and higher qualities of their products and services, and would also contribute to image and brand improvement of European destinations and, therefore, support their long-term competitiveness.

Within this context, we suggest that an innovative decision support system (DSS) for tourists willing to visit Europe can be developed based on a genuine umbrella "Qualité Tourisme" certification/labeling system adopted at the European level.

As we have already proposed in a recent article, several ICT innovative tools can be developed at the destination level in order to provide support for tourists who want to make responsible choices for their vacations [38]. The demand of planning tools for tourists was determined by the need of gaining more time for the vacation. In order to plan a vacation, tourists usually have to spend their time collecting information about the desired destination and the points of attraction they are going to visit. As a result, establishing the desired route at a destination is a time-consuming task for tourists [39] and ideas for addressing this problem arose. It has been stated that "web based decision support applications are excellent aids for tourists who want real support for tourist planning problems" [40]. This so-called "touristic trip planning problem" [41] has been addressed in the last decade by several authors in scientific research papers [40,41,42,43,44,45,46,47,48,49,50], proposing the creation of decision support systems for tourists, both web-based and mobile, generally called "trip planners". These instruments mainly serve to tailor a trip according to the specific needs and interests of a tourist/group of tourists and usually produce trip plans or trip maps. It has been said that these recent developments revolutionized tourism [40]. However, taking into consideration the recent research findings regarding the way of thinking and the real needs of current consumers [9,10,11], such trip planners, mainly conceived for mobile devices, are tailored for the profile of the consumer who belongs to the new generation of "digital natives" and "tech-savvy" people.

In the context presented above, the proposal formulated within this section is highly innovative. Exactly like the first proposal, presented in the previous section of this discussion paper, a DSS for tourists based on an umbrella European certification/labeling system for tourism quality represents a concrete example of how the new Tourism Action Framework adopted by the European Commission can be implemented. Even if such a DSS is mainly conceived for tourists, offering real support for the "quality-sensitive" tourists, it also contributes to raising the competitiveness of tourist destinations willing to implement this kind of ICT instrument, as stated before. However, important steps need to be taken at the European level before such an approach can be transformed into reality. First of all, significant efforts have to be made in order to create the umbrella European certification/labeling system, "Qualité Tourisme". Developing a European trip planner based on "Qualité Tourisme" would definitely provide higher visibility and attractiveness for such a European umbrella certification/labeling system, making it easier to implement.

Both proposals presented above need to be developed and properly implemented in European destinations by specific stakeholders in order to bring a real contribution to the long-term competitiveness of the local area where their importance is acknowledged. As a result, the problem of understanding and assuming roles and responsibilities regarding the implementation of the Tourism Action Framework at the destination level needs further clarification.

3.3. Roles and Responsibilities Regarding the Competitiveness of European Tourist Destinations

As mentioned above, the concept of "tourist destination" is a complex one; the tourist destination is both tangible and intangible, it is a place situated in a specific area, with specific physical and geographical coordinates, but also an intangible socio-cultural entity, mainly represented by the people, traditions, lifestyle, culture, and social interactions within it [51]. At the same time, the tourist destination is an image and an experience which exists in the consumers' minds and souls. Everything a tourist destination represents can be integrated within a unique brand. New important concepts appeared in the tourism literature in this context, such as a destination's personality, spirit or profile [34,52,53]. Taking into consideration all these aspects, tourist destinations are specific entities with specific needs which are not very easy to manage. In order to properly manage a tourist destination, the responsible stakeholders have to understand the specific needs of the destination, correctly prioritize them, and always start from the premises that tourist destinations are living entities with a particular sinusoidal life-cycle. All these concepts enable us to make analogies of tourist destinations and human beings. Therefore, based on Maslow's Hierarchy of Needs [54], the present discussion paper proposes the creation of a pyramidal model called Destination's Hierarchy of Needs, graphically depicted in Figure 3.

Within the Destination's Hierarchy of Needs, each level of specific needs represents, at the same time, a level of responsibilities for the destination's stakeholders. In order to properly fulfill the destination's specific needs, different stakeholders share specific roles. Therefore, a tourist destination's competitiveness highly depends on the degree stakeholders are aware of and assume their roles and responsibilities [55]. In other words, there is a relationship of direct determination between the social capital of a tourist destination and its competitiveness.

There are three main categories of stakeholders who share responsibilities regarding the competitiveness of a destination: public stakeholders, private stakeholders, and the local community [1]. In the context of the discussion of this paper, it is of vital importance to identify those stakeholders who become decision-makers within the process of implementing the European Tourism Action Framework at the destination level. From a theoretical perspective, a Destination Management Organization (DMO) should have such responsibilities and roles as a unique entity which makes and assumes decisions for the entire destination [1,56]. However, at the European level, there is a great diversity and broad typology of tourist destinations and the percentage of European destinations which have a legal and functional DMO is not significant yet, but it is continuously growing. Therefore, local authorities play a crucial role in the process of strategic implementation of the European Action Framework for tourism.

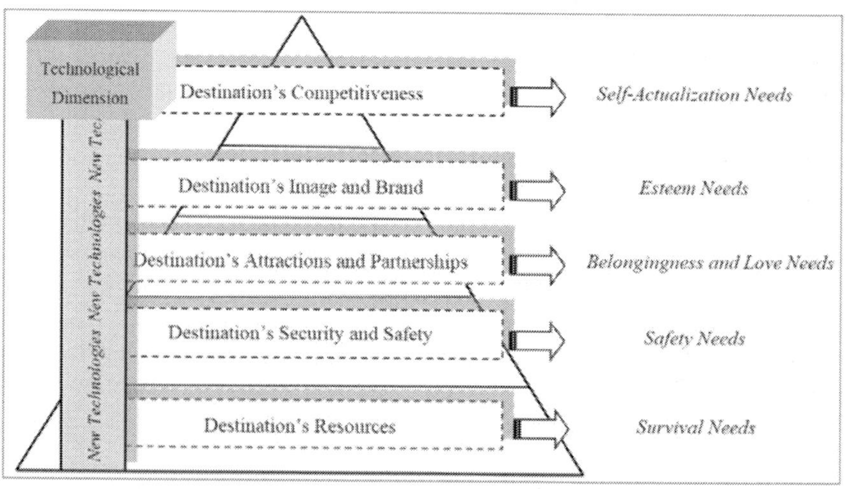

Figure 3. Destination's Hierarchy of Needs.

As presented above, the European perspective on a destination's competitiveness strongly relies on sustainable development and high quality. In the new generational context, we support the idea that local decision-makers at European destinations should focus on identifying ways to make better use of ICT in order to develop the technological strengths of destinations. The present discussion paper introduced two ideas of how innovative ICT tools can be developed in order to better implement the new Tourism Action Framework adopted by the European Commission at the destination level, aiming to increase the competitiveness level. However, there are infinite possibilities to strengthen the technological dimension of the competitiveness for tourist destinations. It should be a focal preoccupation for decision-makers at the destination level because the new generation of tourists are tech-savvy, and they are the "digital natives" [11]; technology is a natural (and vital) part of their lives and their specific profile requires specific strategic measures at the destination level as well [9,10,11]. New technologies can actually be used for satisfying all the levels of a destination's needs, as presented inFigure 3 above. As stated by Law, Leung and Buhalis [47], ICT can bring a real contribution to the dynamic management of information by destination decision-makers, offering them true support in the strategic planning process and, therefore, positively influencing the competitiveness of tourist destinations.

The discussion within this paper focused on the relationship between competitiveness, sustainable development, and innovative ICT solutions at the tourism destination level. The new Tourism Action Framework proposed by the European Commission in 2010 is based on this trinomial. According to European experts from the European Commission, there can be no competitiveness without sustainable development. Short-term competitive advantage no longer exists from a European perspective because this concept proved to have no solid foundation. Only long-term competitiveness is acceptable at the European level, which is based on constant efforts to develop tourism areas and businesses respecting the European sustainable development principles. The new generational context, with the Y generation (and after it, the Z generation) of "digital natives" and "tech-savvy" people entering the workforce and the tourism market, forces tourist destinations and businesses to re-profile according to the new sets of values that characterize these new cohorts. Nowadays, both tourists and tourism employees need ICT and new technologies in order to make decisions; it is a vital part of their lives. In such a context, innovation has become a "must" in all industries, tourism included. Therefore, tourist destinations definitely need to develop innovative solutions based on new technologies in order to enhance long-term competitiveness, as the experts of the European Commission have constantly acknowledged since 2000. With respect to the new generational profile, long-term competitiveness can only be achieved through innovation and proper use of new technologies. Taking into consideration this new social context, it is the ultimate sustainable approach. Forasmuch as the present discussion demonstrated, the relationship between destination competitiveness, sustainable development, innovation and ICT may be considered a symbiotic one and, due to this specific nature, decision-makers of European destinations need to conceive strategies based on this trinomial, sustainable development-innovative ICT solutions competitiveness.

4. CONCLUSIONS

Managing and marketing tourist destinations are very difficult and challenging activities, due to the complexity of the concept of the "tourism destination" itself [1]. European destinations need to be managed in a sustainable way because, due to the high demand for tourism in Europe, travel intensity will increase in the future and there is a risk of partially or completely damaging its resources. All efforts made at the European level regarding the sustainable development of tourism need to be assisted by appropriate information tools and networks.

Important steps have already been made at the European level in the direction of combining sustainability, promotion, and high-quality aspects with ICT with the main aim of offering support to tourism enterprises and destinations in order to gain a competitive advantage. Important related actions are included within the new framework for action established by the European Commission in the most recent Tourism Policy, adopted in 2010. Some of them have already been successfully implemented.

Concepts like "technological strengths of tourist destinations" have only recently started to be used [2] (p. 5). Taking into consideration the technological dimension of the competitiveness, this should be a focal point for European tourist destinations in the near future, finding the best ways to develop technological strengths through innovative approaches based on the main existing assets that are the sustainable development and the high quality of European tourism. Starting with the simple Internet promotion and following with establishing collaborative web platforms and all types of business intelligence tools, the decision-makers at European destinations should find innovative ways to implement the new Tourism Action Framework adopted by the European Commission through ICT applications. Two such examples were formulated within the present discussion paper. The first one introduced the DIMAST Project, which aims to develop a flexible and adaptable DSS for tourist destinations willing to develop in a sustainable way, based on the ETIS Indicators System launched by the European Commission in 2013. The project is in progress, as shown in Figure 1. The second proposal formulated within the present discussion paper presents the idea of developing a trip planner based on the umbrella

European certification/labeling system, "Qualité Tourisme", which will be created in the near future, as stated within the new Tourism Action Framework.

Within the new context created at the European level by the Lisbon Treaty, which introduced a new competence and, therefore, new powers for the European Union to complement the action of Member States in the tourism sector [57], the two proposals formulated within the present discussion paper represent concrete examples of how the new Tourism Action Framework adopted by the European Commission in 2010 can directly contribute to the enhancement and promotion of the long-term competitiveness of European destinations, through making the best use of innovative ICT tools. As it has been emphasized all through the present discussion paper, taking into consideration the present social, economic, and technological context, creating innovative tools based on new technologies is the only path to follow in order to enhance competitiveness. The new generations of "digital natives" and "tech-savvy" people live their lives, make their decisions, and depend on new technologies [11]. In respect of this symbiotic relationship, which characterizes the present social context, the natural way is to develop business intelligence tools tailored to the profile of these new consumers and new employees/managers. The European experts from the European Commission correctly identified this new profile in 2000 [18], and have been proving their commitment to respect it ever since. Therefore, in the most recent Tourism Policy, "innovation" and ICT/new technologies are keywords for competitiveness enhancement. European destinations need to adapt their profile to one which characterizes the new generation in order to gain a long-term competitive advantage.

Ali and Frew [58], in their recent paper, advanced the idea of the great value that social media has for sustainable tourism. In the context of this discussion paper, we support the idea that the European Commission and European destinations should also concentrate on using social media and transforming it into powerful strengths. There is also room for innovation within this field, due to the fact that a new concept appeared: Web 3.0 or the Semantic Web. This is a new direction for future research at the European level.

ACKNOWLEDGMENTS

This paper is supported by the Sectorial Operational Programme Human Resources Development (SOP HRD), financed from the European Social Fund and by the Romanian Government under the Project number PODRU/159/1.5/S/134378.

AUTHOR CONTRIBUTIONS

Ray Iunius, Laura Cismaru and Diana Foris equally contributed to this discussion paper. Respectively, Laura Cismaru and Diana Foris analyzed the research concepts within the most recent Tourism Policies adopted by the European Commission, Ray Iunius and Laura Cismaru formulated and depicted the innovative proposals within the discussion paper, and Diana Foris collected the research resources and systematized the content.

REFERENCES

1. Buhalis, D. Marketing the Competitive Destination of the Future. Available online: http://core.ac.uk/download/pdf/101658.pdf (accessed on 27 Fabruary 2015).
2. Ritchie, J.R.B.; Crouch, G.I. *The Competitive Destination: A Sustainable Tourism Perspective*; CABI Publication: Cambridge, UK, 2003.
3. Vanhove, N. *The Economics of Tourist Destinations*; Butterworth-Heinemann: Oxford, UK, 2005.
4. Kotler, P.; Bowen, J.; Makens, J. *Marketing for Hospitality and Tourism*; Prentice Hall: Upper Saddle River, NJ, USA, 1999.
5. Nicolaescu, L. *Zooming the Image of Romania. Country Branding and Rebranding*; The Academy of Economic Sciences: Bucharest, Romania, 2008.
6. European Commission. Europe, the World's No 1 Tourist Destination—A New Political Framework for Tourism in Europe. Available online: ec.europa.eu/enterprise/sectors/tourism/documents/communications/commission-communication-2010/index_en.htm (accessed on 18 September 2015).
7. Pike, S.D. ToMA as a measure of competitive advantage for short break holiday destinations. *J. Tour. Stud.* **2002**, *13*, 9–19.
8. Crotti, R., Misrahi, T., Eds.; *The Travel & Tourism Competitiveness Report 2015. Growth through Shocks*; World Economic Forum: Geneva, Switzerland, 2015.
9. Cismaru, L.; Iunius, R. Rural tourist destinations and the tourists of Generation Y. In *Romanian Rural Tourism in the Context of Sustainable Development. Present and Future Perspective*; Paduraru,

T., Tacu, G., Ungureanu, D., Talaba, I., Eds.; PIM Publication: Iasi, Romania, 2015.

10. Benckendorff, P.; Moscardo, G.; Pendergast, D. *Tourism and Generation Y*; CAB International: Beijing, China, 2010.
11. Prenksy, M. Digital natives, digital immigrants. *Horizon* **2001**, *9*, 1–6.
12. ECORYS SCS Group. Study on the Competitiveness of the EU Tourism Industry-with Specific Focus on the Accommodation and Tour Operator & Travel Agent Industries. Final Report to Directorate-General Enterprise & Industry. Available online: http://ec.europa.eu/enterprise/sectors/tourism/files/studies/competitiveness/study_on_tourism_competitiveness_2009_en.pdf (accessed on 21 February 2015).
13. Visiteurope. Available online: http://www.visiteurope.com/en/ (accessed on 12 March 2015).
14. European Commission's Virtual Tourism Observatory (VTO). Available online: https://ec.europa.eu/growth/tools-databases/vto/home (accessed on 12 March 2015).
15. eCalypso. Available online: http://www.ecalypso.eu/steep/public/index.jsf (accessed on 11 March 2015).
16. European Commission. Implementation Rolling Plan of Tourism Action Framework. Available online: http://ec.europa.eu/enterprise/sectors/tourism/files/communications/com_implementation_rolling_plan_en.pdf (accessed on 12 March 2015).
17. TOURISMlink. Linking Tourism Professionals with the Digital Market. Available online: http://www.tourismlink.eu/the-project/what-is-tourismlink/ (accessed on 13 March 2015).
18. European Commission. Commission communication to the Council, the European Parliament, the Economic and Social Committee and the Committee of the Regions: Working Together for the Future of European Tourism. Available online: http://aei.pitt.edu/5237/1/5237.pdf (accessed on 18 September 2015).
19. Brătucu, G.; Cismaru, L. The sustainable development of the Euroregions, as tourist macro destinations, through an informatics decision support system based on the European Tourism Indicators System for Sustainable Destinations launched by the EU Commission in 2013. In *Sustainable Economic and Social*

Development of Euroregions and the Cross-border Areas, Proceedings of the International Conference Sustainable Economic and Social Development of the Euroregions and the Cross-border Areas, Balti, Moldova, 27 June 2014; Tehnopress: Iasi, Romania, 2014; pp. 45–61. (In Romanian).

20. DIMAST Homepage. Available online: www.dimast.eu (accessed on 20 March 2015). (In Romanian).
21. Keen, P.G.W. Decision support systems: A research perspective. In *Decision Support Systems: Issues and Challenges*; Fick, G., Sprague, R.H., Eds.; Oxford Pergamon Press: Oxford, UK, 1980.
22. Baggio, R.; Caporarello, L. Decision Support Systems in a Tourism Destination: Literature Survey and Model Building. In Proceedings of the itAIS 2005—2nd Conference of the Italian chapter of AIS (Association for Information Systems), Verona, Italy, 1–2 December 2005; pp. 1–15.
23. Alter, S.L. *Decision Support Systems: Current Practice and Continuing Challenges*; Addison-Wesley series on decision support; Addison-Wesley: Upper Saddle River, NJ, USA, 1980.
24. Buhalis, D.; Costa, C. *Tourism Management Dynamics*; Routledge: London, UK, 2012.
25. Singh, S.P.; Sharma, J.; Singh, P. A Web-based Tourist Decision Support System for Agra City. *Int. J. Instrum. Control Autom (IJICA)* **2011**, *1*, 51–54.
26. Laniado, E.; Cappiello, A.; Cellina, F.; Cerioli, F.; Hernandez, K.; Laghi, A. A decision support system for sustainable tourism: The SFIDA project. In Proceedings of the ENVIROSOFT Conference, Ancona, Italy, 2–4 June 2004.
27. Laniado, E.; Botta, M.; Cappiello, A.; di Guardo, A. Environmental information systems for land planning. In Proceedings of the ENVIROINFO Conference, Geneva, Switzerland, 21–23 October 2004.
28. Balaton Project. Available online: http://www.balatonline.net (accessed on 20 May 2014).
29. SFIDA Homepage. Available online: http://sfida-life.it (accessed on 27 February 2015).
30. Bousset, J.P.; Skuras, D.; Tesitel, J.; Marsat, J.B.; Petrou, A.; Fiallo-Pantziou, E.; Kusova, D.; Bartos, M. A Decision Support System for Integrated Tourism Development: Rethinking Tourism Policies and

Management Strategies. *Tour. Geogr. Int. J. Tour. Space Place Environ.* **2007**, *9*, 387–404.

31. Wober, K.W. Information Supply in Tourism Management by Marketing Decision Support Systems. *Tour. Manag.* **2003**, *24*, 241–255.
32. Austrian National Tourist Office. Available online: http://www.tourmis.info/ (accessed on 12 May 2015).
33. Cismaru, L. European Tools for Managing and Monitoring the Sustainable Development of Tourist Destinations. *Rev. Gen. Manag.* **2015**, *1*, 101–112.
34. Cismaru, L.; Ispas, A. Improving the Profile of European Tourist Destinations through the European Tourism Indicators System. *Bull. Transilv. Univ. Brasov.* **2015**, *8*, 87–95.
35. European Tourism Quality Principles. Available online: http://ec.europa.eu/enterprise/sectors/tourism/quality-label/index_en.htm (accessed on 11 May 2015).
36. European Commission. Proposal for a Council Recommendation on European Tourism Quality Principles. Available online: http://eur-lex.europa.eu/legal-content/EN/TXT/?uri=celex:52014PC0085 (accessed on 18 September 2015).
37. CEPS. Estimated Impacts of Possible Options and Legal Instruments of the Umbrella European Tourism Label for Quality Schemes. Available online: http://ec.europa.eu/enterprise/ newsroom/cf/_getdocument.cfm?doc_id=7655(accessed on 19 August 2015).
38. Iunius, R.; Cismaru, L.; Brătucu, G. The real contribution of a DSS system developed for the Sustainability oriented tourists to the sustainable development of the hospitality industry within the Euroregions. In *Sustainable Economic and Social Development of Euroregions and the Cross-border Areas*, Proceedings of the the International Conference Sustainable Economic and Social Development of the Euroregions and the Cross-border Areas, Balti, Moldova, 27 June 2014; Tehnopress: Iasi, Romania, 2014; pp. 101–113. (In Romanian).
39. Braunschweig, A.; Saladino, D. The Sustainability Certification for Hotels. Available online: http://www.oe-plus.ch/uploaded_files/113871571821.pdf (accessed on 17 August 2015).
40. Souffriau, W.; Vansteenwegen, P. Tourist Trip Planning Functionalities: State-of-the-Art and Future. In Proceedings of the

10th International Workshop on Web Engineering & Tourism, Vienna, Austria, 5–9 July 2010.

41. Portolan, A.; Milicevic, M.; Zubrinic, K. Concept of Mobile Device Integration in Current Travel and Tourism Industry. In Proceedings of the 10th WSEAS International Conference on Applied Computer and Applied Computational Science, Venice, Italy, 8–10 March 2011; pp. 154–159.
42. Garcia, A.; Arbelaitz, O.; Linaza, M.; Vansteenwegen, P.; Souffriau, W. Personalized Tourist Route Generation. In Proceedings of the 10th International Workshop on Web Engineering & Tourism, Vienna, Austria, 5–9 July 2010.
43. Godart, J.M. Combinatorial Optimisation Based Decision Support System for Trip Planning. In *Information and Communicatin Technologies in Tourism*; Buhalis, D., Ed.; Springer-Werlag/Wein: Berlin, Germany, 1999.
44. Goy, A.; Magro, D. STAR: A Smart Tourist Agenda Recommender. In Proceedings of the Configuration Workshop on ECAI, Valencia, Spain, 23–24 August 2004.
45. Kurata, Y. CT-Planner2: More Flexible and Interactive Assistance for Day Tour Planning. In *Information and Communication Technologies in Tourism*, Proceedings of the International Conference, Innsbruck, Austria, 26–28 January 2011.
46. Kurata, Y. Challenges in User-Adaptive Tour Planning Systems. In Proceedings of the AGILE 2009 Workshop on Adaptation in Spatial Communication, Hannover, Germany, 2–5 June 2009.
47. Law, R.; Leung, R.; Buhalis, D. Information Technology Applications in Hospitality and Tourism: A Review of Publications from 2005 to 2007. *J. Travel Tour. Mark.* **2009**, *26*, 599–623.
48. Maervoet, J.; Souffriau, W.; Vansteenwegen, P.; Vaden Berghea, G.; van Oudheusden, D. Tourist Decision Support for Mobile Navigation Systems: A Demonstration. *Appl. Artif. Intell.* **2008**, *22*, 964–985.
49. Maruyama, A.; Shibata, N.; Murata, Y.; Yasumoto, K.; Ito, M. P-tour: A personal navigation system for tourism. In Proceedings of the 11th World Congress on ITS, Nagoya, Japan, 17–22 October 2004; pp. 18–21.
50. Sinleunam, P.; Poolsawad, K.; Tritilanunt, S. A Web-based decision support system for eco-tourism planning: A case study of national park in Chiangmai province. In Proceedings of the Computer Science and Software Engineering (JCSSE) Eighth International Joint

Conference, Nakhon Pathom, Thailand, 11–13 May 2011; pp. 96–101.

51. Reid, R.D.; Bojanic, D.C. *Hospitality Marketing Management*; John Wiley and Sons: Hoboken, NJ, USA, 2009.
52. Del Bianco, P. The Spirit of Place between the Intangible and the Tangible. In Proceedings of the 16th General Assembly and International Scientific Symposium, Quebec City, QC, Canada, 29 September–4 October 2008.
53. Pike, S.; Ryan, C. Destination positioning analysis through a comparison of cognitive, affective and conative perceptions. *J. Travel Res.* **2004**, *42*, 333–342.
54. Maslow, A.H. *Motivation and Personality*, 3rd ed.; Frager, R., Fadiman, J., McReynolds, C., Cox, R., Eds.; Addison Wesley: Upper Saddle River, NJ, USA, 1987.
55. Garau, C. Perspectives on Cultural and Sustainable Rural Tourism in a Smart Region: The Case Study of Marmilla in Sardinia (Italy). *Sustainability* **2015**, *7*, 6412–6434.
56. Pike, S. *Destination Marketing Organizations*; Elsevier: Amsterdam, The Netherland, 2004.
57. European Commission. Consolidated Versions of the Treaty on European Union and the Treaty on the Functioning of the European Union. Available online: http://eur-lex.europa.eu/legal-content/EN/TXT/?uri=celex:12012E/TXT(accessed on 18 September 2015).
58. Ali, A.; Frew, A.J. *Information and Communication Technologies for Sustainable Tourism*; Routledge: London, UK, 2013.

Index

T

U

W

Y